Genetics and Ethics

GENETICS and ETHICS
An Interdisciplinary Study

Edited by
GERARD MAGILL, Ph.D.

Saint Louis University Press
2004

Copyright © 2004 by Saint Louis University Press

All rights reserved. No part of this publication may be reproduced, stored in a retrieval system, or transmitted in any form or by any means—electronic, mechanical, photocopy, recording, or any other—except for brief quotations in printed reviews, without the prior permission of the publisher.

Library of Congress Cataloging-in-Publication Data

Genetics and ethics : an interdisciplinary study/edited by Gerard Magill.—1st ed.
   p. ; cm.
  Includes bibliographical references and indexes.
  ISBN 0-9652929-7-5 (alk. paper)
  1. Genetics—Moral and ethical aspects. 2. Human genetics—Moral and ethical aspects.
  [DNLM: 1. Genetics, Medical—ethics. 2. Ethics, Medical. 3. Genetic Techniques—ethics. QZ 50 G32815 2003] I. Magill, Gerard, 1951–
QH438.7.G445 2004
174.'296042—dc21

2003007578

Printed in the United States of America
07 06 05 04 03 5 4 3 2 1
First edition

In memory of Lawrence, such a cheery brother and kind man,
for endless generosity and *joie de vivre*.

CONTENTS

Preface ix

1. Introduction: The Development of Ethics Discourse on Genetics 1
   Gerard Magill and John Brehany

SECTION I. CONTEXT ISSUES: LAW AND BUSINESS

2. Policy Challenges: Ethical, Legal, and Social Implications of Genetics 23
   Ellen Wright Clayton

3. The Genetics Revolution: Can the Law Cope? 35
   Sheila A. M. McLean

4. Genetics, Investment, and Business: Organizational Ethics for Genomics Companies 53
   Gerard Magill

SECTION II. THEORETICAL ISSUES: CULTURE AND DIGNITY

5. Geneticization: The Sociocultural Impact of Gentechnology 80
   Henk A. M. J. ten Have

6. Genes and Gender: An Egalitarian Analysis 101
   Mary Briody Mahowald

7. Using Human Dignity to Constrain Genetic Research and Development: When It Works and When It Does Not 116
   Jan C. Heller

8. Engineering Our Grace: An Old Idea and New Genetic Technologies 126
   Brent Waters

CONTENTS

SECTION III. PRACTICAL ISSUES: IMPACT AND CHALLENGE

9. The Impact of Genetics and Genomics on Views of Aging   144
   David Schlessinger

10. Genetic Labels and Long-Term-Care Policies: A Winning or Losing Proposition for Patients?   153
    Ruth Purtilo

11. The Challenge of Advances in Genetic Medicine for the Biomedical and Organizational Ethics of Health Care Delivery   164
    Dennis Brodeur

12. The Challenge of Physician Education in Genetics   176
    Carol Bayley

SECTION IV. CLINICAL ISSUES: SCREENING AND THERAPY

13. Pharmacogenetics, Genetic Screening, and Health Care   186
    Ruth Chadwick

14. Ethical Aspects of Prenatal Screening and Diagnosis   197
    Norman M. Ford

15. Gene "Therapy": A Test Case for Research with Children   216
    M. Therese Lysaught

SECTION V. EPILOGUE

16. Science, Ethics, and Policy: Relating Human Genomics to Embryonic Stem-Cell Research and Therapeutic Cloning   253
    Gerard Magill

Select Bibliography   285

Notes on Contributors   343

Index   353

PREFACE

Fifty years ago, in 1953, James D. Watson and Francis Crick announced their discovery of the double helix on the structure of deoxyribose nucleic acid (DNA). Their discovery fueled the genetics revolution and effectively reinvented the science of biology. Over the subsequent five decades, study of the structure of DNA has led to amazing breakthroughs in science, especially in recombinant DNA technology, which has facilitated the diagnosis and treatment of many diseases. And over the past few years, scientists have decoded most of the human genome sequence, providing a comprehensive map for deciphering the genetics of human disease.

These amazing accomplishments in human genetics raise many vexing questions and a plethora of ethical dilemmas that we must address now as science increases its control over the human genome contained in each cell's DNA. This book presents an interdisciplinary study of genetics and ethics by international scholars who explore many of these crucial issues. The authors were invited to present papers at a major international conference on the topic; they subsequently revised and updated their chapters for publication in this collection. The conference was hosted by the Center for Health Care Ethics in the health sciences complex at Saint Louis University in St. Louis and was funded by a grant of $100,000 from Tenet Healthcare Corporation in the United States, whose executives I sincerely thank for their generous support.

The purpose of this collection is interdisciplinary in order to provide a representative sample of the current debates in this fast-changing arena. The simple title *Genetics and Ethics* has been selected to convey a wide-ranging approach that we hope will attract both the general reader and the scholar. To highlight the interdisciplinary nature of ethics discourse on human genetics, the chapters have been organized into five distinct and related sections after an introductory chapter on "The Development of Ethics Discourse on Genetics."

First, the section on *context issues* considers ethical concerns in genetics from the perspectives of law and of business. Second, the section on *theoretical issues* considers ethical concerns in genetics from the perspectives of culture and dignity. Third, the section on *practical issues* considers ethical concerns from the perspectives of the impact and challenge of genetics. Fourth, the section on *clinical issues* considers ethical concerns from the perspectives of screening and therapy in genetics. The final section offers an *epilogue* that relates ethics discourse on human genetics to the emerging policy debate on embryonic stem-cell research and therapeutic cloning.

The organization of these five sections begins with context issues in Section I in order to understand the theoretical issues in Section II; building on the context and theoretical issues, Sections III, IV, and V deal with the practical, clinical, and policy issues that are emerging as critical concerns in human genetics. A brief summary of the chapters in each section may help the reader to understand how the interdisciplinary collection holds together.

Section I, on *context issues*, offers three chapters on ethical concerns in genetics from the perspectives of law and of business. The first chapter is by Professor Ellen Wright Clayton. Her chapter on "Policy Challenges" discusses the ethical, legal, and social implications of genetics. She argues that because genetics tends to affect almost every aspect of our lives, policy has an important influence on the translation of genetic science and technology into the practice of our lives, especially from a perspective that upholds equality. Hence she discusses some of the challenges that policy development will encounter, including in the arena of reproduction and the arena of disease diagnosis and prediction. The next chapter is by Professor Sheila McLean. Her essay on "The Genetics Revolution" discusses whether the law can cope by considering the connection between law and genetics. She analyzes why we should regulate at all by considering the need to develop a more robust view of privacy, especially genetic privacy in an environment that increasingly places confidentiality and genetic information in tension. The new genetics, which emphasizes how interrelated we are, may cause us to revise some of the basic principles assumed by Western society, such as a view of privacy that will fit more comfortably with responsibilities to and for others.

The third chapter in this section is by Professor Gerard Magill. His chapter on "Genetics, Investment, and Business" discusses an approach to organizational ethics for genomics companies. He considers the volatile partnership between medicine and money in genetics by analyzing the accountability of genomics companies to community stewardship of resources and genetics outcomes. This reciprocity can enable genomics companies to realize their corporate integrity by holding in tension the related constituencies of health outcomes, shareholder investment, and business patents.

Section II, on *theoretical issues*, offers four chapters on ethical concerns in genetics from the perspectives of culture and dignity. The first chapter in this section is by Henk A. M. J. ten Have. His chapter on "Geneticization" discusses the sociocultural impact of genetic technology. He considers the original meaning of "geneticization" (as describing interactions between medicine, genetics, society, and culture) in order to reintroduce into ethical discourse forgotten, neglected, or discarded issues in genetics. For example, just as the "medicalization" debate moved the locus of medical power from the physician alone to include the patient, the "geneticization" debate seeks a broad sociocultural terrain for ethics discourse on genetics, perhaps contributing to developing the concept of genetic civilization.

The next chapter is by Professor Mary Briody Mahowald. Her chapter on "Genes and Gender" undertakes an egalitarian analysis that promotes gender equality as a subset of justice in the clinical setting of genetics. In light of gender differences in genetics, based upon empirical differences in the impact of genetics both biological and psychosocial, she applies a view of gender equality to a specific case of misattributed paternity to argue for greater consumer participation in decision-making processes about clinical genetics. This call for participation, especially regarding disadvantaged groups that are significantly affected by advances in genetics, is a means for promoting justice for everyone.

The third chapter is by Jan Heller. His chapter on "Using Human Dignity to Constrain Genetic Research and Development" considers whether the concept of human dignity can guide genetic engineering. He argues that any genetic application that can be used to create human beings undermines the traditionally recognized constraint of human dignity for such applications. He suggests that one legitimate classification of humans, which he refers to as contingent future persons, includes those whose future existence depends on genetic applications yet who cannot be said to have dignity. Although contingent future persons have no dignity to which we can appeal in evaluating genetic applications, we can appeal to the dignity of those persons who will be affected by choices to bring or not to bring contingent future persons into existence.

The last chapter in this section is by Brent Waters. His chapter on "Engineering Our Grace" discusses an old theological idea on human destiny and the need for divine grace in the context of new genetic technologies. If genetic technology leads us to believe we are masters of our destiny and engineers of our grace, we may in fact become more adept at engineering our sin, for example, if children with congenital defects become more stigmatized or suffering becomes increasingly inexplicable. Hence, in the realm of new genetic technologies, as we increase our focus upon safety, efficacy, and access, we must also continue to treasure life as a gift that we hold in trust.

Section III, on *practical issues*, offers four chapters on ethical concerns from the perspectives of the impact and challenge of genetics. The first chapter in this section is by David Schlessinger. His chapter on "The Impact of Genetics and Genomics on Views of Aging" considers a basic conundrum for the genetics of aging: we must wait for several decades to have access to the DNA of successive generations if we are to adopt a traditional approach for such studies in genetics. To resolve this problem, scientists are working on the links between aging and embryology, such as on the shift from "immortal" (totipotent) cells to "mortal" cells, or on tissue formation and regeneration.

The second chapter is by Ruth Purtilo. Her chapter on "Genetic Labels and Long-Term-Care Policies" considers whether these have the impact of being a winning or losing proposition for patients. She argues that because of increasing information about the genetic component in chronic diseases, numerous chronic conditions that were typically met via long-term-care services are now being discussed solely as genetic diseases. She discusses the ethical issues aligned to the use of labels in this situation and recommends resisting changes that may demean the affected person or weaken the patient's relation with the health care setting. The simple shift of category between two less-than-perfect labels raises problems of justice and possible compromises of the ethical standards of health care professionals.

The third chapter in this section is by Dennis Brodeur. His chapter discusses "The Challenges of Advances in Genetic Medicine" for both biomedical and organizational ethics in health care delivery. As an ethics practitioner in a major health system, he considers the ethical challenges that health systems will encounter with new genetic developments, including the shift from a model of infectious disease to a molecular model and recognition of the multifactorial aspect of disease by enhancing the relation between health care delivery and public health.

The final chapter in this section on practical issues is by Carol Bayley. Her chapter on "The Challenge of Physician Education in Genetics" discusses the contribution of attitude, knowledge, and skills to this enterprise. She argues that defensive attitudes of physicians contribute to their downplaying the significance of genetic discoveries, that the increased learning curve in mastering the new knowledge of genetics can be daunting for physicians, and that new skills, such as nondirectiveness and ethical sensitivity to patient needs, are crucial for physicians in our new genetic environment.

Section IV, on *clinical issues*, offers three chapters on ethical concerns from the perspectives of screening and therapy in genetics. The first chapter is by Professor Ruth Chadwick. Her chapter on "Pharmacogenetics, Genetic Screening, and Health Care" discusses the importance of access to information and informed consent in introducing social and political dimensions to the ethical debate. She develops her argument by highlighting the ways in which the ethics of new health technologies such as genetics requires attending to research on bioethics. Focusing her argument upon pharmacogenetic developments, she explains that we need to reconsider ethical principles in this new arena of genetics, with informed consent and genetic screening being obvious examples.

The next chapter in this section on clinical issues is by Norman M. Ford. His chapter on the "Ethical Aspects of Prenatal Screening and Diagnosis" undertakes an analysis of current testing procedures. In his ethical evaluation he considers respect for the life of the fetus, informed consent and counseling, the link between prenatal diagnosis and selective abortion, responsible parenthood, confidentiality and the common good, and sensitivity toward people with disabilities.

The last chapter in this section on clinical issues is by M. Therese Lysaught. Her chapter on "Gene Therapy" considers a test case for research with children. Given the increasing public pressure to enroll children as subjects in clinical trials (and in the wake of the first reported gene therapy death of a patient in 1999), significant ethical issues need attention. She considers the relation between relevant federal and professional guidelines, focusing upon an HIV protocol involving children in human gene transfer research.

Section V, the *epilogue*, offers a final chapter by Gerard Magill. His chapter on "Science, Ethics, and Policy" relates emerging research on human genomics to embryonic stem-cell research and therapeutic cloning. He analyzes the policy incongruity in the United States whereby respect for human life appears to yield bipolar results in human genomics and embryonic stem-cell research. He concludes by arguing for a coherent policy resolution of this incongruity to protect society's interests.

Finally, to assist and encourage further research and study on genetics and ethics, there is a select bibliography and detailed subject and name indices.

Many support staff have assisted in the preparation of this book, and I offer them all heartfelt gratitude. For endless hours researching the bibliography and citations in the book, my thanks to Carmen Schmidt, Alison Miller, and Ethan Brewster as undergraduate students, to John Lomperis, who dedicated a summer's undergraduate internship to the project, and to Alvenio Mozol, a graduate student in the Ph.D. program at our Center for Health Care Ethics. Also, I thank Carol Murphy at Saint Louis University Press for shepherding this project along. Above all, I extend my sincere gratitude to Sue Brewster, my administrative assistant, whose tireless support facilitated the time and energy needed for me to complete this work.

Gerard Magill
St. Louis, 2003

1

Introduction: The Development of Ethics Discourse on Genetics

Gerard Magill, Ph.D. and John Brehany, Ph.D.

THE DAWN of a new millennium brought us a map of the human genome, a multibillion-dollar, international, decade-long effort to sequence the entirety of the human genome. Over the past decade, ethics discourse on genetics has been increasing dramatically as scholars anticipated this milestone. Despite the media hype, it has been difficult to exaggerate the significance of the event. The possibilities for breakthroughs in biotechnology and pharmacogenomics astound the imagination. With these high expectations in molecular medicine and genetic research, there is an increasing need to bridge science, ethics, and policy.

Despite the exposure that genetics has received over the past decade in the media, there remain many concerns about new gene technologies. Of course, there are the buzzword issues, such as stem-cell research and human cloning, that attract considerable time and attention. And there are many exaggerated issues that spin extreme reports ranging from draconian politics (such as eugenics) to doomsday predictions (such as evolutionary genetics). In reality, there is an abundance of serious scholars, from research scientists and clinicians to social scientists and legislators, who seriously engage the ethical discourse that surrounds genetics. This collection of essays seeks to contribute from interdisciplinary and international perspectives to scholarly discourse on ethics and genetics. The Center for Health Care Ethics at Saint Louis University's Health Sciences Center hosted an international conference on Genetics and Ethics. The essays in this collection (except for the introductory and final chapters) are drawn and updated from papers presented at that conference. To set the stage, so to speak, for the subsequent chapters in this collection, this introductory essay provides a summary development of interdisciplinary discourse on genetic research.[1]

History of Genetic Science and Engineering

A genome is the sum total of genetic information contained within the cells of a living being.[2] The human genome (in the nucleus of every cell) is organized in the form of a spiraling double helix: deoxyribonucleic acid (DNA), which resembles a twisting rope ladder with its rungs composed of a series of four bases, adenine, thymine, cytosine, guanine (A, T, C, and G), each of which pairs in a regular manner (A & T; G & C). A gene is a section of DNA containing the instructions for assembling a part of a specific protein or enzyme.[3] Proteins underlie all of a body's structure and functions, ranging from the construction and maintenance of skin, bones, and blood, to producing antibodies to fight infections, to digesting food.[4]

It is hard to believe that the science of genetics is just over a hundred years old. The development of genetic science, and hence ethics discourse about it, occurred through several distinct stages. The first stage of genetic science can be described as classical genetics, a field that encompasses the study of the effects of genes (but not their molecular structure) through patterns of inheritance. Gregor Mendel (1822–1884) is recognized as the founder of classical genetics. His experiments with peas in the 1860s provided science with a basic framework for the laws of heredity. He discovered that our somatic traits are determined by genes (subsequently named as such); these are possessed in pairs and inherited in single copies from parents. He demonstrated that some of these genes can be dominant (when a single gene is present), whereas others can be recessive (where a paired set of genes is required, for example, for a disease to be present).

Over the first half of the twentieth century, Mendel's insights were confirmed, extended, and linked to ongoing investigations of cell biology. T. H. Morgan not only confirmed Mendel's laws, but also showed that chromosomes determined sex and that some genes were carried only on the sex chromosomes. Several of Morgan's students went on to make important contributions. Calvin Bridges showed that chromosomes "cross over" during the formation of gametes, thus creating new genetic combinations, and Alfred Sturtevant constructed in 1913 the first "linkage map" to help trace the presence and approximate location of genes based on a study of patterns of inheritance. Hermann Muller, perhaps Morgan's most famous student, demonstrated in 1927 that radiation introduces mutations into genetic material, making it possible to create and study genetic mutations. Also during this period, key distinctions and terms were clarified, including "genotype" and "phenotype," "heterozygosity" and "homozygosity," and "monogenic" and "polygenic" traits. By 1940, the basic laws of heredity had been worked out, and the second stage of the history of genetics was about to begin.[5] (Incidentally, it is worth noting that dramatic results have emerged recently from the use of DNA to track inheritance).[6]

The second stage of genetic science emerged with the era of molecular genetics, in which the nature, structure, and function of genes were investigated at the molecular level. This era was defined when in 1953 Watson and Crick discovered the structure of DNA (a molecule with the intertwined strands that form a double helix).[7] Watson's and Crick's insight, as momentous as Mendel's, ultimately explained the ability of the genetic code both to self-replicate and to direct protein synthesis. Many discoveries occurred subsequently in the 1950s and 1960s, such as proof of the first chromosome-linked disease (1959) and the relationship between DNA and the amino acid chains that make up proteins (1966).

The third era of genetic science, which built upon the discoveries of molecular genetics, can be described as the era of recombinant DNA (rDNA). While earlier periods dealt with discovering the effects and nature of the genetic code, the era of rDNA was characterized by the development of tools to manipulate DNA. Manipulating genes, whether through copying, sequencing, or inserting them into new organisms, enabled scientists to understand the nature and function of individual genes. Significant tools included the discovery and development of restriction enzymes (naturally occurring DNA cutters that enable scientists to break long strands of DNA into predictable and manageable lengths); recombinant DNA technology (techniques for inserting genes into other organisms, including yeast); sequencing technologies (including the Sanger, Northern blot, and FISH methods of identifying the precise order of nucleotide base pairs of specific strands of DNA); and copying technologies, above all polymerase chain reaction (PCR), a method for quickly making millions of copies of specific fragments of DNA.[8]

Now we find ourselves in the era of the Human Genome Project, which is mapping the molecular landscape for applied genetics. In 1990 the Human Genome Project was initiated by a collaboration between the Wellcome Trust in Britain and the federal government in the United States (the Department of Energy and the National Institutes of Health). To support the Human Genome Project, Congress allocated more than $3 billion over fifteen years to map and sequence the human genome. Other countries also were involved. The government-sponsored project in the United States, led by James D. Watson at first and then by Francis Collins, allocated 3 percent of its funds for the study of the "Ethical, Legal, and Social Implications" (ELSI) of the human genome.[9] The ELSI project was especially concerned with privacy and fairness in using genetic information, with integrating new technologies in clinical environments, and with educational programs on the human genome.[10] More generally, there were clear goals for the Human Genome Project that included the following: mapping the humane genome (and other select organisms); develop mechanisms to analyze this genetic information; and promoting the ELSI program.[11] Because of the success of the Human Genome Project, its completion date for a draft map of the human genome was earlier than anticipated, being presented as a draft on June 26, 2000,[12] and then in its more comprehensive form in February 2001,[13] and finalized in 2003.

In the mapping of the human genome the territory first described by Watson and Crick has been surveyed. Of course, the more we discover about the human genome, the more we realize what we do not know. We now find ourselves on the cusp of an era of applied genetics that will increasingly challenge our models for understanding the human genome as well as previously established approaches to genetic medicine. There are several arenas in which applied genetics will bring significant developments and challenges. One arena is that of diagnosis: there will be an increasing knowledge of genetic causes and predispositions, and the relation between the genotype and the environment will be better understood. Another arena in which applied genetics will have a substantive impact will be the world of pharmaceuticals, especially with regard to the design and testing of individually targeted drug therapies. Yet another and perhaps more controversial arena for applied genetics will be that of genetic engineering—somatic-cell genetic engineering as well as germ-line genetic engineering. Of course, these arenas will need to address the momentous task of integrating new technologies into health care delivery systems.[14] And new biotechnological capabilities that previously seemed impossible are tantalizingly close, such as harnessing DNA to develop computer memories with greater speed and storage capacity than exists currently.[15]

Not surprisingly, regulatory oversight has struggled to keep apace with the unprecedented progress in genetic science and biotechnology. A brief overview of recent policy development in genetic medicine can set a helpful context for understanding the emerging ethics discourse in the field.

History of the Regulatory Debate

The discovery and development of tools to study individual genes and to transfer genes between organisms took place at the end of the 1960s, a time when ethical discourse on the abuses in scientific research was increasing in prominence (as with the Tuskegee syphilis experiments). A regulatory framework for genetic research began at the Asilomar (California) conference for molecular biologists in 1975.[16] The focus of the conference was upon issues related to rDNA, and there was an agreement to restrict experiments with rDNA in order to avoid the creation and release of new organisms.[17] When attorneys spoke at the conference, their discussion of penalties associated with negligence and liability raised the need for appropriate caution. After presenting its consensus statement to the National Institutes of Health, the Recombinant DNA Advisory Committee (RAC), established in 1974, issued guidelines that relied upon peer review for oversight responsibility.[18]

Another threshold for ethics discourse on genetics occurred in the United States during the early 1980s. In 1982 the President's Commission for the Study of Ethical Problems in Medicine and Biomedical and Behavioral Research published its report, *Splicing Life*.[19] The commission invited the Recombinant DNA Advisory Committee to develop an ethical protocol for gene transfer experiments. As a result, in 1984 RAC published "Points to Consider,"[20] emphasizing several basic "points" for genetic research to honor: specifying the medical condition accurately; considering therapeutic alternatives to gene therapy; weighing the potential benefit and harm; selecting research candidates fairly; obtaining consent properly; and maintaining privacy and confidentiality.[21]

Moreover, two important decisions regarding patents and intellectual property were made in the 1980s. In 1980, the Supreme Court upheld the grant of a patent on a genetically engineered microorganism.[22] Then, in 1987, the Patent and Trademark Office ruled that any genetically engineered multicellular organism, including human tissue but not human beings, is potentially patentable.[23] Patents have been applied for and granted on genetically modified organisms and on gene sequences (as long as their function is known), although the legal and ethical validity of these patents are still contested.[24] In parallel to the increasing oversight of genetic engineering by regulatory bodies in the United States, similar developments have occurred in other countries. For example, in Britain the Committee on the Ethics of Gene Therapy, formed in 1989, recommended in 1993 regulating gene therapy by establishing a supervisory body known as the Gene Therapy Advisory Committee (GTAC), working collaboratively with other regulatory agencies.[25]

Contrasting policy positions reflect the regulatory debate on law and ethics for genetic technology.[26] One side argues that the risks of rDNA technology are not inordinate, with biotechnology research promising health benefits and financial rewards. This side treats genetic technology like other technology and therefore seeks to reduce regulations.[27] Another side argues that the human genome is unique and applauds government-industry collaboration and economic competition to influence the policy debate.[28] With such disagreement, it is not surprising to encounter the patchwork nature of regulations on genetic biotechnology.

The patchwork of U.S. regulations on genetics is due in part to the separation of powers that is characteristic of U.S. governance and to the system of federalism in the United States. The effects of federalism and the tensions inherent in it can be seen in the debate over how to regulate the health insurance industry in general and the use of genetic information by that industry in particular.[29] Not surprisingly, aligned closely with concerns about the insurance industry are concerns about the influence of the market upon genetic medicine, including its commercialization and commodification.[30] The current status of U.S. law on genetics results from the fact that legislation at the federal level in the U.S. Congress needs more attention.[31] Occasionally the judiciary has acted, for example by interpreting the Americans with Disabilities Act[32] and the Patent and Trademark Act.[33] Regulation has stemmed largely from the executive branch and federal agencies. In the course of the 1980s and 1990s, various federal agencies claimed oversight of research, testing, application, and marketing of genetics. So these agencies issued and modified many of their policies.[34] The fact that they did not act in concert highlights how much their approaches to regulation and policies can differ.[35] Although many states have passed laws protecting genetic privacy and regulating the use of genetic information, not all states have passed laws nor are all of these laws uniform. There is increasing urgency for adequate regulations to address privacy and confidentiality in the new age of genetic medicine. This patchwork nature of regulation on genetics also occurs internationally.[36]

Alongside the development of science and technology in molecular medicine and accompanying the development of regulatory policy on genetics, a sophisticated body of ethics discourse has emerged. A brief overview of the development of this discourse can shed light on the ethical concerns that are likely to characterize the field of genetics for the foreseeable future.

The Development of Ethics Discourse

The development of ethics discourse on genetics in a sense began during the first part of the twentieth century with the rhetorical imagination of writers such as Haldane and Muller, and especially in Huxley's *Brave New World*, which has never gone out of print.[37] But it was not until after Watson's and Crick's discovery of the structure of DNA that ethics discourse on genetics began in earnest. As the parameters of the field became more established, the early literature was not lacking in controversy, being characterized by what one ethicist described as: "diabolical specters of eugenics and transgenic freaks, threats to human freedom and integrity, and fears of environmental release and uncontrollable plagues."[38] A significant part of this controversial atmosphere was caused by the evident tension between the promise and peril of genetic science. For example, by the end of the 1960s, Joshua Lederberg's optimism (that with enough research funding, "we could do anything we wanted in twenty or thirty years")[39] stood in stark contrast to the call for caution by Marshall Nirenberg, a Nobel prizewinning geneticist (arguing that "when man becomes capable of instructing his own cells, he must refrain from doing so until he has sufficient wisdom to use this knowledge for the benefit of mankind.")[40]

This call for caution in genetic science was developed by many influential voices in genetic ethics. One of the most articulate opponents of radical genetic engineering was Paul Ramsey, who memorably suggested that humans should "not attempt to play God before they had learned how to be men"; and that "after they had learned to be men they would not want to play God."[41] Leon Kass expressed a similar position from a philosophical perspective, arguing that limitless self-modification through genetic engineering might "mark the end of *human* life as we and all humans have known it."[42] And Jeremy Rifkin began his media-savvy opposition to genetic engineering with a book entitled *Who Should Play God* and orchestrated a series of demonstrations and lawsuits to slow the pace of biotechnology development.[43] Not surprisingly, this religious metaphor of "playing God" has continued to recur in the ethics literature on genetics. Examples can be found in the discourse on genetic determinism and human freedom that inquires whether genetic engineering is a form of "playing God" or being a "co-creator" with God,[44] or somehow "connecting God with genetic processes."[45]

After the president's commission published the document *Splicing Life* in 1982 and the RAC published its *Points to Consider* in 1984, gene therapy protocols began to appear. The first gene transfer protocol was submitted in 1988, and the first truly therapeutic human gene transfer protocol took place in 1990. By the turn of the new millennium, there were almost four hundred gene therapy trials in process throughout the world.[46] While there is a near-universal consensus on the propriety of using somatic-cell gene therapy to treat serious diseases, the research difficulties inherent in applying gene transfer technology remain significant.[47] While the scientific complexities cannot be *over*estimated, the financial and other incentives at stake for scientists should not be *under*estimated. Not surprisingly, in the wake of the first death of a gene therapy patient, Jesse Gelsinger, in September 1999, ethical concerns rightly arose about business and economic incentives that might entice researchers to proceed too fast.[48]

At a very basic level, ethics discourse on molecular medicine and genetic biotechnology seeks to differentiate between genetic traits and genetic diseases, a crucial distinction for determining the legitimacy of medical treatment when it is feasible. A helpful fourfold matrix for understanding the emerging discourse on the ethics of genetic therapy is presented by LeRoy Walters and Julie Gage Palmer. Their matrix distinguishes between the two categories of genetic therapy: one category is prevention, treatment, and cure of disease; the other category is enhancement of capabilities or characteristics. These two categories can be interpreted with regard to both somatic-cell gene therapy and germ-line gene therapy.[49] Moreover, ethics discourse also seeks to consider the impact of assessing genetic risks upon social policy for health care in the new world of genetic medicine.[50] At the heart of this discourse is not only the ethical debate on genetic enhancement,[51] including the rhetorical challenge of seeking perfect babies,[52] but also the concern about eugenics that seeks to improve the human gene pool through social policy.[53] These practical concerns are at the forefront of the debate on the new genetics, requiring an insightful combination of relevant ethical principles and emerging medical cases.[54] However, given the pace of developments in molecular genetics, there is a need to caution researchers about rushing to disseminate outcomes. For example, in one study, data suggest that reported applications of genetic research such as genetic testing have deficiencies that might be overcome by closer attention to the reporting principles of epidemiological science.[55]

While there seems to be general agreement that the ethical concerns regarding somatic-cell genetic therapy are akin to the more traditional issues in medical research (consent, burden/benefit, etc.), there is no consensus yet regarding ethical discourse on germ-line gene therapy, whereby genetic changes are passed on to subsequent generations.[56] An influential body of literature on germ-line genetic therapy has been growing since the early 1990s,[57] with many ethicists upholding its validity,[58] based on several perspectives that include the obligation to protect or enhance the health of future generations;[59]

respecting reproductive privacy and autonomy;[60] and the need to influence evolution.[61] Other ethicists present a stance of principled opposition,[62] based on several perspectives that include avoiding experimentation upon human beings without their consent;[63] respect for human nature;[64] and avoiding the danger of eugenics.[65] In the ethical debate about germ-line gene therapy as a form of eugenics,[66] the "slippery slope" argument,[67] as well as the "backdoor" argument,[68] continues to elicit considerable concern, especially with regard to tacit social and political influence.[69]

The new world of biotechnology raises the stakes of old dilemmas that now haunt ethical discourse on genetics. Already, genetic counseling has become a proverbial minefield of ethical challenges as parents seek to grapple with the meaning of *normalcy* and *disability* in the face of genetic risks and probabilities.[70] Some scholars believe we have reached a crucial crossroads in genetic counseling that may lead ethics along a new path ahead.[71] Also, the increasing capacity for genetic testing and screening inevitably raises concerns about employment discrimination, with the concomitant compromise of genetic *privacy*,[72] and persistently draws attention to the need for respect of persons (especially minorities) and accuracy of information.[73] Using genetic information for social purposes will elicit significant controversy both in the policy arena and in normative discourse in ethics—the changing paradigm for the role of insurance in genetic medicine (for example, with regard to Alzheimer's disease) provides a good example of the need for policy and ethics debate.[74] Internationally, the ethical debate on testing and screening has focused upon protecting privacy and freedom while promoting responsibility and counseling.[75] Moreover, the role of prenatal genetic testing for women,[76] whether of fetuses (which can be aborted) or embryos created in the petri dish (which can be discarded), provokes challenging questions about the meaning of *human life* and *deformity*.[77] And as concerns about access, equality, and justice continue to guide ethics discourse generically on these issues,[78] there is an increasing sensitivity to the role of gender justice in genetics, especially from feminist perspectives.[79]

Religious perspectives have also had a broad influence upon the development of ethical discourse on genetics over recent decades. One of the first religious conferences on genetics was sponsored by Gustavus Adolphus College in 1965.[80] Paul Ramsey gave a lecture at this conference which was later reprinted as the first chapter in his famous book, *Fabricated Man*. Long before the government-funded Human Genome Project began, influential religious scholars initiated a dynamic debate on genetics and ethics in both the Protestant[81] and the Catholic traditions.[82] In the 1990s, the government-funded ELSI research program on the human genome generated significant religious interest in genetics and ethics.[83]

Increasingly, religious denominational teachings have been issued to provide ethical guidelines for genetic biotechnology,[84] including advocacy to ban new developments (such as the call for a ban on gene patenting in 1995).[85] Many of the Protestant denominations in North America have sponsored documents on genetics issues, including the Episcopal Church, the Presbyterian Church, the Reformed Church in America, the Southern Baptist Convention, the United Church of Christ, the United Methodist Church, and the United Church of Canada.[86] A large number of religious scholars have contributed to the ethical discourse on genetics,[87] with an obvious increase in the 1990s after the beginning of the government-funded Human Genome Project.[88] Religious scholars continue to worry that ethics discourse on genetics can lose sight of the larger picture of human bodiliness and the relations predicated upon it: the result is that subtle threats to human personhood may not be noticed until substantial damage is done.[89] The opportunity and threat of genetic engineering have increased a sense of urgency, not only for religion and policy to engage each other,[90] but also for science and politics to embrace social values and the common good[91] and for policy on genetics to engage a global perspective.[92]

Hence there are many general and practical ethical issues that continue to provoke dynamic discourse about genetics. However, there have also been very specific topics that have caught the imagination of the media as well as the attention of ethics scholars. Obvious examples are cloning and embryonic stem-cell research. Ethical discourse on cloning began seriously after the successful cloning of a mammal, the now-famous Dolly (which died in February 2003), by somatic-cell nuclear transfer at the Roslin Institute in Edinburgh, Scotland, in the spring of 1997. Since then, the development of cloning techniques for other animals has abounded.[93] Predictably, this sparked a debate over whether and how this technology would be applied to humans[94] — a topic that is pursued further in the final chapter (the epilogue) in this collection.[95]

In conclusion, this introductory chapter has provided a brief overview of the current landscape in genetic science and biotechnology, of the regulatory debate that surrounds this exciting terrain, and of the development of ethics discourse on genetics over recent decades.

NOTES

1. For a complementary account, see the section on human genomics in Gerard Magill, "The Ethics Weave in Human Genomics, Embryonic Stem Cell Research, and Therapeutic Cloning: Promoting and Protecting Society's Interests," *Albany Law Review* 65:3 (2002):701–28, at 704–16.

2. See D. Peter Snustad, Michael J. Simmons, and John B. Jenkins, *Principles of Genetics* (New York: John Wiley and Sons, 1997), 180–379. For a clear introduction for the layperson to the function of DNA and genes, see LeRoy Walters and Julie Gage Palmer, *The Ethics of Human Gene Therapy* (New York: Oxford University Press, 1997), Chap. 1.

3. See Snustad et al., *Principles of Genetics*, 349–62, for the history of definitions of genes.

4. Department of Energy, Human Genome Program, *Primer on Molecular Genetics* (Washington, D.C.: U.S. Department of Energy 1992), 7.

5. For historical materials, see Colin Tudge, *The Engineer in the Garden: Genes and Genetics. From the Idea of Heredity to the Creation of Life* (New York: Hill and Wang, 1993); Robert P. Wagner, "Understanding Inheritance: An Introduction to Classical and Molecular Genetics," *Los Alamos Science* 20 (1992).

6. Nicholas Wade, "DNA Backs a Tribe's Tradition of Early Descent from the Jews," *New York Times* (May 9, 1999):A1 (referring to the Lemba, a Bantu-speaking people of southern Africa).

7. J. D. Watson and F. H. C. Crick, "Molecular Structure of Nucleic Acids: A Structure for Deoxyribose Nucleic Acid," *Nature* 171 (1953):737–38; also see James D. Watson, "The Human Genome Project: Past, Present, and Future," *Science* 248 (1990):44–48.

8. Wagner, "Understanding Inheritance," 38–39, 44, 49–63. See also Snustad, Simmons, and Jenkins, *Principles of Genetics*, 462–500.

9. Eric M. Meslin, Elizabeth J. Thompson, and Joy T. Boyer, "The Ethical, Legal, and Social Implications Research Program at the National Genome Research Institute," *Kennedy Institute of Ethics Journal* 7:3 (1997):291–98.

10. See Francis Collins, "Shattuck Lecture—Medical and Societal Consequences of the Human Genome Project," *New England Journal of Medicine* 341:1 (1999):28–37. Also see the chapter on "Social, Legal, and Ethical Implications of Genetic Testing" in Institute of Medicine, *Assessing Genetic Risks. Implications for Health and Social Policy* (Washington, D.C.: National Academy Press, 1994), 247–89.

11. Fadi F. Haddad, et al., "The Human Genome Project: A Dream Becoming a Reality," *Surgery* 125:6 (1999):575–80.

12. Arielle Emmett, "The Human Genome," *The Scientist* 14:15 (2000):1, 17–19.

13. Francis Collins, et al., "Initial Sequencing and Analysis of the Human Genome," *Nature* 409 (2001):860–921; and Craig Venter, et al., "The Sequence of the Human Genome," *Science* 291:5507 (2001):1304–51; N. Wade, "Human Genome is Complete," *New York Times* (April 15, 2003):D1.

14. See Joan Stephenson, "As Discoveries Unfold, a New Urgency to Bring Genetic Literacy to Physicians," *Journal of the American Medical Association* 278:15 (1997):1225–26; Francis S. Collins, "Preparing Health Professionals for the Genetic Revolution," *Journal of the American Medical Association* 278:15 (1997):1285–86; Joan Stephenson, "Group Drafts Core Curriculum for 'What Docs Need to Know about Genetics,'" *Journal of the American Medical Association* 279:10 (1998):735–36; John Bell, "The New Genetics: The New Genetics in Clinical Practice," *British Medical Journal* 316:7131 (1998):618–20; Ann Louise Kinmonth, et al., "The New Genetics: Implications for Clinical Services in Britain and the United States," *British Medical Journal* 316:7133 (1998):767–70; Thomas H. Murray, Mark A. Rothstein, and Robert J. Murray, Jr., eds., *The Human Genome Project and the Future of Health Care* (Bloomington and Indianapolis, Ind.: Indiana University Press, 1996).

15. Andrew Pollack, "Researchers Harness DNA for Tiny Motors that Could Widen the Use of Genetic Code," *New York Times* (August 10, 2000):C5.

16. Robert Pollack, *Signs of Life: The Language and Meanings of DNA* (Boston: Houghton Mifflin, 1994).

17. See Robert S. Schwartz, MD, Editorial Comment, *New England Journal of Medicine* (1999):244.

18. Clifford Grobstein, *A Double Image of the Double Helix: The Recombinant-DNA Debate* (San Francisco: W. H. Freeman, 1979), 24–25, 28–29; 33–34, 65–73.

19. President's Commission for the Study of Ethical Problems in Medicine and Biomedical and Behavioral Research, *Splicing Life: The Social and Ethical Issues of Genetic Engineering with Human Beings* (Washington, D.C.: U.S. Government Printing Office, 1982).

20. See Donald S. Fredrickson, *The Recombinant DNA Controversy, A Memoir: Science, Politics, and Public Interest, 1974–1981* (Washington, D.C.: ASM Press, 2001).

21. Walters and Palmer, *The Ethics of Gene Therapy*, chap. 3; LeRoy Walters, "Human Gene Therapy: Ethics and Public Policy," *Human Gene Therapy* 2 (1991):115–22. See also Eric J. Juengst, "The NIH 'Points to Consider' and the Limits of Human Gene Therapy," *Human Gene Therapy* 1 (1990):425–33.

22. *Diamond v. Chakrabarty*, 447 U.S. 303 (1980).

23. U.S. Patent and Trademark Office, *Animals—Patentability* (Washington, D.C.: U.S. Government Printing Office, 1987).

24. See, for example, the articles in *Cambridge Quarterly of Healthcare Ethics* 7:4 (1998):417–35; Andrew Pollack, "Genentech Trial on Patent Ends with Jury Deadlocked," *New York Times* (June 3, 1999):C10.

25. See N. C. Nevin, "Experience of Gene Therapy in the United Kingdom," *Annals of the New York Academy of Sciences* 862 (1998):184–87.

26. One of the most influential early works was by George J. Annas and Elias Sherman, *Gene Mapping: Using Law and Ethics as Guides* (New York: Oxford University Press, 1992).

27. See, for example, Henry I. Miller, *Policy Controversy in Biotechnology: An Insider's View* (Austin, Tex.: R.G. Landes Co. 1997).

28. See Susan Wright, *Molecular Politics: Developing American and British Regulatory Policy for Genetic Engineering 1972–1982* (Chicago: University of Chicago Press, 1994); Wright, "The Social Warp of Science: Writing the History of Genetic Engineering Policy," *Science, Technology, and Human Values* 18:1 (1993):79–101.

29. See Eric Mills Holmes, "Solving the Insurance/Genetic/Fair/Unfair Discrimination Dilemma in Light of the Human Genome Project," *Kentucky Law Journal* 85 (1996–1997):503; Jennifer M. Jendusa, "Pandora's Box Exposed: Untangling the Web of the Double Helix in Light of Insurance and Managed Care," *DePaul Law Review* 49 (1999):161.

30. For example, see eight helpful chapters on the "Commercialization of Genetic Information" in *Genetic Information: Acquisition, Access, and Control*, ed. Alison K. Thompson and Ruth F. Chadwick (New York: Kluwer, 1999), 85–174.

31. Phillip R. Reilly, "Laws to Regulate the Use of Genetic Information," in *Genetic Secrets: Protecting Privacy and Confidentiality in the Genetic Era*, ed. Mark A. Rothstein (New Haven, Conn.: Yale University Press, 1997), 369–91.

32. Brian R. Gin, "Genetic Discrimination, Huntington's Disease and the Americans with Disabilities Act," *Columbia Law Review* 97 (1997):1406.

33. U.S. Patent and Trademark Office, *Animals—Patentability*. Also see the discussion in Margaret Clark, "This Little Piggy Went to Market: Xenotransplantation," *Journal of Law, Medicine, & Ethics* 27:2 (1999):143–44, nn.82–91.

34. For a detailed explanation of the historical, legal, scientific, and political issues involved in this regulation, see Henry I. Miller, *Policy Controversy in Biotechnology*, 79–154.

35. Michael J. Malinowski and Maureen A. O'Rourke, "A False Start? The Impact of Federal Policy on the Genotechnology Industry," *Yale Journal on Regulation* 13 (1996):163; Miller, *Policy Controversy in Biotechnology*.

36. Rothstein, *Genetic Secrets*; Wright, *Molecular Politics*; and Phillip R. Reilly, "Laws to Regulate the Use of Genetic Information," 369–91.

37. Aldous Huxley, *Brave New World* (London: Chatto and Windus, 1932); J. B. S. Haldane, *Daedalus: Or, Science and the Future* (New York: E.P. Dutton, 1924); Hermann J. Muller, *Out of the Night: A Biologist's View of the Future* (New York: Vanguard Press, 1935). For a recent study that exploits Huxley's title in the modern context of genetic science, see Bryan Appleyard, *Brave New Worlds: Staying Human in the Genetic Future* (New York: Viking, 1998).

38. M. Therese Lysaught, "Commentary: Reconstructing Genetic Research as Research," *Journal of Law, Medicine, & Ethics* 26 (1998):48–54, at 51.

39. Joshua Lederberg, "Hearings, Commission on Government Operations, U.S. Senate, Joint Res. 145, Establishment of a National Commission on Health, Safety, and Society" (Mar. 7–28, Apr. 2, 1968), 67, cited in Joseph Fletcher, *The Ethics of Genetic Control: Ending Reproductive Roulette* (Garden City, N.Y.: Anchor Books, 1974), 23 n. 24.

40. Marshall Nirenberg, "Will Society Be Prepared?" *Science* 157 (1967):633.

41. Paul Ramsey, *Fabricated Man. The Ethics of Genetic Control* (New Haven, Conn.: Yale University Press, 1970), 138, 151.

42. Leon Kass, "New Beginnings in Human Life," in *The New Genetics and the Future of Man*, ed. M. Hamilton (Grand Rapids, Eerdmans, 1972), 61.

43. Ted Howard and Jeremy Rifkin, *Who Should Play God? The Artificial Creation of Life and What It Means for the Future of the Human Race* (New York: Delacorte Press, 1977).

44. See, for example, Ted Peters, *Playing God? Genetic Determinism and Human Freedom* (New York: Routledge, 1997); also see a much earlier debate in President Carter's commission in the late 1970s that inquired into genetics and was subsequently published as the President's Commission, *Splicing Life*.

45. Ronald Cole-Turner and Brent Waters, *Pastoral Genetics: Theology and Care at the Beginning of Life* (Cleveland, Ohio: Pilgrim Press, 1996), 71.

46. Theodore Friedmann, "Milestones and Events in the Early Development of Human Gene Therapy," *Molecular Genetic Medicine* 3 (1993):1–32.

47. Larry R. Churchill, et al., "Genetic Research as Therapy: Implications of 'Gene Therapy' for Informed Consent," *Journal of Law, Medicine, & Ethics* 26 (1998):38–47; M. Therese Lysaught, "Commentary: Reconstructing Genetic Research as Research," 48–54.

48. See Rick Weiss and Deborah Nelson, "Gene Therapy's Troubling Crossroads: A Death Raises Questions of Ethics, Profit, Science," *Washington Post* (Dec. 31, 1999):A3; LeRoy Walters, "The Oversight of Human Gene Transfer Research," *Kennedy Institute of Ethics Journal* 10 (2000):171–74; Joan Stephenson, "Studies Illuminate Cause of Fatal Reaction in Gene-Therapy Trial," *Journal of the American Medical Association* 285 (2001):2570.

49. Walters and Palmer, *The Ethics of Human Gene Therapy*; chap. 2 addresses somatic cells, chap. 3 addresses germ-line cells, and chap. 4 addresses enhancement for both somatic and germ-line cells.

50. See Institute of Medicine, *Assessing Genetic Risks*.

51. See, for example, Walters and Palmer, *The Ethics of Human Gene Therapy*; E. T. Juengst, "Can Enhancement Be Distinguished from Prevention in Genetic Medicine?" *Journal of Medicine and Philosophy* 22:2 (1997):125–42; Roger L. Shinn, *The New Genetics: Challenges for Science, Faith, and Politics* (London: Moyer Bell, 1996); Hastings Center, "Genetic Grammar: 'Health,' 'Illness,' and the Human Genome Project," *Hastings Center Report* 22 (1992); David Suzuki and Peter Knudtson, *Genethics: The Ethics of Genetic Engineering* (Cambridge, Mass.: Harvard University Press, 1990).

52. See, for example, Glenn McGee, *The Perfect Baby: Parenthood in the New World of Cloning and Genetics* (New York: Rowman and Littlefield, 1997).

53. See, for example, Ted Peters, ed., *Genetics: Issues of Social Justice* (Cleveland, Ill.: Pilgrim Press, 1998); Daniel Wikler, "Can We Learn from Eugenics?" in *Genetic Information*, ed. Thompson and Chadwick, 1–16.

54. For a helpful example of a study that addresses the combination of ethical principles with emerging cases in genetic medicine, see Bernard Gert, et al., *Morality and the New Genetics* (Boston: Jones and Bartlett, 1996).

55. Sidney T. Bogardus Jr., John Concato, and Alvan R. Feinstein, "Clinical Epidemiological Quality in Molecular Genetic Research: The Need for Methodological Standards," *Journal of the American Medical Association* 281:20 (1999):1919–26.

56. In the United States, in contrast to other industrialized nations, germ-line genetic therapy (GLGE) is not forbidden. Rather, the RAC, since 1990, has refused to consider any proposals for its use. RAC, "Points to Consider," Fed. Reg. 7444 (1990).

57. John C. Fletcher, "Evolution of Ethical Debate about Human Gene Therapy," *Human Gene Therapy* 1 (1990):55–68; John C. Fletcher and W. French Anderson, "Germ-Line Gene Therapy: A New Stage of the Debate," *Law, Medicine, and Healthcare* 20:1–2 (1992):27–39; Eric T. Juengst, "Germ-Line Gene Therapy: Back to Basics," *Journal of Medicine and Philosophy* 19 (1991):587–92.

58. H. Tristam Engelhardt Jr., "Persons and Humans: Refashioning Ourselves in a Better Image and Likeness," *Zygon* 19:3 (1984):281–96; Engelhardt, "Human Nature Technologically Revisited," *Social Philosophy and Policy* 8:1 (1990):180–91; Engelhardt, "Germ-Line Genetic Engineering and Moral Diversity: Moral Controversies in a Post-Christian World," *Social Philosophy and Policy* 14 (1996):47–62; Heta Hayry, "How to Assess the Consequences of Genetic Engineering," in *Ethics and Biotechnology*, ed. Anthony Dyson and John Harris (New York: Routledge, 1994), 144–56.

59. John Harris, "Is Genetic Engineering a Form of Eugenics?" *Bioethics* 7:2/3 (1993):178–87; Harris, *Clones, Genes and Immortality: Ethics and the Genetic Revolution* (Oxford: Oxford University Press, 1998); Bernard D. Davis, "Germ-Line Therapy: Evolutionary and Moral Considerations," *Human Gene Therapy* 3 (1992):361–63.

60. Robert M. Cook-Deegan, "Human Gene Therapy and Congress," *Human Gene Therapy* 1 (1990):163–70; Cook-Deegan, "Germ-Line Gene Therapy: Keep the Window Open a Crack," *Politics and the Life Sciences* 13:2 (1994):217–20; William Gardner, "Can Human Genetic Enhancement Be Prohibited?" *Journal of Medicine and Philosophy* 20:1 (1995):65–84.

61. Bernard Haering, *Ethics of Manipulation* (New York: Seabury Press, 1975); Fletcher, *The Ethics of Genetic Control*; Williard Gaylin, "Fooling with Mother Nature," *Hastings Center Report* (1990):17–21.

62. Andrea L. Bonnicksen, "National and International Approaches to Human Germ-Line Gene Therapy," *Politics and the Life Sciences* 13:1 (1994):39–49; Maurice de Wachter, "Ethical Aspects of Human Germ-Line Gene Therapy," *Bioethics* 7 (1993):166–77 and "The European Convention on Bioethics," *Hastings Center Report* 27:1 (January 1997):13–23.

63. Council for Responsible Genetics, Human Genetics Committee, "Position Paper on Human Germ Line Manipulation," *Human Gene Therapy* 4 (1993):35–37.

64. Hans Jonas, *The Imperative of Responsibility: In Search of an Ethics for the Technological Age* (Chicago: University of Chicago Press, 1984); Leon R. Kass, *Towards a More Natural Science* (New York: The Free Press, 1985); Enquete Commission, "A Report from Germany—An Extract from *Prospects and Risks of Gene Technology: The Report of the Enquete Commission to the Bundestag of the Federal Republic of Germany*," *Bioethics* 2 (1988):256-63; P. S. Greenspan, "Free Will and the Genome Project," *Philosophy and Public Affairs* 22:1 (1993):31-43; Daniel Callahan, "The Moral Career of Genetic Engineering," *Hastings Center Report* 9 (1979):9; Callahan, "Manipulating Human Life: Is There No End to It?" in *Medicine Unbound: The Human Body and the Limits of Medical Intervention*, eds. Robert H. Blank and Andrea L. Bonnicksen (New York: Columbia University Press, 1994), 18-31; "Can Nature Serve as a Moral Guide?" *Hastings Center Report* (Nov.-Dec. 1996):21-22.

65. Edward M. Berger and Bernard M. Gert, "Genetic Disorders and the Ethical Status of Germ-line Gene Therapy," *Journal of Medicine and Philosophy* 19 (1991):667-83.

66. John Harris, "Is Gene Therapy a Form of Eugenics?"178-87.

67. Emmanuel Agius, "Germ-Line Cells: Our Responsibilities for Future Generations," in *Ethics in the Natural Sciences*, ed. Dietmar Mieth and Jacques Pohier (Edinburgh: T. and T. Clark Ltd., 1989).

68. Troy Duster, *Backdoor to Eugenics* (New York: Routledge, 1990).

69. Diane B. Paul, "Eugenic Anxieties, Social Realities and Political Choices," in *Are Genes Us? The Social Consequences of the New Genetics*, ed. Carl F. Cranor (New Brunswick, N.J.: Rutgers University Press, 1994), 142-54; Paul, "Is Human Genetics Disguised Eugenics?" in *Genes and Human Self-Knowledge: Historical and Philosophical Reflections on Modern Genetics*, ed. Robert F. Weir, Susan C. Lawrence and Evan Faces (Iowa City: University of Iowa Press, 1994), 67-83; Paul, *Controlling Human Heredity, 1865 to the Present* (Atlantic Highlands, N.J.: Humanities Press, 1995).

70. An influential work in the early 1990s was by D. M. Bartels, B. S. LeRoy, and A. L. Caplan, *Prescribing Our Future: Ethical Challenges in Genetic Counseling* (New York: De Gruyter, 1993); Maureen Junker-Kenny and Lisa Sowle Cahill, *The Ethics of Genetic Engineering*, 2 (1998) *Concilium* (Maryknoll, NY: Orbis Books, 1998); Helga Kuhse, "Preventing Genetic Impairments: Does It Discriminate against People with Disabilities?" in Thompson and Chadwick, *Genetic Information* (1999), 17-30.

71. Glenn McGee and Monica Arruda, "A Crossroads in Genetic Counseling and Ethics," *Cambridge Quarterly in Health Care Ethics* 7:1 (1998):97-100.

72. For example, John F. Kilner, et al., *Genetic Ethics: Do the Ends Justify the Genes?* (Grand Rapids, Mich.: Eerdmans, 1997); Rothstein, *Genetic Secrets*; Peters, *Genetics: Issues of Social Justice*; American Society of Human Genetics, "ASHG Statement: Professional Disclosure of Familial Genetic Information," *American Journal of Human Genetics* 62:2 (1998):474–83.

73. For example, Dorothy C. Wertz and Robin Gregg, "Genetics Services in a Social, Ethical and Policy Context: A Collaboration between Consumers and Providers," *Journal of Medical Ethics* 26 (2000):261–65; also see Dorothy C. Wertz, "Society and the Not-So-New Genetics: What Are We Afraid Of? Some Predictions from a Social Scientist," *Journal of Contemporary Health Law and Policy* 13 (1997):299–346; Wertz, "The Difficulty of Recruiting Minorities to Studies of Ethics and Values in Genetics," *Community Genetics* 1 (1998):175–79.

74. For the policy debate, see British Medical Association, *Human Genetics: Choice and Responsibility* (New York: Oxford University Press, 1998), especially chaps. 1 and 8; for the ethics debate, see the doctoral dissertation of R. H. M. V. Hoedemaekers, *Normative Determinants of Genetic Screening and Testing: An Examination of Values, Concepts and Processes Influencing the Moral Debate* (Nijmegen, Netherlands: University of Nijmegen, 1998); for the impact of the debate upon insurance, see five valuable chapters on "Genetics and Insurance" in Thompson and Chadwick, *Genetic Information* (1999), 31–84. For an application of the debate to Alzheimer's disease, see Stephen G. Post and Peter J. Whitehouse, eds., *Genetic Testing for Alzheimer's Disease: Ethical and Clinical Issues* (Baltimore, Md.: John Hopkins University Press, 1998).

75. For example, President's Commission for the Study of Ethical Problems in Medicine and Biomedical and Behavioral Research, *Screening and Counseling for Genetic Conditions* (Washington D.C.: Government Printing Office, 1983); Nuffield Council on Bioethics, *Genetic Screening: Ethical Issues* (London: Nuffield Council, 1993); British Medical Association, *Human Genetics*; World Health Organization, *Proposed International Guidelines on Ethical Issues in Medical Genetics and Genetic Services* (Geneva, Switzerland: WHO, 1998); Ruth Chadwick, Henk A. M. J. ten Have, et al., "Genetic Screening and Ethics: European Perspectives," *Journal of Medicine and Philosophy* 23:3 (1998):255–73; Ruth Chadwick, Henk A. M. J. ten Have, et al., *The Ethics of Genetic Screening* (New York: Kluwer, 1999).

76. For example, Karen H. Rothenberg and Elizabeth J. Thompson, eds., *Women and Prenatal Testing: Facing the Challenges of Genetic Technology* (Columbus, Ohio: Ohio State University Press, 1994).

77. See, for example, Roger A. Willer, ed., *Genetic Testing and Screening: Critical Engagement at the Intersection of Faith and Science* (Minneapolis, Minn.: Kirk House, 1998).

78. Maxwell J. Mehlman and Jeffrey R. Botkin, *Access to the Genome: The Challenge of Equality* (Washington, D.C.: Georgetown University Press, 1998).

79. Mary Briody Mahowald, *Genes, Women, and Equality* (New York: Oxford University Press, 2000).

80. John D. Rolansky, ed., *Genetics and the Future of Man: A Discussion of the Nobel Conference Organized by Gustavus Adolphus College, St. Peter, Minnesota 1965* (New York: Appleton-Century-Crofts, 1966).

81. For example, Ramsey, *Fabricated Man*; Joseph Fletcher, "Ethical Aspects of Genetic Controls: Designed Genetic Changes in Man," *New England Journal of Medicine* 285:14 (1971):776–83; Fletcher, "Indicators of Humanhood: A Tentative Profile of Man," *Hastings Center Report* 2:5 (1972):1–4; Fletcher, "Medicine and the Nature of Man," in *The Teaching of Medical Ethics* (Hastings-on-Hudson, N.Y.: Hastings Center, 1973), 47–58; Fletcher, *The Ethics of Genetic Control*.

82. For example, Karl Rahner, "The Experiment with Man," in *Theological Investigations IX*, Graham Harrison trans. (New York: Herder and Herder, 1972), 205–24; "The Problem of Genetic Manipulation," in *Theological Investigations IX*, Graham Harrison trans. (New York: Herder and Herder, 1972), 225–51; Bernard Haering, *Ethics of Manipulation* (New York: Seabury Press, 1975).

83. See Cynthia S. W. Crysdale, "Christian Responses to the Human Genome Project," *Religious Studies Review* 26:3 (2000):236–42, at 237. For a recent review of religious literature on reproductive technology on genetic medicine, see A. R. Chapman, "Ethics and Human Genetics," *Annual of the Society of Christian Ethics* 18 (1998):293–303. Also see Peters, *Genetics: Issues of Social Justice*; and J. Robert Nelson, *On the New Frontiers of Genetics and Religion* (Grand Rapids, Mich.: Eerdmans 1994).

84. For example, Catholic teaching can be found in many of the statements, such as: Pope John Paul II, "Biological Research and Human Dignity," *Origins* 12 (1982):342–43; "Ethics of Genetic Manipulation," (Address by Pope John Paul II to Members of the World Medical Association) *Origins* 13 (1983):385; 387–89; Congregation for the Doctrine of the Faith, *Instruction on Respect for Human Life in Its Origin and on the Dignity of Procreation: Replies to Certain Questions of the Day* (Donum Vitae), Vatican City: Vatican Polyglot Press, 1987.

85. Richard Stone, "Religious Leaders Oppose Patenting Genes and Animals," *Science* 268 (May 26, 1995):1126. See also Mark J. Hanson,

"Religious Voices in Biotechnology: The Case of Gene Patenting," *Hastings Center Report* 27:6 (1997):S1.

86. For a discussion of these documents and statements, see Roger Shinn, "Genetics, Ethics and Theology," in Peters, *Genetics: Issues of Social Justice*, 122-43; Chapman, "Ethics and Human Genetics," 294-96.

87. Some influential examples are: James M. Gustafson, "Genetic Engineering and the Normative View of the Human," in Preston N. Williams ed., *Ethical Issues in Biology and Medicine: Proceedings of a Symposium on the Identity and Dignity of Man* (Cambridge, Mass.: Schenkman Pub. Co., 1973), 46-58; Ronald Cole-Turner, "Is Genetic Engineering Co-Creation?" *Theology Today* 44 (1987):338-49; Cole-Turner, "Genetic Engineering: Our Role in Creation," in John M. Mangum ed., *The New Faith-Science Debate: Probing Cosmology, Technology and Theology* (Minneapolis: Augsburg Press, 1989); Thomas A. Shannon, *What Are They Saying about Genetic Engineering?* (New York: Paulist Press, 1985).

88. Ronald Cole-Turner, "Religion and the Human Genome," *Journal of Religion and Health* 31 (1992):161-73; Cole-Turner *The New Genesis: Theology and the Genetic Revolution* (Westminster, U.K.: John Knox Press, 1993); Cole-Turner and Waters, *Pastoral Genetics*; James M. Gustafson, "Genetic Therapy: Ethical and Religious Reflections," *Journal of Contemporary Health Law and Policy* 8 (1992):183-200; Gustafson, "A Christian Perspective on Genetic Engineering," *Human Gene Therapy* 5 (1994):747-54; Ann Lammers and Ted Peters, "Genethics: Implications of the Human Genome Project," *Christian Century* 107 (1990):868-71; Ted Peters, "'Playing God' and Germline Intervention," *Journal of Medicine and Philosophy* 20 (1995):365-86; Peters, *For the Love of Children: Genetic Technology and the Future of the Family* (Louisville, Ky.: Westminster John Knox Press 1996); Peters, *Playing God?*; Peters, *Genetics: Issues of Social Justice*; Roger Shinn, *The New Genetics*; Shinn, "Genetics, Ethics and Theology: The Ecumenical Discussion," in Peters, *Genetics: Issues of Social Justice*, 122-43; Shinn, "Genetic Research and the Elusive Body," in Lisa S. Cahill and Margaret A. Farley, eds., *Embodiment, Morality and Medicine* (Boston: Kluwer Academic Publishers, 1995), 59-73; Shinn, "Christian Perspectives on the Human Body," *Theological Studies* 55 (1994):330-46; Thomas A. Shannon, "Genetics, Ethics and Theology: The Roman Catholic Discussion," in Peters, *Genetics: Issues of Social Justice*, 144-79; Shannon, "Ethical Issues in Genetics," in *Theological Studies* 60 (1999):111-23; Shannon, *Made in Whose Image? Genetic Engineering and Christian Ethics* (Amherst, N.Y.: Humanity Books, 2000).

89. James F. Keenan, "What Is Morally New in Genetic Manipulation?" *Human Gene Therapy* 1 (1990):289-98; Keenan, "Genetic Research and the

Elusive Body," in L.S. Cahill and M.A. Farley, eds., *Embodiment, Morality and Medicine* (Boston: Kluwer Academic Publishers, 1995), 59–73; Keenan, "Christian Perspectives on the Human Body," *Theological Studies* 55 (1994):330–46.

90. See Shinn, *The New Genetics*.

91. See, for example, Daniel J. Kevles and LeRoy Hood, *The Code of Codes: Scientific and Social Issues in the Human Genome Project* (Boston: Harvard University Press, 1992); Robert Cook-Degan, *The Gene Wars: Science, Politics, and the Human Genome* (Norton, 1994); Lisa Sowle Cahill, "The Genome Project: More than a Medical Milestone," *America* 183:4 (2000):7–13; Walters and Palmer, *The Ethics of Human Gene Therapy*, chap. 5.

92. See, for example, Kilner, *Genetic Ethics*; Junker-Kenny and Cahill, *The Ethics of Genetic Engineering*.

93. Early examples varied from cloning cows (see *New York Times*, December 9, 1998:A1), to embryo splitting of rhesus monkeys (see *New York Times*, January 14, 2000:A13), to cloning pigs for human organ transplants (see Ron Winslow, "Scientists Clone Pigs, Lifting Prospects of Replacement Organs for Humans," *Wall Street Journal*, August 11, 2000:A6).

94. See Gina Kolata, "Human Cloning: Yesterday's Never Is Today's Why Not?" *New York Times* (Dec. 2, 1997); Leon R. Kass and James Q. Wilson, *The Ethics of Human Cloning* (Washington, D.C.: AEI Press, 1998).

95. Gerard Magill, "Science, Ethics, and Policy: Relating Human Genomics with Embryonic Stem-Cell Research and Therapeutic Cloning," Chapter 16 in this volume.

2

Policy Challenges: Ethical, Legal, and Social Implications of Genetics

Ellen Wright Clayton, MD., J.D.

IN THIS COLLECTION, you will read a great deal about the promises and perils of our growing understanding of the contributions of genetics to human health and well-being. How this scientific information will actually affect individuals is influenced by many factors—some obvious ones like the structure of the health care system and the ability of health care professionals to understand this information and incorporate it into their practices, as well as some more general factors such as the individual finances, cultural norms, and religious beliefs. My comments will focus on the role of policy in influencing the translation of genetics into practice, focusing particularly on some challenges that we will face in the process of policy development.

What I mean by policy in this area is the ways in which various potential actors act or fail to act to affect the use of genetic information. Certainly, the most obvious place to find genetics policy is in the laws, regulations, and directives promulgated by our governmental entities. It is important to recognize, however, that these official pronouncements represent only a small part of the picture. One also has to consider those actions that could have been taken but were not, whether because the actor did not think of it or because the actor made a deliberate decision not to act. Many bills dealing with genetics, for example, have been introduced into legislatures but failed to pass; other issues presented by genetics appear never to have come onto the legislative radar screen. Put another way, policy is the product not only of what people and institutions do but also of what they do not do. Policy, in this view, is also the product of the unwritten practices of governmental entities and of the explicit and unspoken actions of numerous actors in society, including third-party payers, health care professionals and institutions, and employers. Part of the difficulty of understanding policy and its development, then, is that one has to consider both action and inaction and that so much of the process occurs outside the public eye.

So what are our policies about genetics? Here one gets different views depending on one's level of focus. If one looks at genetics per se, the following picture emerges. Genetics first emerged on the policy agenda in the first half of this century in the context of eugenics.[1] When we look back, we tend to focus on the Nazi atrocities during World War II and our own history of eugenic sterilization, but we need also to remember that eugenic notions were responsible for many of the restrictive immigration policies adopted around the turn of the last century. Although those days are officially behind us, it is difficult to underestimate the impact of this history and its unsavory sequelae on our current responses to the new genetics.

Genetics and Reproduction

Interestingly, one of the next topics on the policy agenda was again in the area of reproductive genetics, with the development of the ability to detect carriers for Tay-Sachs disease and sickle cell disease in the late sixties and early seventies.[2] Carriers for recessive disorders like these are not sick themselves, but if they have children with other carriers, there is a one-in-four chance with each pregnancy that the child will be affected. The experience with sickle cell carrier testing was particularly telling. Congress enacted a series of laws in the early 1970s providing funds to study and "prevent" a series of inherited diseases. Sickle cell disease was the first, but others, such as Cooley's anemia (now called thalassemia) soon followed. These programs were often actively sought by the affected communities, including, interestingly enough, the Black Panthers, and were sometimes supported more broadly as a way of making up for past wrongs.

During the same period, a number of states, generally in the South, enacted laws requiring carrier screening for sickle cell disease. Here as well, the motives of the legislators were apparently not all bad, but the consequences of these disease-specific laws and programs were largely negative. It was quickly recognized that the public did not understand the inheritance of sickle cell disease and the minimal health impact of being a carrier. As a result, many carriers suffered unjust discrimination. It was also quickly recognized that, at least in the 1970s, finding out that one was a carrier for sickle cell disease was not useful for the individual. Prenatal diagnosis for that disease was not possible at that time, and even had it been technically possible, it probably would not have been readily available in the health care system and probably would not have been acceptable to many members of the population at risk. It did not help that Linus Pauling, a Nobel laureate, recommended that carriers for sickle cell disease be tattooed with an S on their foreheads.[3] It did not help that the basic population at risk— African-Americans—were and still are victims of serious discrimination in our society. The result was that these programs were quickly characterized as eugenic in nature, and these laws were rapidly repealed or at least were no longer enforced. State-mandated carrier screening was widely seen as a flop.

Within a few years, a different trend emerged in which parents who had children with genetic disorders sued their physicians, alleging that had they been informed about their risk, they would have prevented the birth of these children. For the most part, these litigants were successful in their claims, obtaining at least some damages.[4] In this way, litigants effectively "pulled" reproductive genetic testing into use where the "push" of state-mandated screening had failed. To be sure, some states resisted this pull, either because their courts rejected these claims[5] or more often, as in Missouri,[6] because their legislatures passed laws forbidding these claims, but the basic landscape was permanently changed, and reproductive genetic testing became part of routine prenatal care.

Looking, then, just at the fate of publicly mandated carrier screening and at the impact of so-called wrongful birth litigation, one might conclude that the result is that decisions about using genetic information in reproductive planning are largely a matter of personal choice. A different picture comes into focus, however, when one takes a broader view.[7] First of all, many third-party payers do not pay for genetic testing. The most dramatic fact is that almost half of women in this country could not act on the results of reproductive genetic testing even if they got it because they do not have access to abortion. For many, the barrier is the ban on using federal funds to pay for abortion;

for others, the barrier is the distance many women live from centers that provide these services; for still others, the barrier is the fact that health care institutions and professionals refuse, often for religious reasons, to provide these services. I point this out not necessarily to condemn these barriers but to make clear that we cannot accurately describe our policies regarding access to genetics without looking at whether people can actually obtain or use their results.

Another important barrier to access is the fact that health care professionals, individually and collectively, make decisions about the appropriate use of reproductive genetic testing. Most geneticists, for example, refuse to provide prenatal diagnosis and selective abortion for the sole purpose of sex selection[8] or to prevent the birth of children who will have "relatively minor" disorders. Many refuse to use these technologies to detect the presence in fetuses of mutations associated with late-onset disorders such as Huntington's disease or breast and ovarian cancer.[9] Many of these positions and their supporting arguments have been published, where they can be subjected to scrutiny and debate, but the fact remains that the individual clinician decides in any particular case whether or not to offer prenatal diagnosis. Women whose providers say no to their request for services are left to "shop around" to the extent that they can to try to find a willing physician.

A larger view may also lead us to ask why people, and more particularly women, choose to use reproductive genetic testing. Doubtless their reasons are complex, but selfishness is rarely the only motive. At least some of their reasons are the result of broadly based social policies. The fact that it is so difficult in our society to obtain the full range of services needed to optimize the life of a child with special needs is surely one factor. The fact that so many middle- and upper-middle-class women work outside the home and would have a hard time meeting the additional demands of child and medical care of a child with special needs is another. Here, I say middle-and upper-middle-class because poorer women have always engaged in wage work and have always faced these dilemmas, albeit out of the public eye. Another factor is social stigma. I have seen people ask mothers of children with Down syndrome why they did not "do something" to prevent that. The fact that a major cause of divorce in our society is illness in a child is another factor. Major determinants of the distribution of wealth in our society are marriage and men's greater earning power, and a woman may be reluctant to risk having a child with a major disorder for fear of losing her spouse and the poverty that divorce would mean for her and her children. The fact is that life is easier in many ways, particularly for women, if children are healthy.

The timing of reproductive genetic testing raises policy issues as well. Most of these technologies, such as amniocentesis, chorionic villus sampling, and ultrasound, have to be performed during pregnancy because they assess the health of the developing fetus. One important test, however—carrier screening—could be done prior to conception. If it were, couples who were found to be at risk would have additional options to avoid having affected children, including using gamete donation and foregoing childbearing altogether. Most people, however, wait until conception to be tested. In a recent group of studies of carrier screening for cystic fibrosis (CF) funded by the National Institutes of Health (NIH), for example, far less than half of those people offered free carrier screening prior to pregnancy accepted, while the uptake during pregnancy was as high as 80 to 90 percent. Explicitly relying on these data about differential uptake, the members of the consensus conference convened by the NIH to decide how best to incorporate CF carrier screening into clinical practice recommended that carrier screening should be offered as part of prenatal care.[10] The conferees did not ask why different rates of utilization before and after conception were observed or whether it would be a good idea to try to shift carrier screening to the period prior to conception. There might, however, be good reasons to try.

One might argue, for example, that it would be a good thing for adults to think about the health of their prospective children prior to conception. They could try to stop smoking and cut down or eliminate alcohol consumption, and women could take folic acid to reduce the risk that their unborn child would develop a neural tube defect. Carrier screening could be a part of and a foundation for that effort to improve preconception care. One might also observe that confining testing to pregnancy disproportionately and adversely affects women. In any event, screening during pregnancy inevitably raises the difficult question of abortion; gamete donation and contraception, options available prior to pregnancy, while difficult themselves, are less vexed for many people. Instead, the conferees seemed to make the mistake of believing that data inevitably define the appropriate course of action, failing to recognize that data at most can only inform policy choices and that policy choices always require the consideration of values as well. In an era in which we appropriately and increasingly demand data to inform decision-making and to ensure that we get the outcomes we seek, it is critical that we be ever mindful that data, while often necessary, are never sufficient bases for action. The unfortunate result of what I believe to be the consensus conference's flawed conclusion is that efforts to conduct carrier screening prior to conception will be stunted.

Genetics and the Diagnosis and Prediction of Disease

Next I will turn my attention to some of the dilemmas posed by testing that is done to detect genetic factors that determine or predict the health and/or characteristics primarily of the person who is being tested. What do I mean to include here? Everything from a sweat test to see if a child who is failing to thrive and has foul-smelling stools and recurrent pneumonia has cystic fibrosis, to an iron saturation test to see if a person has hereditary hemochromatosis, to a fasting lipoprotein phenotype to see if a person is susceptible to coronary artery disease, to a test for a mutation in BRCA1 in a woman who has many relatives with early-onset breast and ovarian cancer, to apoE testing which may give clues about predispositions for cardiovascular disease and Alzheimer's disease.

I picked the examples above to illustrate a number of points. Most information about genetic risk factors is and will continue to be derived from personal and family history, physical examination, and laboratory testing of specific analytes such as cholesterol or iron saturation. Very little genetic information is derived from direct testing of DNA.

In addition, much testing is done to diagnose the etiology of symptoms that are already present. Direct mutation testing will play a growing role in the diagnostic armamentarium, which will at worst raise some questions of defining whether or not a person has a particular disease. To illustrate the questions that could arise, it is now known that certain mutations in the CF gene cause only male infertility.[11] Is it appropriate to say that people who have these mutations have CF?

Testing to identify people who are at risk to develop disease later on is also not new. We are interested in detecting hypertension not because people who are hypertensive are currently ill but because we know that uncontrolled hypertension is an important risk factor for cardiovascular disease. We are interested in finding adults with elevated levels of cholesterol for the same reason. We already understand that some of the features we look for reveal simply an increased possibility of later disease—not everyone who has the feature will get sick. Here as well, mild hypertension is a good example. We also understand that for a much smaller number of characteristics, such as having a serum cholesterol level of 900, the person who has it will surely become ill if they live long enough. We call the former characteristics predisposing and the latter predictive. We examine environmental factors for the same reason. We know, for example, that a person who worked with asbestos for years has a much higher risk of developing mesothelioma than does a person who has not had such an exposure.

The same phenomena hold true for mutations as well. Having a mutation in BRCA1 greatly increases the likelihood that a woman will develop breast or ovarian cancer some time during her lifetime, but she is not doomed to get it. By contrast, it appears that virtually everyone who has a large enough expansion of trinucleotide repeats in the gene for Huntington's disease will be affected. So what then is so new, or put another way, so threatening about presymptomatic genetic testing? It cannot be the fact that one who has been shown to have a mutation cannot somehow become "mutation-free," because other predictors of disease are similarly unchangeable. As evidence of the latter, a person rarely loses the label "hypertensive" even if his blood pressure is subsequently controlled and the risk of later disease is thereby averted.

The more likely reasons for concern about genetics come from both in and outside of medicine. Because our understanding of genetics is so new, we often simply do not know what to do to decrease the risk that a person who has a predisposing mutation will become ill. We can do nothing, for example, to prevent or even delay the onset of HD in a person who has the mutation. The problem of the lag between the ability to diagnose and the ability to intervene effectively is not unique to genetics, and while the gap usually narrows, we cannot count on it to do so in a predictable way. For example, we have made only modest progress in decreasing the morbidity and mortality of sickle cell disease even though the exact nature of the biochemical and genetic defect that causes that disorder has been known for decades.[12]

The larger problem is that bad things happen to people in our society if it is known that they currently or will one day need expensive health care, regardless of whether the medical intervention will prevent or cure the disease and regardless of the etiology of the disease. We know that a person who is predisposed to develop colon cancer can greatly decrease his or her risk by undergoing routine colonoscopy and, in some cases, colectomy. We know that a person who develops HD will require costly care, much of which may be borne outside the health care system. The issue is that third-party payers, whether they be traditional insurance companies or employers (the ultimate payers in many cases) do not want to pay for this care. Beyond the issue of expensive health insurance, employers, moreover, do not want to have sick people in their workforce for another reason, namely that sick people disrupt the flow of work.

One policy question is what sorts of limits can be placed on the entirely understandable desire of employers to hire only cheap, healthy, efficient workers. Do not get me wrong, I am a strong advocate of the importance of making accommodations for workers who are ill and who have ill family members and for workers with disabilities. I believe that we are a better society when we are more inclusive and when we attend to *all* the needs of individuals. We want people to get the health care they need either for prevention or for treatment. At the same time, we need to acknowledge that making these accommodations and providing this coverage entail costs for employers, even though we may justify them as a "cost of doing business" just as we do the protection of the environment or the payment of taxes. We also need to admit that the Americans with Disability Act[13] and the Family and Medical Leave Act,[14] while they help, are entirely unequal to the task of providing employees with adequate protection. Employees almost never win those lawsuits. Moreover, the prevalence of self-insurance in employer-provided coverage means that employers have the exact information they need to discriminate against potentially costly employees. Even though employers are required to keep health information separate from employment records, recent surveys indicate that many of them use health information to make employment decisions.[15] Even the most aggressive health care information privacy legislation does nothing to solve this issue. Here as well, the greater understanding of genetics did not cause these problems; it only caused their magnitude to increase.

A policy question that results from our current system of third party payment is how much inequality in access to the benefits of the new genetics we are going to tolerate. Let us take a case in point. It is a matter of received wisdom that 30 percent of patients who receive a prescription will benefit, 30 percent will not take it, 30 percent will get no benefit, and 10 percent will be harmed. The promise of pharmacogenetics is that understanding the genes that influence how people respond to drugs could help us eliminate the last 10 to 40 percent. Yet I, as a primary care provider for children whose care is paid for, if at all, by TennCare, recognize that not everyone will get access to this sort of targeted therapy. My patients almost surely will not. We are never going to have a single standard of care in this country, but one of the questions that we must face is what basic level of access must exist. The fact that the pharmaceutical industry relies to a large extent on the results of publicly funded research in developing products makes the issue of whether it is just to deny access to drugs to some of the public all the more pressing.

Let me demonstrate yet another dilemma. We know that lipid-lowering drugs can decrease the risk of coronary artery disease in people with elevated cholesterol levels. We also know that weight control, exercise, and diet can decrease this risk as well. As a society, we appear to love drugs, particularly if they allow us to continue our current behaviors. Focusing on genetics may encourage us toward more medical prevention and intervention. The question is how far we ought to go in that direction or whether we should encourage healthier behaviors and improving environmental risk factors.

Genetics and Equality

I would like to close by saying a few words about the challenges presented by genetics to our level of commitment to equality. It seems to be an intrinsic part of human nature for one person to want to demonstrate that he or she is "better" in some way than others. Yet what we have seen in this century is a growing sense that, as a matter of public policy, some things that distinguish us from each other ought not to "matter" or "count" as a basis for advantage or disadvantage. Thus we say in our laws against discrimination that race and sex and disability ought to be irrelevant for some purposes. Of course, similar ideas are an important part of the Christian tradition, with notions that all people are equal in God's eyes and hence are equally entitled to respect. But we also know that we honor equality, to the extent that we do, more by our words than by our deeds. And at least at the policy level, recognizing human dignity is not the only force behind efforts to limit unwarranted discrimination. Rather, one important reason we say that certain characteristics are irrelevant is that these characteristics are poor proxies for what we actually care about. We cannot truthfully say, for example, that all blacks and all women are inherently unable to obtain the skills needed to hold leadership positions, or unable to understand complex concepts. We have come to realize that, on a purely utilitarian level, excluding all women and all blacks harms society by eliminating potentially effective leaders and contributors.

Our understanding of genetics could both promote and undermine our commitments to equality. We now know that we share almost more than we can imagine with our fellows. Any two people share more than 99.99 percent of their DNA. In fact, we share a lot of our DNA with all living creatures.[16] Moreover, we know that there is more genetic variation within racial and ethnic groups than there is between groups. My DNA is likely to be more different from that of another white Anglo-Saxon Protestant than that of either of us is likely to differ from that of a person of Asian or African descent. One would hope that these facts of shared heritage would do much to promote respect for each other.

On the other hand, we will certainly find mutations that affect, sometimes profoundly, characteristics that we seem to believe are legitimate reasons for treating one person differently from another, in particular behavior and inborn ability to learn. It is inconceivable that behavior and intelligence are untouched by genes. Of course, these attributes are almost always affected by numerous social factors as well, most notably socioeconomic status, but at the end of the day, there will be definable differences in genetic endowment as well. We cannot escape this knowledge simply by deciding not to do that type of research; the apoE story reveals that a single gene can affect both body and mind. The points I want to drive home are that genetics is inextricably involved with every aspect of our lives, and that we will be unable to avoid the issue of genetics and behavior completely. Perhaps the most difficult policy challenge for us will be whether we will treat these differences as reasons to treat some individuals worse than others, à la Huxley's *Brave New World*, or whether we will continue our fledgling efforts to level the playing field for all people.

Acknowledgments

This project and article was supported in part by funding from 1 R01 HG01974-01.

Notes

1. Troy Duster, *Backdoor to Eugenics* (New York: Routledge, 1990); Daniel J. Kevles, *In the Name of Eugenics: Genetics and the Uses of Human Heredity* (New York: Penguin, 1985).

2. Philip Reilly, *Genetics, Law, and Social Policy* (Cambridge, Mass.: Harvard University Press, 1977).

3. Linus Pauling, "Reflections on the New Biology," *UCLA Law Review* 15 (1968):268–72.

4. Ellen Wright Clayton, "What the Law Says about Reproductive Genetic Testing and What It Doesn't," in *Women and Prenatal Testing: Facing the Challenges of Genetic Technology*, ed. Karen Rothenberg and Elizabeth Thomson (Columbus, Ohio: 1994), 131–78.

5. *Etkind v. Suarez*, 519 S.E.2d 210 (Ga. 1999).

6. Mo. Ann. Stat. §188.130 (1998).

7. Ellen Wright Clayton, "What the Law Says about Reproductive Genetic Testing and What It Doesn't," 131–78.

8. Dorothy Wertz and John C. Fletcher, "Ethical and Social Issues in Prenatal Sex Selection: A Survey of Geneticists in 37 Nations," *Social Science and Medicine* 46:2 (1998):255–73.

9. Sherman Elias and George C. Annas, "Generic Consent for Genetic Screening," *New England Journal of Medicine* 330:22 (1994):1611–13.

10. "Genetic Testing for Cystic Fibrosis," NIH Consensus Statement, *Online* (April 14–16, 1997), 15:4 (1997).

11. T. Dork, B. Dworniczak, C. Aulehla-Scholz, et al., "Distinct Spectrum of CFTR Gene Mutations in Congenital Absence of Vas Deferens," *Human Genetics* 100:3–4 (1997):365–77.

12. P. A. Lane, "Sickle Cell Anemia," *Pediatric Clinics of North America* 43:3 (1996):639–64.

13. 42 U.S.C. §§12100 et seq. (1999).

14. 29 U.S.C. §2601 et seq. (1999).

15. Department of Labor, Department of Health and Human Services, Equal Employment Opportunity Commission, Department of Justice, "Genetic Information and the Workplace," January 20, 1998, http://www.dol.gov/asp/programs/history/herman/reports/genetics.htm.

16. Kelly Owens and Mary-Claire King, "Genomic Views of Human History," *Science* 286 (1999):451–53.

3

The Genetics Revolution: Can the Law Cope?

Sheila A. M. McLean, LL.B, M.Litt., Ph.D.

INTRODUCTION

In the second half of the nineteenth century, what has been described as a "conceptual revolution"[1] occurred in modern medical science, as disease states (and increasingly personal characteristics) came to be viewed in terms of biochemistry. Now, according to the World Medical Association, "[a] new revolution is happening . . . which locates in the gene the instructions for all the biochemical processes in the body's cells."[2] From Watson's and Crick's groundbreaking work on human DNA and the subsequent investigations of their distinguished colleagues throughout the world came a new picture of the human being.

It is immediately apparent that the new genetics is more than just a scientific phenomenon. It has consequences that go far beyond the interests of the scientist and will inevitably have moral, social, political, and legal implications. Sometimes referred to as the "Holy Grail of modern biology,"[3] the information gained from the Human Genome Project will have profound effects on us all as individuals, on our world collectively, and on our very perception of what it is to be a human being. Even although some doubt its current therapeutic potential, and even in the face of the current failures of gene therapy, hope remains high that in the future, cures for common conditions such as cancer will be found as a result of the genetics revolution. For the moment, however, it has been suggested that: "It could be said of molecular biology that, insofar as human genetics is concerned, it has gained a scientific empire but not yet found its real clinical role."[4]

The Genome Project, then, is the new face of human genetics. But it must be said that the history of genetics is not a pretty one. In the early parts of this century regimes as disparate as the United States and Nazi Germany made extensive use of elementary genetic information to enforce politically, and with the backing of the law, program that resulted in the compulsory sterilization of those thought to be genetically unfit to reproduce.[5] The scientists involved in genetics have taken a long time to throw off this history, but they remain well aware of the power of genetic knowledge for good and for bad. Indeed, recognition of the potential nonscientific consequences of the Genome Project was obvious from the beginning of the program itself. As the *Human Genome News* reports:

> In a significant departure from other previous large-scale scientific undertakings, the U.S. Human Genome Project specifically supports studies on ethical, legal, and social issues (ELSI) that can arise from the increasing availability of genetic information about individuals and populations. The decision to establish the DOE ELSI program ... was spurred by the early realization that while human genome research itself does not pose any new ethical questions, use of the research data could raise very challenging issues.[6]

Society, too, should bear in mind the potential of advances in knowledge—should remember that knowledge is seldom value-free and that once acquired, it will almost inevitably be used.

The motivation of those involved in seeking remedies for our ills is generally—and rightly—assumed to be benign, and the fact that this is so may blind us to the potentially uncontrollable problems posed by this seemingly beneficent intervention. Equally, medical and scientific knowledge are now so far ahead of our understanding that we may feel disenfranchised from tackling them. Yet it is essential that an informed public, and if necessary a directive legal system, demand answers to difficult questions, particularly when the progress being made has such fundamental importance to us all. Scientists, therefore, carry a major responsibility to inform citizens of both what they are doing and what it may mean. Whereas some are sanguine about the implications of the new genetics, suggesting that "the widespread fear of genetics cannot be justified,"[7] others note that: "Scientists are at least no better and no worse than the society of which they are members, and criticism of science is part of legitimate social critique."[8]

Perhaps as important as engaging in this critique is the need to decide what controls, if any, we wish to see in place. The Genome Project has the capacity to challenge many of the principles that have been central to our social, cultural, and legal traditions. Just as early genetic information became a powerful political tool, we should not be complaisant about our capacity to react and use this more sophisticated information in any more responsible a manner in the twenty-first century.

However, the aim of this discussion is not scaremongering, nor is it a Luddite approach to scientific advance. The new genetics will almost certainly do much good,[9] and the information gained from the Genome Project has the potential to transform health, to enhance quality of life, to promote longevity, and to facilitate personally appropriate reproductive choices. For the moment the knowledge we have is just that—knowledge—but now that the genome is effectively mapped, its real value will lie not solely in the possession of that information but in the uses to which that knowledge can be put.

It is central to this discussion that account must be taken, ethically and legally, of the values inherent in genetic information and the implications of the uses to which it may be put. In their commentary on the draft Genetic Privacy Bill, Annas, Glantz, and Roche made the point that "[t]he DNA molecule contains information about one's probable medical future, and this information is written in a code that is currently being broken at a rapid pace."[10] Moreover, they concluded that: "Current legal protections for medical information, tissue samples, and DNA samples are inadequate to protect genetic privacy."[11]

In couching the debate in terms of the protection of privacy, Annas and his colleagues mirror and bolster the values that have traditionally underpinned Western societies. Communities based on individualism, autonomy, and personal integrity have good reason to focus on privacy as a concept of critical importance in maintaining political, legal, and ethical equilibrium. Pellegrino says:

> The dominant characteristics of Western science, ethics, and politics are mutually supportive. Western science is empirical and experimental, pursuing objectivity and quantification of experience. Ultimately it attempts to control nature to the greatest extent possible. Western ethics is analytical, rationalistic, dialectical, and often secular in spirit. Western politics is liberal, democratic, and individualistic and governed by law.

Western science, ethics, and politics provide an environment that gives rise to and sustains the use of complex medical technologies. As a result, it is difficult to divorce medical knowledge and the benefits it offers from the Western cultural and ethical milieu that support and sustain them.[12]

Although rating only a brief mention in this quotation, law nonetheless plays a central part in reflecting, shaping, and reinforcing the values of the culture from which it stems and over which it has control. The response of the law to the genetics revolution, therefore, will play as important a part in contemporary society as it did in the shameful abuse of individuals in the early part of this century—it is hoped, however, that this time its influence will be for the good. But the question remains of whether the Western legal tradition is equipped to manage the issues raised by the new genetics. More subtly, perhaps, an investigation may be needed into whether the values—as well as the structure—of the law are appropriate to regulate the future that will be shaped by the completion of the genome project and the application of the knowledge derived from it. Brahams suggests that: "Society will expect the law to protect its wider ideals and, in particular, the individual citizen from the excesses of over-enthusiastic doctors and scientists, greedy corporations and immoral profiteers and manipulators. The law will have to balance the need for future research against the need to protect society from its dangers and evils."[13]

She also notes—significantly—that whereas medicine and science have their sights set firmly on the future, the law "traditionally looks mainly to the past to interpret the legal position of futuristic techniques."[14] Hence the values that have historically informed and that still underpin current law will likely play an important part in regulation and control—a part as significant as the available mechanisms themselves. For example, and the Annas Bill is but one example, protection of the individual from unwarranted intrusion into private decision-making has substantially depended on development of the kinds of privacy rights that form the central plank of Annas's and others' attempts to devise a modern legal response to the genetics revolution. In the United States, of course, the development of a privacy right from the case of *Roe v. Wade* has been, although not uncontroversial, nonetheless a significant protector of reproductive and other choice.[15] In the United Kingdom, a right to respect for private and family life (which may or may not be a direct equivalent of a privacy right) was only built into the law by the passing of the Human Rights Act 1998, which finally incorporates the European Convention on Human Rights into U.K. law. However, protection of the individual was not absent before this; rather it derived from a variety of other concepts closely linked to privacy.

LAW AND GENETICS—WHERE ARE THEY CONNECTED?

The range of issues that demonstrate the connection between genetics and the law is immense. From obvious situations such as employment and insurance to more esoteric concerns about fetal status and reproductive choice; from "black letter" law such as intellectual property to concepts of personal responsibility and accountability, few aspects of our daily life will escape the impact of the genetics revolution, and few areas of law will remain untouched.[16] It should be immediately apparent from the kinds of issues briefly mentioned above that societal and legal abdication of responsibility to total scientific/clinical self-regulation is not an option. The tradition (in the United Kingdom at least) of handing over significant responsibilities to professionals such as doctors has long been the subject of critical assessment. The genetics revolution seems likely to increase disquiet, already widely expressed, about the transference of decisions that are essentially about human rights to any one discreet professional group. As Kennedy has noted, "the scope of the alleged unique competence of the doctor has become as wide, as imprecise, and as flexible as the meaning given to the notions of health and ill-health."[17] Like it or not, the genetics revolution may sound the death knell of extensive professional self-regulation. The privileges of professionalism seem to be under attack.

WHY REGULATE?

It has already been suggested that there is—indeed must be—a role for the law in controlling technological advance. Kinderlerer and Longley turn this suggestion into a positive rather than an apparently negative claim. As they say:

> Whilst scientists and physicians are capable of identifying the possibilities which may advance therapeutic procedures, they are no better equipped than anyone else to identify the social and moral issues that might result from their use. Only society can ultimately decide the degree of importance to be attached to the benefits, hazards and impact of these. Such fundamental decisions within an area of inherent uncertainty can only be made within a properly constituted regulatory framework. Only law can provide the mechanisms that can facilitate effectively the negotiation between fact and value and the taking of necessary decisions about the nature and direction of scarce resources for both research and policy. Regulation also carries with it a number of other advantages. By ensuring that controversial developments and the practices of scientists and clinicians are properly authorised and monitored a potent regulatory framework can assist public acceptance of "cutting edge" techniques. . . . . Besides being instrumental in the policy process law thus promotes legitimacy and accountability.[18]

However—and this is the structural point—in few societies are the coherent mechanisms available that would permit the law to act as facilitator between the scientists and the community. Just as with assisted reproduction, it is all too common that regulatory control is spread through a variety of legal and quasi-legal sources. Black, for example, noted that in the United Kingdom at the time of writing, a wide range of sources sought to provide regulation, and while agreeing with Kinderlerer and Longley that regulation by law is important, found that in addition to practice notes, ethical codes, international agreements, circulars, and nonlegal rules, there were: "In the UK, at national level alone, . . . over eleven different bodies involved in the regulation of some aspect of genetic technology."[19] In the United States also, bodies with supervisory or regulatory functions range from the Food and Drug Administration (FDA) to the large companies that hope to benefit economically from the Genome Project and its spinoffs, producing a varied and diverse form of regulation that potentially has no common core principles.

The inadequacy of the law referred to by Annas and his colleagues pertains in the United Kingdom also. In this sense, the inadequacy is structural rather than based on principle, although this will be returned to later. However, no matter on which principles the law is based, it seems likely that its impact—and its importance—are threatened by the lack of a consistent and coherent structure. Of course, many feel that the law is useful only in response to presenting problems. The British Medical Association, for example, believes that reactive law is preferable to proactive,[20] and the association's view is supported by a number of other commentators.[21]

However, this view arguably takes an unnecessarily narrow view both of the types of law available and of their potential content. Law may be coherent while at the same time presenting in different forms—for example, legislation, thought by many to be too heavy-handed to deal with the subtle and complex issues raised by areas such as genetics, is by no means the only vehicle available to regulators. Indeed, it is areas such as family law, reproductive law, and issues of confidentiality that will form a significant part of the need for a regulatory framework, and these are traditionally common-law areas in many jurisdictions. Hence to require the law to take a substantially or solely reactive role is to mistake the scope of legal control and to lack an imaginative approach to regulation.

However, no matter how it is structured, there remains one fundamental matter of principle that will have profound effects both on the contribution of law and on the individuals subject to it. As was noted earlier, much of the current emphasis in law reform in respect of genetics has been on the development or strengthening of genetic privacy. As has also been noted, privacy is a concept entirely in tune with liberal Western democracies and their emphasis on individualism. The movement towards genetic privacy legislation is strong in the United States. In the United Kingdom, too, the House of Commons Select Committee on Science and Technology endorsed the view that genetic privacy should be absolute in most, if not all, circumstances.[22]

Although privacy is an important weapon in the armory that protects individuals, it will require to be increasingly robust in the face of the new genetics. Indeed, the new genetics arguably challenges the very principles upon which Western democracies rest, in particular in respect of individualism.[23]

Self-evidently, what is different about genetics is its capacity to inform about others. A substantial number of commentators have already taken up this issue. Suter, for example, says that: "When genetic testing of one person can benefit another family member, privacy and autonomy interests of the former may collide with the relative's interest in protecting her health or planning her future."[24]

Equally, the adoption of the UNESCO *Declaration of the Responsibilities of the Present Generations towards Future Generations* adds credence to the concept of "intergenerational justice."[25] The declaration unequivocally states that: "The present generations have the responsibility of ensuring that the needs and interests of future generations are fully safeguarded."[26] No exception is made in the interests of providing a privacy right for current generations. As Fletcher and Wertz note: "The completion of the human genome project will provide a basis for acting on a moral obligation for *future* generations, a claim that has appeared weak in the past. A generation *with* such knowledge who neglected to use it to minimize the risks in reproduction could hardly be said to respect the requirements of intergenerational justice."[27] And Macer says: "A common feature of many issues raised by the human genome project data is that we need to consider the effects of knowledge and technology on future generations. We have a responsibility to future generations. The beneficiaries and those at risk may not yet exist. In the sense of benefits and risks, it is their genome project more than ours. We have an obligation to the future based on the principle of justice."[28]

Arguably, therefore, depending on the weight given to intergenerational justice, the privacy right expounded in *Roe v. Wade*[29] and presaged in cases such as *Skinner v. Oklahoma*[30] and *Griswold v. Connecticut*[31] risks revision. The Universal Declaration of Human Rights and the European Convention on Human Rights also guarantee rights to respect for private and family life. As is usual, they are of course subject to exceptions, which may not be unimportant for the development of the law and genetics. The European Convention, for example, in Article 8(2) permits derogation from the right to respect for private and family life in situations such as "for the protection of the rights and freedoms of others."

Of course, it is not just our obligations to future generations that may be affected by genetic capacity. It is arguable that — in the near future, if not the present — there may be increased pressure on individuals to make disclosure of their genetic condition in order to protect others' freedoms — perhaps even lives. As the gap between diagnostic and therapeutic capacity narrows, the reasoning applied in the U.S. case of *Tarasoff v. Regents of the University of California*[32] to genetics — namely, a duty to warn in the face of a real threat — might become an integral part of the management of genetic information. Simple adherence to the historical—some might say, atomizing—traditions of privacy rights may ultimately prove inadequate in providing an explanation as to why breaches of individual privacy might not reasonably fall within the internationally accepted exceptions to the general rule.

One further issue also arises that may shake adherence to Western legal tradition—namely the extent, already mentioned, to which genes are shared. As the Danish Council of Ethics notes:

> Genes are a fundamental element of a person's biology. The unique genetic blend is an essential contributing factor in making a person what and who he is. It thus influences the factors which belong to the personal sphere (including disease, physical development, pain, suffering and death). On the other hand, genes are "public domain," shared with other people: parents, children, siblings, etc. When disease is genetically conditioned, it creates a special kinship with other people.[33]

The problems that may arise from the peculiar nature of genes can perhaps best be explored by a hypothetical but by no means extreme example. One facet of individual privacy is the traditional commitment of medicine to maintaining confidentiality. Confidentiality is widely agreed to be a concept of vital importance to the individual and reflects the uniqueness of each person. What the Genome Project will do is to enable all of us to identify our place in the gene pool. But in so doing, we also identify the place of others. At the same time, genetic information might be thought to be the most confidential information of all. If my genetic makeup is what makes me the person I am, then I might reasonably argue that this information is mine and mine alone. I might claim a right to have this information maintained in the utmost confidentiality, even if it does affect others about whom I care. After all, I have no legal obligation to rescue and only an arguable moral one. The fact that my knowledge about myself affects or may affect others might be said to be irrelevant viewed from this perspective.

However, we must ask: Where individuals can be identified as being specifically at risk, does or should this affect any duty to disclose or the right to confidentiality? This question has a peculiar resonance in terms of the Human Genome Project, because there is no way round the fact that my genetic characteristics are intimately linked to those of others in my family. In other words, I may be argued to be under a moral obligation to waive my own privacy right rather than ignore risk to other members, actual or potential, of my own family.

And whatever this does to my moral obligation to share information, there are others also affected by possessing the knowledge. As I have said, I may have no duty to tell anyone else, but what of my doctor? Where can he or she look for guidance on this? My doctor might subscribe to an ethical commitment to the individual patient—me—but at the same time run the risk of being tackled by a variety of other laws, for example, those concerning negligence or professional discipline if he or she protects my confidentiality by not disclosing information that is relevant to other, identifiable individuals to whom he or she also owes a duty of care.[34] And, of course, disclosure of this information may lead to professional and/or legal sanction through the laws on confidentiality.

For example, if my brother is also a patient of my doctor, my doctor owes to him the same duty of care as he owes to me. If the doctor discovers that I have the gene for Huntington's disease, and is at the same time attempting to resolve the problems of an apparent lack of fertility in my brother's relationship, does my doctor have an obligation under the laws of negligence to inform my brother of the risks to which his future children might be exposed? If my brother's wife is already pregnant, should the doctor disclose the risk in the interests of permitting the prospective parents the opportunity of seeking screening and termination if necessary? And if the doctor does not, can he or she be sued subsequently?

The Annas Bill, and probably the House of Commons Report, would appear specifically to rule out the doctor transmitting information to my family; but, although this may supply an answer to the conundrum posed, it is an answer with which many may be unhappy. Yet the answer must be logically right if privacy is the overwhelming "good" in dealing with genetic information. Privacy, after all, is an absolute value and takes no account of the potential harm that might flow from its dominance. Absolute adherence to privacy has already been threatened by the management of HIV and AIDS and to a lesser extent by the management of other medical and psychological conditions. However, it is from genetics that it faces its sternest test. Particularly when diagnostic and therapeutic potential converge, or at least become more closely aligned, what now may seem like an unreasonable demand—namely the sharing of information traditionally regarded as private—may well come to seem not only reasonable but responsible and desirable.

This is a major dilemma confronting our management of the new genetics. Structure is important in terms of regulation, but the value systems that inform that structure take logical precedence. As the Danish Council of Ethics puts it: "it can be said that a decisive stand on the new challenges must be based on the help motive and respect for the individual. The question, in other words, is: Must the principle purpose of applying human genetics be formulated in terms of the gain for the common good or in terms of the individual?"[35]

A New Legal Order?

Ultimately, the "new genetics" may require revision of the principles upon which Western society is based. The rich world values concepts such as autonomy and—as we have already seen—notions such as privacy. But the genetics revolution shows starkly how interrelated we are, essentially limiting our individuality. The deontological ethics that traditionally inform the Western world and to which notions of human rights are integral are arguably challenged by the recognition of people as intimately connected. Hence it may be necessary to reevaluate the central platform of Western society to take account of what we share with our immediate and wider community. Will a more communitarian philosophy be the outcome—a philosophy based on how we are alike rather than how we are different, how we are responsible to others as well as to ourselves?

Whatever modifications in approach may or may not be needed, they will generally be achieved at the political level and often through the law. However, as J. Windeyer noted in the Australian case of Mount Isa Mines v. Pusey,[36] the law is often described as being "in the rear and limping" in relation to the march of medical (including *ex hypothesi* genetic) progress.[37] Certainly it seems unlikely that nonscientists will ever be in advance of or even up to tempo with our scientific colleagues in terms of knowledge. Nonetheless, an effort to close the gap must be made in the face of the greatest scientific and human challenge to date.

A cursory look at most legal systems would, however, suggest that there are few, if any, relevant laws in place to protect the individual against the unwarranted or unwanted use of genetic information, and that information is increasingly available and likely to be used. This is true despite the fact that legislative response to genetics has been, in certain areas at least, unusually rapid. In the United States, for example, there has been considerable legislative interest in the implications of genetics for both employment and insurance. Despite this activity in the area of employment, however, Rothstein notes that: "Since enactment in 1983, Minnesota is the only state with a statute that limits an employer's access to an employee's non-job-related medical information."[38] And in the United Kingdom it has been argued that "law is underdeveloped with regard to workplace genetic testing."[39] In addition, The *Human Genome News* reported that: "A 1995 Harris poll revealed that over 85 percent of Americans are concerned about insurers or employers having access to their genetic information. Another study showed that many high-risk people refuse to take advantage of new genetic tests for fear of losing their jobs or their insurance."[40]

In terms of insurance, although few insurers request genetic information,[41] if it can be shown that this information is relevant to actuarial calculations, then it is plausible to imagine that this position will not continue for long. In the United Kingdom, the basis for not requiring genetic tests to be undertaken rests on a voluntary code of practice drafted by the Association of British Insurers (ABI), which represents over 96 percent of the business of U.K. insurers.[42] Although not requiring tests to be taken, the industry reserves the right—in light of the nature of the insurance contract—to seek access to genetic information already held, with one limited exception.[43]

In the United States, Reilly reported that: "As of April 1997, at least 15 states had enacted genetic privacy laws. More than 75 similar bills are pending in more than 30 states . . . and several federal bills have been introduced into the 105th Congress."[44] Worldwide, Austria has currently legislated to ban both insurers and employers from access to genetic test results, and Australia has a voluntary agreement that genetic tests should not be requested by insurers and that tests will not be used for underwriting purposes. The position in New Zealand is similar to that in the United Kingdom, and in Norway legislation prohibits obtaining, handling, or using genetic test information.[45]

The variations in approach may reflect cultural differences but they also show the relative novelty of such directive intervention. Although historically the industries have been constrained by law in many ways, it is relatively recently that legislators have had to consider a more head-on approach to industries whose underwriting or employment practices have been largely—although not entirely—a matter for themselves. Equally, our traditional reliance on the professional self-regulation of groups such as doctors will no longer suffice to meet the challenges posed by our previously unimagined power. Arguably, therefore, privacy laws may no longer be the most effective way of protecting the individual because they seem vulnerable to attack by other principles that we also value.

Perhaps one of the most significant and far-reaching consequences of the Human Genome Project may be the challenge to scrutinize closely our own value systems. Seldom if ever before have communities confronted such a wealth of sensitive information with such a plethora of uses. To be sure, it may all turn out well, and the advances in science may prove to be only beneficial. However, neither history nor common sense would allow us to dwell in the oasis of that particular aspiration for long. Our history of negative eugenics, coupled with the current race for scientific discovery and intellectual property rights over these discoveries, does not lead this writer, at least, to have confidence that all will inevitably be well. This is not an argument against progress but is rather a plea for informed debate and rational legal response before it is too late. As Dan Kevles said: "Given that changes in individual attitudes inevitably affect the scope of institutional action, both public and private, history surely teaches that serious attention is owed to the warnings, however shrill they may sometimes be, of the dissenters from the eugenic revival."[46]

There is a role for us all to play through our democratic and legal processes. And it is not just a role that we should play—it is a role that we must play. The genetic truth about mankind is unfolding before our eyes, its promises and pitfalls emerging from the shadows, yet our eyes do not see or are at best myopic. Decisions that will affect us all are being taken, yet we are not involved. As Franklin says: "One of the most important aspects of modern technological practice is that it allows the control of people in ways that make the control invisible. No longer does Big Brother blare out of loud speakers. Big Brother barely beeps today. The invisibility of control ought to concern us very profoundly."[47]

Structures to which we have clung for protection for centuries are inadvertently at risk, yet we passively wait to succumb. The Human Genome Project has the potential to be of enormous benefit to communities worldwide. Even allowing for cultural differences, it is expected to facilitate and enhance quality and quantity of life. But it will also confront us with hitherto unthinkable conundrums. We cannot blame the scientists for leaving us behind if we are not prepared to grasp the nettle now. We must question and we must challenge, for the way forward is not obvious or simple. However, we must also resist the temptation of the "knee-jerk" reaction, which is all too often the response of politicians ill-versed in the ethical and social issues of specific scientific developments, and of certain areas of the mass media.

The picture is presented clearly by Tom Wilkie, formerly science editor of *The Independent* newspaper in the United Kingdom and a scientist himself:

> Like the sword of justice, technology has a double edge. It can be used for ill or for good. The technologies that will follow, that have already followed, from the advances in human genetics can alleviate human suffering and misery. But we, the laity and not the professionals, must master the new knowledge and its technical application. The only way to do so is to transcend the geneticists' approach—to remember that human beings are more than just vehicles for transmitting genetic information from one generation to the next, and that human life is more than just the expression of some computer program written in the biochemical language of DNA.[48]

It is here that the role of society and the law become critical. It is a paradox, not unnoticed by commentators on human rights, that the State is most often the organization that seeks to limit human rights and yet it is to the State—generally as represented by the law—that we must look for protection of these very same rights.[49] But there is a further question that the Human Genome Project is beginning to raise, and that is whether the legal and social principles with which we have apparently been comfortable over many years are sufficient to meet the new challenges of science. Gostin acknowledges that "[p]ermitting the Human Genome Initiative to proceed unabated will have costs in personal privacy,"[50] and concludes that: "While careful security safeguards will not provide complete privacy, the public should be assured that genomic information will be treated in an orderly and respectful manner and that individual claims of control over those data will be adjudicated fairly."[51]

Perhaps this begins to point to the way forward. Concern about privacy and acceptance that privacy may also include responsibility to others may provide a newly formed basis from which to address and perhaps answer the profound changes that the genetics revolution will bring about. What seems clear is that privacy as an atomizing concept, one that separates us from others, may need to be redefined and reinforced if it is to encapsulate the intricacies of the reality of the new genetics. In the long run, the real challenge may be to assert control over information, regulating its external use rather than denying the validity of that use. Clearly, adherence to privacy rights already provides control, but it has been the thrust of this argument that the notion of what is *appropriate* control may have to be rethought in view of our obligations to others. Hence resolution of the problems may demand closer scrutiny of the content and breadth of the concept of control as well as describing the kinds of rights and interests we may legitimately claim as part of both a close and a distant community. If so, and if we are able to accommodate both the maintenance and the relaxation of control, individual privacy rights may yet sit comfortably with responsibilities to and for others and their well-being, as well as maximizing self-interest. Once we unpick the control issue, the best form of regulation — informed by principle — may become more obvious and more attainable.

Notes

1. Preamble, World Medical Association Declaration on the Human Genome Project, adopted at its 44th Assembly in 1992.

2. Ibid.

3. T. Wilkie, *Perilous Knowledge: The Human Genome Project and Its Implications* (London, Faber and Faber, 1993), 1.

4. J. Davis, "Ethical Issues," in M. P. Ryan, et al., "Genetic Testing for Familial Hypertrophic Cardiomyopathy in Newborn Infants," *British Medical Journal* 310 (1992):858.

5. For further discussion, see D. Meyers, *The Human Body and the Law* (Edinburgh, U.K.: Edinburgh University Press, 1971); S. A. M. McLean, "The Right to Reproduce," in *Human Rights: From Rhetoric to Reality*, ed. Campbell, et al. (Oxford: Basil Blackwell, 1986).

6. *Human Genome News* 4:2 (1992):1.

7. J. Maddox, "New Genetics Means No New Ethics," *Nature* 364 (1993):97.

8. J. Schmidtke, "Who Owns the Human Genome? Ethical and Legal Aspects," *Journal of Pharmacy and Pharmacology* 44:supplement 1 (1992):205–10.

9. For example, Wilkie, note 3, says: "The new genetical anatomy will transform medicine and mitigate suffering in the twenty-first century," at 1.

10. George J. Annas, Leonard H. Glantz, and Patricia A. Roche, "The Genetic Privacy Act and Commentary," Health Law Department; Boston University School of Public Health, 1995; available from http://www.ornl.gov/TechResources/Human_Genome/resource/privacy/privacy1.html, sec. 2(a)(1).

11. Ibid., at 2(a)(7).

12. E. D. Pellegrino, "Intersections of Western Biomedical Ethics and World Culture: Problematic and Possibility," *Cambridge Quarterly of Healthcare Ethics* 3 (1992):191–92.

13. D. Brahams, "Human Genetic Information: The Legal Implications," in *Human Genetic Information: Science, Law and Ethics*, Ciba Foundation Symposium 149 (Chichester: John Wiley and Sons, 1990), 117.

14. Ibid., at 112.

15. 410 US 113 (1973).

16. For a discussion of many of these areas, see R. Brownsword, W. R. Cornish and M. Llewelyn, *Law and Human Genetics: Regulating a Revolution* (Oxford: Hart Publishing, 1998).

17. I. Kennedy, *Treat Me Right: Essays in Medical Law and Ethics* (Oxford: Clarendon Press, 1988, repr. 1994), 23.

18. J. Kinderlerer and D. Longley, "Human Genetics: The New Panacea?" in Brownsword, et al., note 16, at 28.

19. J. Black, "Negotiating the Genetic Revolution," in Brownsword, et al., note 16, p. 29. This number has been somewhat reduced by the creation of the Human Genetics Commission.

20. British Medical Association, *Our Genetic Future: The Science and Ethics of Genetic Technology* (Oxford: Oxford University Press, 1992), 137: "At first sight anticipation of risks seems far superior to the trial and error of reactive legislation, but it has serious limitations arising from the difficulty of evaluating the future consequences of a previously untried event."

21. See, for example, J. B. P. Garcia, "The Law and the New Challenges Posed by Genetic Science. Presentation of a Meeting," *Law and Human Genome Review* 4 (1996):202. "The law, it must be said, is not renowned for anticipating problems. Acting 'after the event' is not always a bad thing. . . . . Acting in haste can be a bad thing also."

22. House of Commons Science and Technology Committee, Third Report, Session 1994–95, *Human Genetics: The Science and Its Consequences* (London: HMSO, 1994–1995), 41–1, at para 224, lxxiv, "In our view, the fundamental question is not about genetic information per se, but about personal privacy."

23. For a discussion of its impact on the doctor/patient relationship, see S. A. M. McLean, "The New Genetics: A Challenge to Clinical Values?" *Proceedings of the Royal College of Physicians, Edinburgh* 26 (1996):41.

24. M. S. Suter, "Whose Genes Are These Anyway? Familial Conflicts over Access to Genetic Information," *Michigan Law Review* 9 (1993):1855.

25. UNESCO, *Declaration of the Responsibilities of the Present Generations towards Future Generations* (Paris, 1997).

26. Ibid., Article 1.

27. J. C. Fletcher and D. C. Wertz, "An International Code of Ethics in Medical Genetics before the Human Genome Is Mapped," in *Genetics, Ethics and Human Values: Human Genome Mapping, Genetic Screening and Therapy*, ed. Z. Bankowski and A. Capro, *xxiv CIOMS Round Table Conference* (1991), 103.

28. D. Macer, "Whose Genome Project?" *Bioethics* 5:3 (1991):209.

29. Note 15.

30. 316 US 535 (1942).

31. 318 US 479 (1965).

32. 529 P 2d 55 (Cal, 1974); on appeal 551 P 2d 334 (Cal, 1976).

33. Danish Council of Ethics, *Ethics and Mapping of the Human Genome* (Copenhagen: Danish Council of Ethics, 1993), 61–62.

34. See McLean, note 23.

35. Danish Council of Ethics, note 33, p. 64.

36. (1970) 125 CLR 383.

37. Ibid., at 395.

38. M. Rothstein, "Protecting Genetic Privacy by Permitting Employer Access Only to Job-Related Employee Medical Information: Analysis of a Unique Minnesota Law," *American Journal of Law & Medicine* XXIV:4 (1998):400–1.

39. P. Gannon and C. Villiers, "Genetic Testing and Employee Protection," *Medical Law International* 4 (1999):40.

40. *Human Genome News* 9 (1998):1–2.

41. Annas, Glantz, and Roche, note 10, sec. 112(f).

42. The ABI Web site and Code of Practice is available from http://www.abi.org.uk.

43. The exception is where the life insurance policy is for less than 100,000 pounds sterling and is attached to a mortgage.

44. P. R. Reilly, "Fear of Discrimination Drives Legislative Interest," *Human Genome News* 8 (1997):3–4.

45. Further information about a number of countries can be found at http://www.geneticsinsuranceforum.org.uk/.

46. D. J. Kevles, *In the Name of Eugenics: Genetics and the Uses of Human Heredity* (Harmondsworth, U.K.: Penguin, 1985), 299.

47. U. Franklin, "New Threats to Human Rights Through Science and Technology—The Need for Standards," in *Human Rights in the Twenty-First Century*, ed. K.E. Mahoney and P. Mahoney (Dordrecht, Netherlands: Kluwer Academic Publishers, 1993), 734.

48. Wilkie, note 3, at xi.

49. For further discussion, see Campbell et al., note 5.

50. L. O. Gostin, "Genetic Privacy," *Journal of Law, Medicine, & Ethics* 23:4 (1985):328.

51. Ibid.

4

Genetics, Investment, and Business: Organizational Ethics for Genomics Companies

Gerard Magill, Ph.D.

ORGANIZATIONAL ETHICS can enable genomics companies to maintain corporate integrity in the market economy of biotechnology, where community stewardship must balance health outcomes, shareholder investment, and business patents. The accelerating pace of scientific breakthroughs and technological developments among genomics companies is breathtaking. Genetic science and technology are changing the face of health care and creating controversial opportunities, as illustrated by the media coverage of the policy and ethics debate on whether government should fund or permit stem-cell research.[1] Genetics is even contributing to aesthetics through new opportunities for art and advertisement.[2] And more imaginatively, gene technology provides the opportunity for what some refer to as directed evolution as a design paradigm for all areas of biotechnology.[3] Moreover, gene research outside earth's one-gravity environment (courtesy of NASA's space shuttle) is already demonstrating promising results—evidently, microgravity influences gene expression, thereby demonstrating for the first time that a physical environment can effect such changes.[4]

For clarity in this paper, genetics and genomics need to be distinguished. Genetics usually means the study of gene inheritance, the passing on of traits and mutations down the generations. Genomics refers to the systematic identification and analysis of human genes and their functions (for example, seeking the molecular pathway between a mutation and tumor) as well as the information systems to process and interpret that data.[5]

Market forces are exacerbating the strenuous competition that fuels this remarkably promising research. However, the healthy competition that has always characterized academic pursuit of medical discovery now has a new edge sharpened by the financial pursuit of profit. Genomics companies have enhanced the connection between intellectual inquiry for health promotion and investment return for shareholder profit. This new partnership between medicine and money threatens to eschew the traditionally collaborative pursuit of research and therapies—and it is a very volatile partnership that runs the danger of measuring success more in fiscal productivity than health outcomes. This paper considers whether market forces are unethically mining the human genome, pursuing profitable seams to reap a bonanza for shareholders, as feared by some,[6] rather than contributing to medical science or population health.

Genomics companies are in a race in the genetics revolution. While the race occurs on the running track of medical genetics, the prize is financial gain: "a race of entrepreneurs to cash in on the genetics code."[7] This booming business of genomics needs to be held accountable to ethics discourse.

The purpose of this paper is to consider the ethical implications of this volatile partnership between medicine and money in human genomics. Because this partnership reflects how genomics companies function as organizations, the paper focuses upon organizational ethics. For the purposes of this analysis, "organizational ethics" refers to the integration of the values of the genomics companies with appropriate community stewardship of relevant resources and anticipated outcomes. The thesis is that organizational ethics can enable genomics companies to maintain their corporate integrity in the market economy of the biotechnology and biopharmaceutical industries. The means of promoting ethical integrity will be through community stewardship that helps genomics companies to realize their basic values of health outcomes, shareholder investment, and business patents. This first part of the paper considers these values to understand what they entail.

The Values of Genomics Companies

There are three basic values that genomics companies seem to share: promoting genetics health outcomes; enhancing shareholder investment; and seeking business patents. The reciprocity between medicine and money in genomics companies entails a keen balance between these values. When the balance is accomplished, it is likely that genomics research can work honorably and productively with market pressures. When a balance is out of alignment, it seems likely that financial gain can threaten to commodify medical pursuits. A fundamental question, then, is whether genomics can have ethical integrity in a market environment or whether companies will simply turn genomics into a commercial commodity.

1. Gene-Based Health Outcomes

Health outcomes of genomics research are multifaceted. The potential range of the impact of genomics includes considering several related aspects of gene research: the goals of gene sequencing and annotation; the changing paradigms for medicine; and developments in disease diagnosis and treatment.

1.A. Gene Sequencing and Annotation

Genomics refers to the science of the human genome, including sequencing the DNA in human genes. This involves sequencing the three billion letters in the human genetic code. The process of annotation in sequencing involves deciphering a gene's function, including the identification of sequence variation within genes, and entering the information into databases. Early in 2001, two competing groups published what was effectively a comprehensive draft of the human genome, identifying the 30,000 or so genes (not the 100,000 that many projected) in the human species.[8]

Single nucleotide polymorphism (SNP) technology transformed this genetic research because it facilitated scanning large areas of human DNA samples. The purpose is to detect subtle genetic variations specific to each individual, and to study factors that make them prone to disease. These small genetic differences located along the DNA molecule function as landmarks. And the landmarks enable researchers to sift through genetic material from thousands of individuals to uncover genes that predispose people to disease. Many genetic diseases are associated with SNPs, such as cystic fibrosis and muscular dystrophy.[9] The significance of this SNP-map technology should not be underestimated. For example, the technology enabled the human genome project to reach its goal much earlier than originally anticipated; also the technology is also likely to unveil the biological/molecular basis underlying all human variability.

1.B. Changing Paradigms for Medicine

The potential significance of human genomics is that the discoveries of science and developments of technology in genetics may change the focus of medicine from treatment to prevention. That is, instead of the focus on diagnosing and treating diseases of individuals (the so-called rescue paradigm of curative medicine), the human genome project may enable medical practice to focus on predicting and preventing diseases among populations and communities. Certainly genetic medicine will provide a welcome early-warning system that should enhance health promotion.[10] Even prior to mapping the human genome, genetic testing led to many clinical interventions that save lives.[11] And it is likely that genetic testing will change the current nature of health insurance[12] and probably even create a new wave of court rulings on discrimination.[13]

In this regard, there is emerging a clearer understanding of genetic influences on environmentally associated diseases, as indicated in the Environmental Genome Project (EGP) initiated by the National Institute of Environmental Health Sciences (NIEHS). The purpose is simply to develop disease-prevention strategies and programs of intervention. Specifically, the EGP will study common sequence variations in genes (known as genetic polymorphisms) that are not necessarily associated with an increased risk of disease. It will focus upon allelic variants with frequencies of more than 1 percentage in a population.[14]

1.C. Disease Diagnosis and Treatment

Genomics research places our society on the cusp of major breakthroughs for the treatment of gene-caused diseases. Sometimes gene technology enhances our capacity for better treatment by reliable early diagnosis. For example, the hereditary disease hemochromatosis is the most common genetic disorder in white populations. It is a genetic disorder causing iron overload that can lead to early death. The disease is easily treated by simple phlebotomy (removing blood regularly, thereby reducing iron levels) after gene diagnosis. Of course, this capacity for early diagnosis raises the question of widespread population-based screening, especially regarding the relevance of age and phenotypical expression, to detect and treat the disease effectively.[15] Also, genomics research continues to provide new opportunities for enhancement purposes, for example to assist the growth of human hair for baldness.[16]

Perhaps the most significant discoveries in genomics research occur with the development of gene-based drugs. There is a fundamental difference between biotechnology drugs and the more traditional chemotherapeutic agents that have been developed over recent decades. That is, biotechnology drugs are more narrowly targeted and they affect only the pathological cells.[17]

Some examples of gene-related drug research include the following. Genzyme Corp. (Cambridge, Massachusetts) has developed the drug Cerezyme for Gaucher's disease, and Genentech Inc. (San Francisco, California) has developed Activase to treat blood clots for heart attack victims. Another breakthrough in the Alzheimer's field occurred based on a set of experiments involving over eight thousand repetitions at the biotechnology company Amgen Inc. (Thousand Oaks, California).[18] The anticipated development of such gene-based drugs is being compared to equivalent breakthroughs with AIDS research in the mid-1980s—the discovery that HIV needs the replication of a protease (an enzyme that cuts protein), thereby leading to protease inhibitors, which radically enhanced AIDS treatment. The similarity to isolating the HIV protease enzyme is that beta-secretase, the enzyme linked to Alzheimer's, is also a protease. This discovery occurred, beginning in 1997, by seeking the gene that directs the relevant cells to make beta-secretase. The discovery of the gene enabled scientists to make the enzyme and show that it snips a protein linked to the brain cells, thereby releasing toxic fragments in the brain, forming amyloid plaques and so contributing to Alzheimer's disease. However, science continues to clarify whether amyloid buildup is the primary cause of Alzheimer's or a major consequence of it. Nonetheless, the task now for genomics companies is to develop a drug directed to this molecular target in order to block the beta-secretase enzyme.

Research on gene-based drugs has also raised expectations for the diagnosis and treatment of cancers. For example, recent discoveries in genomics are causing many to consider a more precise cancer classification system that focuses upon molecular variations in tumors in order to diagnose and treat cancer in the future.[19] Moreover, new drugs on the market are targeting mutant genes that regulate tumor growth. Some examples of drug companies targeting cancer-causing genes include the following. Herceptin from Genentech targets the gene HER-2, which plays a significant role in one third of breast cancers; Gentech has attributed its recent revenue growth in large part to the sales of Herceptin.[20] AstraZeneca PLC (United Kingdom) and Pfizer Inc. (New York) are researching drugs for another gene, the EGF receptor, that will help against breast, prostate, lung, hand, and neck cancers. Other companies, such as Merck & Co. (Whitehouse Station, New Jersey) and Bristol-Myer Squibb Co. (New York) are researching drugs against another gene, RAS, to treat colon, lung, and pancreatic cancers.[21]

Another example of research on gene-based drugs involves research by Glaxo Wellcome PLC on three common diseases that have proven remarkably difficult to treat: adult-onset diabetes, migraine headaches, and the skin ailment psoriasis.[22] To accomplish its objectives, Glaxo has joined with nine other drug companies using a new technology that tracks disease-causing genes: SNP technology. Other companies using SNP technology are also making significant discoveries in genetics. For example, Genset has been researching the disease-causing gene for schizophrenia, and Johnson & Johnson (New Brunswick, New Jersey) is licensed to use the discovery for the development of new drugs, which will take several years. Johnson & Johnson has also worked with a company from Montreal, Algene Biotechnologies Corp., which focused upon an isolated Canadian community to collect samples from forty schizophrenic patients traced to a common ancestor. Eli Lilly & Co. has accomplished success with its new drug for schizophrenia called Zyprexa. Other rivals researching schizophrenia include Rhone-Poulenc SA and also Roche Holding, with has a partnership with the Icelandic company Decode Genetics Inc. Because current schizophrenia drugs only alleviate symptoms, the race for a gene-based cure is becoming increasingly competitive.[23]

However, there are also dark clouds on the horizon of genomics research. Despite the upbeat scenario of gene research, with high hopes for new drugs and technology, a wind of caution wafts over genome companies. It is unlikely that many significant gene-therapy drugs will be widely available in the immediate future. This is because of the technological difficulties involved, especially with regard to effective vectors for gene therapy. An example is the Swiss drug-maker

Sandoz AG, which merged with its Swiss rival Ciba-Geigy AG in 1996 to form Novartis AG. In 1995 Sandoz bought the U.S. biotechnology company Genetic Therapy Inc. (Gaithersburg, Maryland) for $300 million and in 1997 acquired SyStemix Inc. for $548 million. In a subsequent move, all but one of the programs on human gene testing were constrained at GTI and SyStemix. Given its original investment of approximately $848 million to acquire these biotechnology companies, the corporation became more cautious, because hundreds of human studies conducted around the world offered disappointing results on gene therapies. In 1998 Novartis abandoned its flagship gene-therapy project for the brain cancer glioblastoma after poor results from studies since 1990, including a large-scale trial with 200 patients beginning in 1996. Novartis postponed new human gene–related tests until technical issues were resolved.[24]

In sum, the hope for human genomics is that gene-based health outcomes will result from several related phenomena that include the sequencing and annotation processes in the Human Genome Project; a concomitant shift in medical paradigms that will give primacy to detection and prevention; and new drugs that will provide effective gene diagnosis and therapy. However, to accomplish these health outcomes, genomics companies need to raise enormous capital investment for research and development. Understanding this relation between anticipated health outcomes and antecedent shareholder investment provides the business context for developing organizational ethics for genomics companies.

2. Shareholder Investment

The impact of shareholder investment upon genomics companies cannot be underestimated. While attempting to protect and enhance this investment, these companies are faced with intense competition that puts them at financial risk, which they manage by mergers and collaborations. This business scenario clarifies the context for understanding why these companies must be responsive to shareholder investment.

2.A. Competition

Genomics companies function in an adversarial system of intense competition in two main areas: first, between the public and private sectors of research; and second, among fellow companies whose rivalry for success is measured by the size of potential profits for shareholders.

There is a continuing competition between the private research by genomics companies and the public research of government-funded projects. The public consortium in the United States and United Kingdom, which began its work in 1990, was financed by the National Institutes of Health (NIH; Bethesda, Maryland) and the Wellcome Trust (London). Subsequently, the investments of genomics companies in an effort to sequence the human genome led to a concomitant increase in government funding to maintain a competitive edge. Some of the toughest competitors in the private sector are former personnel of the NIH. For example, J. Craig Venter left NIH to form the Celera Genomics Group (Rockville, Maryland) with Perker Elmer Corporation, the largest manufacturer of genetic sequencing machines; he was also the founder and chairman of the board of the Institute for Genomics Research (Rockville, Maryland). He began his work at Celera on the human genome in May 1998, selling much of the emerging data to subscribers, thereby attracting several blue-chip subscribers such as Novartis AG, Amgen, and Pharmacia & Upjohn.[25] Another example is William Haseltine, M.D., chairman of Human Genome Sciences Inc. (Rockville, Maryland). When he was a professor with the Dana-Farber Cancer Institute, he isolated the HIV protease, which after nine years led to the protease-inhibitor drugs.[26]

Biotechnology and genomics companies are funded by venture capitalists and private investors, by investment banks and the stock market, and by large pharmaceutical companies. So the days of government funding leading most of the basic research are over. Hence genomics companies must remain responsive to their varied investment constituencies as they address the multifaceted environment of financial risk.

2.B. Financial Risk

The financial risk for genomics companies is extraordinarily high. Previously, it took about ten years at a cost of $350 million for a new drug to reach the market, with a 99 percent failure rate. The process of gene-based drug development ordinarily involves these steps: first, the gene must be tracked and identified; second, scientists must identify the function of the chemical protein that the gene produces; third, there is a determination of the role the protein plays in causing a specific disease; and only then can drug-makers begin to develop a drug that addresses the disease pathway. Despite this pessimistic ratio, by the turn of the millennium genomics companies represented a sustained growth sector for the biotechnology investment markets hoping for major breakthroughs.[27]

However, the experience of Novartis provides a high-profile example of the financial risk entailed by genome research. After investing approximately $848 million to acquire GTI and SyStemix to pursue its gene research programs, Novartis curtailed its gene therapy programs. Although this amount is only a fraction of the $2 billion annual health care research budget at Novartis, the results caused the company to adopt a more cautious approach to gene drug testing.[28] Even large companies, such as Amgen and Genentech, have difficulty competing in size with the major pharmaceutical companies. In a Hambrecht and Quist report for February 1999, 53 percent of the biotechnology companies surveyed had only two years or less of cash available and 30 percent had one year or less. This financial situation places genomics companies in a hostile landscape where prospering requires success in developing drugs and information databases.[29]

The staggering amounts of money involved in research for gene-based drugs demonstrate the financial risk that genomics companies face. A few examples illustrate this concern. The market capitalization of Millennium Pharmaceuticals Inc. (Cambridge, Massachusetts) exceeded two billion dollars, with revenues in 1998 of $134 million. Similarly, the market capitalization of Human Genome Sciences was just over one and a half billion dollars, with revenues in 1998 of $30 million. And a similar picture emerged with regard to the genomic companies that focused upon selling information; the market capitalization of Incyte Pharmaceuticals Inc. (Palo Alto, California) was $713 million, with revenues in 1998 of $135 million, and the market capitalization of Celera was $650 million, with revenues in 1998 of $6 million.[30]

In this fast-moving business environment of genomics companies, the cost and availability of technology often determine who will be the successful competitor. For example, Celera was the first company to gain access to the automated gene-sequencing machines (from PE Biosystems Group, which belonged to the PE Corporation). These function by forcing fragments of DNA through hair-fine tubes in which the sequence of the bases (the four units of DNA) are read electronically. In 1999 (with approximately 300 such devices at a cost of

$300,000 each) Celera dwarfed the largest public genome center, the MIT laboratory which had only 120 similar devices. Most other genomics companies had only a few of these devices; Celera was completing a billion base pairs every month or less than a month.[31]

In this market environment, the concern whether one company may become dominant (akin to Microsoft in computing software) tends to spawn abundant mergers and collaborations.

2.C. Mergers and Collaboration

One of the more prominent mergers occurred when the biotechnology company Genzyme bought the gene therapy company Cell Genesys Inc. for approximately $350 million in the form of a stock swap.[32] And one of the more prominent collaborations in the biotechnology industry was between Onyx Pharmaceuticals Inc. (Richmond, California) and Warner-Lambert Company. Their agreement was to work together to develop and market a new cancer drug. The reason for the arrangement was interesting because the deal came at a critical moment for Onyx's survival. Warner agreed to pay $155 million (in a combined cash payment and equity investment) to Onyx for access to its therapy using a genetically modified virus to kill cancer cells. The drug was referred to as ONYX-015, an adenovirus that was genetically altered to replicate in cells lacking the p53 gene, a tumor suppressor. Phase 3 trials (the last phase before the Food and Drug Administration considers a new drug for market approval) were planned, but Onyx did not have the resources to fund the trials. The agreement included Warner providing $40 million for the Phase 3 trials. The payoff, so to speak, was immense: Warner agreed to pay more than $100 million to Onyx when it successfully reached specific milestones such as demonstrating positive results or receiving marketing approval. Although Warner and Onyx will share profits equally in North America, Warner will commercialize the products for the rest of the world, paying only a royalty on net sales to Onyx.[33]

Another example of high-profile collaboration was Millennium Pharmaceuticals securing financing of over one billion dollars from companies including Roche Holding AG, Eli Lilly & Co., American Home Products, Pfizer, and Bayer AG as partners. Millennium also partnered with Monsanto for agricultural products, an arena that sparked an intense international backlash over the biotechnology crops and food industry (opponents claiming that the biotechnology companies treat people as guinea pigs).[34]

Briefly, the importance of shareholder investment for genomics companies emerges from recognizing a competitive business environment with colossal financial risks that constantly spawn mergers and collaborative ventures to maximize market position. As genomic companies seek to be responsive to their investments, they pursue business patents vigorously, thereby raising one of the greatest concerns from the perspective of organizational ethics. Patents raise the problem of exclusivity, in contrast to government-sponsored research which is openly available in the public domain for all.

3. Business Patents

Seeking business patents constitutes a third value of genomics companies. These companies try to protect their research investment by seeking patents and licensing agreements whereby intellectual property rights (a negative right to exclude others) can be claimed for a period of two decades. By acquiring patents they establish an assured basis to fund their subsequent research while returning profit to their shareholders. Traditionally, in the United States patents were recorded only for inventions. But in a landmark case in 1980 the U.S. Supreme Court ruled (with four of the nine judges dissenting) in *Diamond v. Chakrabarty* to permit a patent for genetically engineered microorganisms, specifically a bacteria strain that was altered genetically to cause oil deterioration. With this radical extension of patent law as a protection of intellectual property rights, patents can be granted for gene-sequencing discoveries.[35] Also, in 1987 the U.S. Patent and Trademark Office permitted patenting for genetically altered animals.[36] Hence the ethical debate on the legitimacy of intellectual property rights for gene discoveries needs to be attentive to the function of patents, and nuanced ethical discourse on patents needs to focus on the main goals of genomics investment, including how these goals are aligned with developments in computer science and artificial intelligence technology.

3.A. Investment Goals of Genomic Companies

Genomics research and investment are built upon combined two goals that shape current science: discovering and distributing new drugs; and discovering and disseminating new genetic information to others such as biopharmaceutical corporations.

First, the focus upon developing gene-based drugs can be illustrated by two successful genomics companies, Millennium Pharmaceuticals and Human Genome Sciences. For example, Millennium Pharmaceuticals develops small-molecule drugs, proteins and antibodies, and gene-based diagnostic tests; it also provides genetic targets for drug discovery by its partners. Human Genome Sciences also develops proprietary proteins, antibodies, and gene-therapy drugs and provides genetic targets for drug discovery by its partners. One such drug is to protect blood-forming cells from chemotherapy toxicity (currently being tested on breast and ovarian cancers); another such drug is a wound-healing agent; and a third therapy involves injecting a gene into a diseased area to grow new blood vessels—the gene, named VEGF-2 and being tested as a drug, is a product of Vascular Genetics Inc., which is a joint venture with Human Genome Sciences and others; early results indicate growth of new blood vessels in 60 percent of the previously blocked site.[37]

However, the unanticipated death of an otherwise healthy eighteen-year-old Arizona teenager in a trial in September 1999 signaled a warning bell for gene drug therapy.[38] The patient, Jesse Gelsinger, died suddenly of liver and other organ failure while being treated experimentally for a rare genetic liver disorder (a genetically caused enzyme deficiency). He was receiving by injection a genetically modified virus (an adenovirus) to deliver high doses of a therapeutic gene to his liver. Researchers were using the virus to insert the p53 gene, which is often defective in cancer tumors, to determine if the gene can slow or stop unwanted cell division.[39] An overreaction of his immune system led to respiratory, liver, and kidney failure, causing his death within a few days. The family eventually settled a lawsuit out of court.[40]

Second, in contrast to developing drugs, other genomics companies focus upon disseminating gene-related information and plan only to sell data necessary for identifying potential drug targets. These companies reap profits by selling access to information via their proprietary databases of genetic information rather than by discovering and patenting disease genes. Examples of these companies are Incyte Pharmaceuticals and Celera Genomics Group. Incyte provides databases of gene sequences and analytical software via subscription; it also manufactures and sells drug discovery tools. Celera provides genomics and biological information; it also plans to offer analytical software and promises complete genomes of human and other species. This dissemination of information technology also occurs via the Internet. For example, Pangea Systems Inc. (Oakland, California) created the Web site Doubletwist.com (a play on the double helix structure of DNA) to make online research in genetic biotechnology readily available to amateurs and professionals alike. The company hopes the Web site will provide a competitive edge in the hot pursuit for business in bioinformatics.[41] In this scenario, some genome companies have created substantial operating income from their sales without being burdened with significant outlays on drug development.

Not surprisingly, in this quest for gene-based drugs and disseminating gene-related information, the contribution of bioinformatics is crucial. The increasing role for artificial intelligence in modern gene technology highlights the urgency of seeking appropriate protections for genomics discoveries by filing patents.

3.B. Genomics and Bioinformatics

Developing gene-based drugs and designing databases for the dissemination of genetic information rely significantly upon the interface between molecular science and computer science, with its artificial intelligence research capacity. Government-funded projects store this electronic data in GenBank, the gene-sequencing repository maintained as an online service by the National Center for Biotechnology Information (NCBI) at the NIH. An example of a crucial breakthrough in bioinformatics is pattern recognition to discover genes and to determine what they do. Pattern recognition uses algorithms to detect hidden structures or patterns in molecular data. The National Human Genome Research Institute's genome informatics branch was spending $20 million annually to support and develop this area as it mapped the human genome. The process of deciphering a gene's function and entering it into a database is called annotation, which became a complementary goal of the Human Genome Project. The Sanger Center (London) and the European Bioinformatics Institute (EBI)—part of the multinational European Molecular Biology Laboratory (Hinxton, United Kingdom)—wanted to ensure that the first draft of the human genome would have annotation attached. This project, called

EnsEMBL, used gene-finding algorithms to provide a map of the location of the genes and the Basic Local Alignment Search Tool (BLAST) program, designed by the National Human Genome Research Institute, to determine the gene's function. Interestingly, this capacity for pattern analysis was at the heart of Celera's plans to compete with the Human Genome Project. Pattern-recognition technology generates specific devices called DNA chips that apply pattern-recognition software to gene expression. The gene's expression level refers to the number of copies of protein it makes at specific times. The DNA chips measure the expression levels of thousands of genes simultaneously. This capacity shifts genetic technology from focusing upon one gene at a time to interpreting thousands of genes simultaneously, thereby detecting the general pattern of gene expression.[42]

Artificial intelligence technology, especially with regard to pattern recognition in genome bioinformatics, generates concomitant discoveries that genomic companies need to protect. Their preferred instrument of protection for these discoveries, in both genetic information and gene-based drugs, is to file a patent that protects their intellectual property rights.

3.C. Intellectual Property Rights

In the race to develop gene-based drugs, accruing patent portfolios is crucial. Genomics companies try to keep specific genetic information proprietary. The more commercially interesting information lies in the DNA variations that are very common. These variations determine the gene performance and help to determine susceptibility to disease. It is by compiling these common DNA variations that genomic companies plan to be commercially viable by filing patents.[43] For example, Genzyme bought Cell Genesys (mentioned previously) with over 200 patents and 330 patent applications pending.[44]

Patents encourage confidence among investors that there is likely to be long-term profit from investment. Patents try to justify the huge up-front expenditures that are necessary for developing new genomic technologies and therapies. Hence genomics companies use patents to protect the billions of dollars they have invested in basic research. If genomics companies could not profit from their investment by protecting their discoveries via patents, it seems unlikely that the torrent of information being produced by these companies would continue.

There is little doubt that the research investment by genomics companies has shown how slow the previous academic model of government-funded research was by comparison. In this sense, patents can clearly support the vigor of research.

However, these patents raise one of the greatest concerns about genomics companies from the perspective of ethics. Patents have a symbolic meaning that projects a crucial question when medicine and money become so entangled in genomics research. The question is whether genomics can have ethical integrity in a market environment, or whether these companies will simply turn genomics into a commercial commodity. Put another way, gene patents raise the ethical question about the justice of claiming intellectual property rights for something not invented but discovered. To do so implies making property claims on human nature. In this regard it seems indispensable to recognize genetic information as social information if we are to create appropriate social policy for the nation in general and regulation for genomics companies in particular.

4. General Conclusion

In a real sense, then, gene patents symbolically depict the tension between medical research and market pressure that causes ethical concerns for many. The quest by genomics companies for patents represents the integration of their corporate values of promoting health outcomes while enhancing shareholder profit. Of course, establishing a balance between medical research and the financial markets is not easy. The debate on patents helps to ascertain whether this balance is accomplished. An appropriate balance may enable genomics research to work honorably with market pressures. The lack of such a balance may cause financial gain to commodify medical pursuits. So patents represent the fulcrum of this balance.

This chapter argues that organizational ethics can help to establish an appropriate balance for genomics companies. That is, organizational ethics can enable genomics companies to maintain their corporate integrity in the volatile market economy of the biotechnology and biopharmaceutical industries. The suggested means of promoting ethical integrity is through community stewardship as genomic companies try to realize their values of promoting genetics health outcomes, enhancing shareholder investment, and seeking business patents. This final section of the paper considers the ethical questions that community stewardship requires for genomics companies to realize their values in a manner that maintains their corporate integrity.

4.A. Community Stewardship

The ethics of stewardship entails a responsible use of limited resources to enable a specific entity to flourish with regard to its legitimate mission or identity. For example, a nation must steward its resources to allow its citizens to flourish in a manner that respects their individuality and commonality. A health care organization must steward its resources to sustain its continued provision of quality care in a manner that is cost-effective for the communities it serves. Similarly, the ethics of stewardship can provide appropriate guidance for genomics companies to use their resources for community benefit in a manner that assures their continuance in a competitive business environment.

Previously, patents were identified as representing the mechanism that enables genomics companies to balance promoting health outcomes and seeking financial gain. Simply, patents represent a more general tension between company profits and the public's interest in genomic research to curtail corporate greed with regard to what might be the very foundation of human life. But community stewardship calls genomics companies to social accountability in this regard. After all, the purpose of stewardship in this exciting landscape is to develop genetic medicine that will enhance human flourishing. Of course, community stewardship must also recognize the scientific and commercial context of genomics companies. The ethics of stewardship can enable genomics companies to maintain their balance with integrity if they can resolve basic dilemmas posed by patents for the community. These issues deal especially with shielding research data and being accountable to community oversight.

4.B. Shielding Research Data

One of the most difficult ethical problems created by genomics companies is shielding research data from public knowledge and from scientific scrutiny in order to maximize potential profits through developing new gene-based drugs and tests. The ethical dilemma addresses the legitimacy of shielding scientific information that is crucial for human flourishing. The dilemma involves the meaning of human rights for individuals and society with regard to protecting the human nature of each person and our species. In religious discourse, the dilemma also raises the metaphor of inappropriately playing God, even though some argue that there are no distinctly religious grounds against patenting human DNA.[45] Religious discourse seeks to have an impact on the entire field of genetic research and technology.[46]

In this sense, patents to enhance profits can harm both scientific and medical research through a threefold legacy: withholding information that can help the public; restricting access to data; and preventing effective competition. For example, withholding genetic information from government or university researchers to protect the company's investment can be construed as stalling scientific progress.[47] Of course, this trend extends beyond genetics, though it is especially pertinent therein. The problem, of course, is that when patents are licensed exclusively to clinical laboratories (with accompanying patent or licensing fees), other scientists may be prevented from performing the patented tests.[48] On the other hand, the assumption that scientific research results are shared readily does not bear close scrutiny. Some data suggest a delay period ordinarily of six months before academics disseminate research data publicly, often to protect the commercial value of results.[49]

Nonetheless, there is a fundamental dilemma that gene patents raise: as the product of nature, may genes become private property to be mined for profit, or should they be permanently available for the public interest? Even Thomas Jefferson, who as secretary of state supervised the drafting of the first patent laws in the United States, opposed patent monopolies that were contrary to the public interest.[50] So gene patents raise basic questions—about individual rights over one's own genes—that entail significant social policy concerns.[51] Until we resolve the debate on gene patents, it seems likely that antagonisms and lawsuits between genomics companies and universities will occur, such as in the patent suit *Regents of the University of California v. Genentech Inc.* The dispute was about the sale by Genentech of a human growth hormone product called Protropin; the University of California claimed infringement on one of its patents. The two-decade-long dispute (it began in 1978) became complicated because of the claim that another Genentech growth hormone product called Nutropin infringed the same patent. The university sought $500 million in royalties and interest, with the possibility of $1.5 billion in damages.[52]

Perhaps there is a way around future standoffs between industry and the academy, especially when gene patents are involved. It may be possible that free access to basic scientific information is compatible with commercial interests. An example is the collaboration, albeit limited, between the academy and industry to voluntarily provide unlimited access to a specific SNP database as a referent for individual companies to develop gene-based drugs.[53]

4.C. Community Oversight

In addition to resolving the current standoff between corporate and community interests with regard to gene patents, there are other basic concerns regarding community oversight of crucial aspects of genomic research. This appeal to community oversight is intended to bring genetic discourse more into the public square of common discourse.[54] Specifically, community oversight seeks to foster more interaction between genetic experts and the broader community in a dialogical manner that integrates effective communication of scientific and public discourse.[55] In other words, community oversight raises a basic question of hermeneutics with regard to fostering effective conversation on values for public life.[56] With regard to community oversight of gene research, there are several particular issues that might be considered.

First, researchers in genomics companies are beginning to alter traditional conventions for reporting scientific discovery, such as reporting progress on disease-causing genes before publishing the data in the traditional form of a peer-reviewed scientific journal. This constitutes a significant shift away from scientific and medical research conventions. Of course, the rationale for this shift reflects the requirements of patent law in the United States. That is, patent attorneys advise researchers to limit what they reveal prior to applying for a patent because the relevant laws require that the content should not be revealed in "prior art" (including reports, abstracts, and published papers).[57] But this narrow focus on patent law requirements can reverse the traditional culture of open scholarly discourse.

The ethics of stewardship seeks community accountability, including the scientific community and the general public. This accountability requires the dissemination of sound data. Previously, traditional conventions were useful for weeding out unreliable discoveries. Only the most serious reasons would seem to legitimize breaking this convention in the new landscape of genetic medicine. Breaking a convention in this manner is somewhat reminiscent of another breach of convention after cloning the Scottish ewe Dolly.[58] In that situation, the scientist published the scientific data prior to the usual duplication of the experiment, though relevant duplication occurred with other animals after a year. This call to proper stewardship of data does not mean a company must reveal everything in the public context of scientific discourse. Clearly, stewardship also respects a company's need to restrict access to data for commercial purposes. In such cases, new information that cannot be publicly justified should not be conveyed in a public context. Presenting scientific results without supporting data can create expectations in the general public that may be deceptive and may thereby undermine confidence in scientific integrity.

A second concern regarding community accountability deals with secrecy that muffles government-funded research. There is a growing trend of genomics companies to establish partnerships with academic research teams where increased funding comes with a hefty condition—the constraint of access to research data as a protection against competitors. This trend inverts the more traditional approach whereby drug companies applied their technology to the biological discoveries made by government-funded laboratories—discoveries available in the public domain to everyone. This problem reflects a significant shift in how basic research is undertaken in genomics today. During the 1990s the tactics of pharmacology changed significantly. Previously, basic research occurred in NIH-funded academic laboratories; pharmaceutical companies then applied the research as they developed effective drugs. However, in the new paradigm, biotechnology companies undertake much of the basic research in a commercially funded manner.[59]

The ethics of stewardship must be attentive to this fundamental shift in paradigm that reconfigures the relation between the public and private realms in genetic research. In the previous paradigm, discoveries funded by government research were available to everyone. From the perspective of community stewardship, there needs to be accountability for access to and use of government-funded research, including use by genomics companies. Stewardship requires social accountability of information gleaned from government-funded research, especially when it has an immense impact on the medical condition of millions of future patients.

A third concern for community accountability deals with competency, accuracy, and oversight of genetic testing. Perhaps our society is not quite ready for the genetics revolution, especially when we consider the impact it may have on a system of health insurance that excludes a large segment of the population.[60] Nonetheless, there must be social accountability with regard to the competency of those who give the tests, the accuracy of the tests, and the entire community's oversight of the quality of genomics technology.

A fourth concern deals with the problem of commodification. This problem is well recognized in the managed care debate, where patients are seen as consumers. This debate revolves around the tension between altruistic service as a perceived tradition of health professionals and self-interest as a perceived hallmark of the for-profit sector of health care. As genomic companies pursue vigorous research to develop their patent portfolios, there emerges an increasing commodification of health-related outcomes. In this context, commodification can be described as "an economically valued good within a relationship of exchange," thereby making the relationship an instrumental rather than an intrinsic value.[61] Insofar as commodification occurs in degrees according to a variety of contexts, it seems that is inevitable in health care to some extent. The critical question is whether commodification of genes, especially through the mechanism of patenting, is ethically inappropriate.

Summary Remarks

Genetic engineering provides a dominion over human nature that has never been equaled. Genomics companies hold the hereditary blueprints of human life. Hence genomic commerce requires the utmost vigilance as a remarkably powerful economic enterprise that explores the depths of our biological condition and will be able to influence virtually every disease imaginable.

This chapter argues that community stewardship can enable genomics companies to accomplish their values in a manner that maintains their corporate integrity while avoiding the danger of reducing genomics to being a commercial commodity. Genomics raises many new ethical problems. Community stewardship must be at the root of resolving them. Generally speaking, this appeal to the ethics of stewardship requires a community accountability that necessarily involves the inclusion of all the relevant constituencies in debates about what ought to occur in genomics research.

Organizational ethics for genomics companies eventually leads to a public appeal for education about the amazing possibilities on the horizon. Such education must address the many specific problems such as privacy and confidentiality, disclosure of results, insurance and discrimination, risk-benefit analysis, and so on.[62] But underlying all of the specific questions will be an ongoing curiosity about the meaning of being alive when humans can harness a proverbial holy grail, the molecular basis of life. In other words, effective education about emerging possibilities for gene manipulation and modification must become the hallmark of community stewardship of genomics research. Ethical discourse may not occur exhaustively prior to scientific discovery and technological breakthroughs. And it may very well be inappropriate to burden the pace of scientific research with endless public debate that may seldom reach consensus. Yet prudence dictates that public scrutiny rather than investment profits should determine the future pathways for genomics research, especially when the evolutionary future of our species is at stake.

NOTES

1. National Bioethics Advisory Commission, *Ethical Issues in Human Stem Cell Research*, 3 vols. (Rockville, MD: Government Printing Office, 1999-2000); Department of Health and Human Services, "National Institute of Health Guidelines for Research Using Human Pluripotent Stem Cells," 65 *Federal Register* 65:166 (2000):51975-81. Also see http://www.nih.gov/news/stemcell/stemcellguidelines.htm; Kevin W. Wildes, "The Stem Cell Report," *America* (1999):12-16; Cahill, L. S., "The New Biotech World Order. Symposium: Human Primordial Stem Cells," *Hastings Center Report* 29:2 (1999):30-48; A. R. Chapman et al., *Stem Cell Research and Applications: Monitoring the Frontiers of Biomedical Research* (Washington D.C.: American Association for the Advancement of Science and Institute for

Civil Society, 1999); John C. Fletcher, "The National Bioethics Advisory Commission's Report on Stem Cell Research: A Review," *ASBH Exchange* 3:1 (2000):8–11; President George Bush, "Address on Federal Funding of Embryonic Stem Cell Research," *Origins* 31 (2001):213–15.

2. Gary Schneider and John Noble Wilford, "Inside Out: Genetic Self-Portrait," *New York Times* Magazine (October 17, 1999):107–13, and the one-page advert on "Genetic Engineering," in *New York Times* (October 11, 1999):A11.

3. Ellen Licking, "Evolution on Fast-Forward: Letting Nature Do the Heavy Lifting, Gene Scientists Are Creating Super New Materials," *Business Week* (September 27, 1999):140–42, at 142.

4. A. J. S. Rayl, "Microgravity and Gene Expression: Early Results Point to Relationship," *The Scientist* 13:18 (1999):1, 8–9.

5. Joe Flower, "Genomics vs Genetics," *Health Forum Journal* 42:4 (1999):61; also see Karl A. Thiel, "The Millennium Minuet," *Forbes ASAP* (May 31, 1999):80–81.

6. Antonio Regaldo, "Mining the Genome," *Technology Review* (September/October 1999):55–63.

7. Lawrence M. Fisher, "The Race to Cash in on the Genetic Code," *New York Times*, "Money & Business" (Sunday, August 29, 1999): sec. 3, pp. 1 and 12, at 12. The information in the first part of this paper is developed from this article. Material from other sources is indicated accordingly in the subsequent notes.

8. Francis Collins, et al., "Initial Sequencing and Analysis of the Human Genome," *Nature* 409 (2001):860–921; Craig Venter, et al., "The Sequence of the Human Genome," *Science* 291:5507 (2001):1304–1351.

9. George Johnson, "Gene Chip Technology," *St. Louis Post-Dispatch* (October 22, 1999):A16.

10. Robert Langreth, "Early Warnings. Genetic Tests Will Allow People to Take Steps to Delay—or Prevent—the Occurrence of Disease," *Wall Street Journal* (October 18, 1999):R7.

11. For example, see William Allen, "Children's Case Shows How Quickly Genetic Discoveries Start Saving Lives," *St. Louis Post-Dispatch* (September 20, 1999):A1, A5. In this story, two children were diagnosed with pulmonary surfactant protein B deficiency which leads to death unless a lung transplant occurs. The year before their birth, researchers identified the genetic mutation causing the deficiency and developed a rapid screening test to detect it and thereby plan for lung transplants for the children.

12. For example, Mark A. Hall and Stephen S. Rich, "Genetic Privacy Laws and Patients' Fear of Discrimination by Health Insurers: The View from Genetic Counselors," *Journal of Law, Medicine & Ethics* 28 (2000):245–57;

Mark A. Hall and Stephen S. Rich, "The Impact of Genetic Discrimination Laws Restricting Health Insurers' Use of Genetic Information," *American Journal of Human Genetics* 66 (2000):293–307; also see Nancy Ann Jeffrey, "A Change in Policy. Genetic Testing Threatens to Fundamentally Alter the Whole Notion of Insurance," *Wall Street Journal* (October 18,1999):R15.

13. For example, Lori Andrews, the director of the Institute for Science, Law, and Technology at Illinois Institute of Technology in Chicago described how a husband persuaded a South Carolina court to require mandatory testing for Huntington's disease, a fatal condition, for the wife during divorce proceedings as the spouses contested custody for their one child; the mother chose to flee, thereby forfeiting custody, rather than submit to genetic testing; see Rhonda L. Rundle, "What Should I Do? The More We Know about Genes, the More Difficult Questions We Will Face," *Wall Street Journal* (October 18, 1999):R16.

14. Richard R. Sharp and J. Carl Barrett, "The Environmental Genome Project and Bioethics," *Kennedy Institute of Ethics Journal* 9:2 (1999):175–222, at 176–78, where the three main objectives of the EGP are described.

15. John K. Olynyk, et al., "A Population-Based Study of the Clinical Expression of the Hemochromatosis Gene," *New England Journal of Medicine* 341:10 (1999):718–32; and Anthony S. Tavill (editorial), "Clinical Implications of the Hemochromatosis Gene," *New England Journal of Medicine* 341:10 (1999):755–57. Also see Alan E. Guttmacher, M.D., "Genetic Medicine in the Next Five Years," *Health Progress* 80:5 (1999):43–44.

16. Nicholas Wade, "Of Mice, Men and a Gene that Jump-Starts Follicles," *New York Times* (October 5, 1999):D1.

17. Joe Flower, "The Way It Is, Is Not The Way It Will Be," *Health Forum Journal* 42:4 (1999):16–27, at 20.

18. Gina Kolata, "Scientists Find Enzyme Linked to Alzheimer's," *New York Times* (October 22, 1999):A1, A20.

19. Joan Stephenson, "Human Genome Studies Expected to Revolutionize Cancer Classification," *Journal of the American Medical Association* 282:10 (1999):927–28.

20. Lawrence M. Fisher, "Roche Cancels Bond Offering on Genentech. But New Stock Issue Will Stay on Track," *New York Times* (October 21, 1999):C6. Also see Ralph T. King, "Roche to Unload as Much as 17 percent of Genentech," *Wall Street Journal* (October 11, 1999):B2. On September 2, 1998, the Advisory Committee of the Food and Drug Administration unanimously recommended approval of the new drug Herceptin for women with advanced breast cancer (*New York Times*, September 3, 1998).

21. Ron Winslow, Robert Langreth, Michael Waldholz, "Disease by Disease," *Wall Street Journal* (October 18, 1999):R4.

22. Michael Waldholz and Elyse Tanouye, "Going Public. Glaxo to Report It's Closing In on Genes Linked to 3 Diseases," *Wall Street Journal* (October 19, 1999):A1 and A12.

23. Stephen D. Moore, "Johnson & Johnson, Genset Find Gene That Could Lead to Schizophrenia Cure," *Wall Street Journal* (October 19, 1999):B4. See a previous report on the pace of progress, Stephen D. Moore, "Gene Hunters Forge Uneasy Ties in Search for a Cure," *Wall Street Journal* (August 25, 1999):B1, B8.

24. Robert Langreth and Stephen D. Moore, "Delivery Shortfall. Gene Therapy, Touted as a Breakthrough, Bogs Down in Details," *Wall Street Journal* (October 27, 1999):A1, A6.

25. Marilynn Larkin, "J. Craig Venter: Sequencing Genomes His Way," *Lancet* 353:9171 (1999):2218.

26. Joe Flower, "A Brave New Medicine: A Conversation with William Haseltine," *Health Forum Journal* 42:4 (1999):28–30, 61, at 30; Joe Flower, "The Promise of Genomics," *Health Forum Journal* 42:4 (1999):62–63, at 62.

27. Cynthia Robbins-Roth, "Magic Bullets: The Breakthroughs, the Business, and the People of Biotechnology," *Forbes ASAP* (May 31, 1999):42–43.

28. Langreth and Moore, "Delivery Shortfall."

29. Karen Southwick, "Slow Growth. Why Biotech Companies Need Time to Mature," *Forbes ASAP* (May 31, 1999):72–73.

30. Lawrence M. Fisher, "The Race to Cash in on the Genetic Code."

31. Nicholas Wade, "Rivals Reach Milestones in Genome Race," *New York Times* (October 26, 1999):D3.

32. "Genzyme Corp," *Wall Street Journal* (October 19, 1999):A12.

33. Lawrence M. Fisher, "Warner-Lambert Signs Up a Partner to Work on Cancer Drugs," *New York Times* (October 19, 1999):C6; also see Associated Press report, "Warner-Lambert Profit Rises," *New York Times* (October 19, 1999):C20.

34. Lucette Lagnado, "Group Sows Seeds of Revolt against Genetically Altered Foods in the U.S.," *Wall Street Journal* (October 12, 1999):B1, B4; also see Michael Pollan, "Playing God in the Garden," *New York Times Magazine* (October 25, 1998):44–51; Charles C. Mann, "Biotech Goes Wild. Genetic Engineering Will Be Essential to Feed the World's Billions," *Technology Review* (July/August, 1999):36–43; Robert Steyer, "Biotech Battle Opens Way for Test-Makers," *St. Louis Post-Dispatch* (September 19, 1999):A8; Robert Steyer, "Monsanto Stock Takes a Beating Amid Biotech

Worries," *St. Louis Post-Dispatch* (October 3, 1999):C3; Bill Lambrecht, "Monsanto Appears Ready to Join Public Debate on Genetically Engineered Food," *St. Louis Post-Dispatch* (October 3, 1999):A2; Bill Lambrecht, "Genetic Research on Plants Steams Ahead Here, Despite Concerns in Europe," *St. Louis Post-Dispatch* (October 10, 1999):A1; Steve Stecklow, "Foodstuff: Genetically Modified on the Label Means . . . Well, It's Hard to Say," *Wall Street Journal* (October 26, 1999):A1, A14.

35. *Diamond v. Chakrabarty*, 447 U.S. 303 (1980). Also see Jonathan King and Doreen Stabinsky, "Patents on Cells, Genes, and Organisms Undermine the Exchange of Scientific Ideas," *Chronicle of Higher Education* (February 5, 1999):B6–B7.

36. This patent was for the Harvard oncomouse. See M. Cathleen Kaveny, "Jurisprudence and Genetics," *Theological Studies* 60 (1999):135–147, at 144–47.

37. Lawrence M. Fisher, "A Vitalizing Gene, Straight to the Heart," *New York Times* (August 29, 1999):12.

38. "Patient Dies in Gene-Therapy Trial at University of Pennsylvania Medical Center," *Chronicle of Higher Education* (October 1, 1999):A23; also see Nicholas Wade, "With a Death, Advocates of Gene Therapy Express Concerns for Future of the Field," *New York Times* (September 30, 1999):A20; Robert Langreth, "Gene Therapy Is Dealt Setback by the FDA," *Wall Street Journal* (October 11, 1999):B1.

39. LeRoy Walters, "The Oversight of Human Gene Transfer Research," *Kennedy Institute of Ethics Journal* 10 (2000):171–74; Joan Stephenson, "Studies Illuminate Cause of Fatal Reaction in Gene-Therapy Trial," *Journal of the American Medical Association* 285 (2001):2570.

40. "Family Settles Suit over Teen's Death During Gene Therapy," *Wall Street Journal* (November, 6, 2000):B6.

41. Lawrence M. Fisher, "Surfing the Human Genome: Data Bases of Genetic Code Are Moving to the Web," *New York Times* (September 20, 1999):C1, C6.

42. Antonio Regaldo, "Mining the Genome," *Technology Review* (September/October, 1999):55–63, at 58. Other companies that specialize in "pattern-recognition" software include: Bioreason (Santa Fe, New Mexico), Compugen (Tel Aviv, Israel), Lion Bioscience (Heidelberg, Germany), Neomorphic (Berkeley, California), Partek (St. Peter's, Missouri), Silicon Genetics (San Carlos, California), and Silicon Graphics (Mountain View, California); see Regaldo, ibid., 59; Christine Gorman, "Drugs by Design," *Time* (January 11, 1999):79–83, at 83.

43. Nicholas Wade, "Rivals Reach Milestones in Genome Race."

44. "Genzyme Corp," *Wall Street Journal*.

45. Ronald Cole-Turner, "Religion and Gene Patenting," *Science* 270 (1995):52; also see James J. Walter, "Theological Issues in Genetics," *Theological Studies* 60 (1999):124–34, at 132–34. For a more comprehensive analysis, see Mark Hanson, "Religious Voices in Biotechnology: The Case of Gene Patenting," *Hastings Center Report* 27 (1997):1–21.

46. See John Thavis, "Vatican Experts OK Plant, Animal Genetic Engineering," *St. Louis Review* (October 22, 1999):2.

47. Marlene Cimmons and Paul Jacobs, "Biotech Battlefield: Profits vs. Public," *Los Angeles Times* (February 21, 1999):A1.

48. Steve Bunk, "Researchers Feel Threatened by Disease Gene Patents," *The Scientist* (October 11, 1999):7; also see J. F. Merz, "Disease Gene Patents: Overcoming Unethical Constraints on Clinical Laboratory Medicine," *Clinical Chemistry* 45 (1999):324–30.

49. David Blumenthal, et al., "Withholding Research Results in Academic Life Sciences: Evidence from a National Survey of Faculty," *Journal of the American Medical Association* 277:15 (1997):1224–28. See David Blumenthal, "Relationships between Academic Institutions and Industry in the Life Sciences — An Industry Survey," *New England Journal of Medicine* 334:6 (1996):368–73.

50. Jonathan King and Doreen Stabinsky, "Patents on Cells, Genes, and Organisms Undermine the Exchange of Scientific Ideas," *Chronicle of Higher Education* (February 5, 1999):B6–B7.

51. David Resnik, "The Morality of Human Gene Patents," *Kennedy Institute of Ethics Journal* 7 (1997):43–61; also see, M. Cathleen Kaveny, "Genetics and the Future of American Law and Policy," *Concilium* 225 (1998):69–70.

52. Goldie Blumenstyk, "U. Of California Patent Suit Puts Biotech Powerhouse Under Microscope," *Chronicle of Higher Education* (August 6, 1999):A45–A46.

53. Editorial, "Case for free access to fundamental data," *Lancet* 353:9166 (May 22, 1999):1721, referring to the *SNP Consortium* discussed in *Lancet* (May 15, 1999):1684.

54. This discussion is intended to be broader than the community review model of Morris W. Foster. See, Morris W. Foster, et al., "Communal Discourse as a Supplement to Informed Consent for Genetic Research," *Nature Genetics* 17 (1997):277–79; and Morris W. Foster, et al., "The Role of Community Review in Evaluating the Collective Risks of Human Genetic Variation Research," *American Journal of Human Genetics* 64 (1999):1719–27.

55. For example, Renato Schibecci, et al., "Genetic medicine: An Experiment in Community-Expert Interaction," *Journal of Medical Ethics* 25 (1999):335–39.

56. For example, Gerard Magill and Marie D. Hoff, "Public Conversation on Values," in Gerard Magill and Marie D. Hoff, *Values and Public Life: An Interdisciplinary Study* (Lanham, New York: University Press of America, 1995), 1–28.

57. Jonathan King and Doreen Stabinsky, "Patents on Cells, Genes, and Organisms Undermine the Exchange of Scientific Ideas."

58. Nicholas Wade, "Cloner of a Sheep Moves to Persuade the Skeptics," *New York Times* (February 28, 1998):A6, referring to the publication of the data by the scientific journal *Nature* which was defended by the editor, Dr. Philip Campbell. Interestingly, Dr. Benjamin Lewin, who was editor of a competing journal, *Cell*, a leading journal of molecular biology, said he probably would have published Dr. Ian Wilmut's paper with the caveat of requiring the authors to mention that they could not *prove* Dolly was derived from an adult sheep cell.

59. Joe Flower, "The Way It Is, Is Not the Way It Will Be."

60. Arthur Caplan, "To Succeed Biotech Will Have to Answer Many Vexing Ethical Questions," *Forbes ASAP* (May 31, 1999):83–84.

61. Mark Hanson, "Biotechnology and Commodification within Health Care," *Journal of Medicine and Philosophy* 24 (1999):267–87, at 267–68.

62. Thomas A. Shannon, James J. Walter, and M. Cathleen Kaveny, "Ethical, Theological, and Legal Issues in Genetics," *Theological Studies* 60 (1999):109–147. For a challenging view upon appropriate limits of privacy, see Allen Wilcox, et al., "Genetic Determinism and the Over-Protection of Human Subjects," *Nature Genetics* 21 (1999):362. On disclosure of results and the implications for nonparticipants, see Susan M. Wolf, "Beyond 'Genetic Determinism': Toward the Broader Harm of Geneticism," *Journal of Law, Medicine, & Ethics* 23 (1995):345–53.

5

Geneticization: The Sociocultural Impact of Gentechnology

Henk A. M. J. ten Have, M.D., Ph.D.

IN PRESENT-DAY CULTURE, genetic issues have a strong public representation. The media are frequently drawing our attention to genetic discoveries. These reports seem to follow a particular pattern: the gene mutation associated with a particular disease has been identified; an effective therapy for the disease is lacking, and it is not probable that the discovery itself will enhance the development of therapeutic possibilities. However, the discovery will lead to more or less accurate prediction of the presence and prognosis of the disease when individuals are tested, even long before the onset of symptoms.

An example of such a reporting pattern occurred when the genetics research group at Nijmegen University announced that they had isolated the genetic structure responsible for Steinert's disease. This is one of the most prevalent muscular dystrophies, occurring in 1 out of 8,000 people. It is characterized by weakness of limb muscles and facial muscles. Onset is very variable, but for many patients it may be as late as the fourth decade. No cure is available as yet. The discovery was published in a scientific journal but also widely reported in the media. In newspaper interviews the researchers pointed out the practical implications. Although this scientific breakthrough will not lead to any therapy, the researchers argued in *De Telegraaf*, a national newspaper, that "it makes it possible to detect with hundred percent certainty the disease long before the onset of the first symptoms."[1] Also the severity of the disease can be prognosticated from the extent of the abnormalities of the DNA.

Public representations of genetic research, following the above pattern, generate a complex set of questions regarding the impact and value of genetic knowledge. The ethical debate however, tends to focus on the impact of genetics at the level of individuals. The emphasis on the proper management of information by individual citizen has a tendency within moral debate to neglect the social dimension of genetic information. Clarification of the right to know and the right not to know, although valuable in itself as a possible way to empower individual persons, also needs to elucidate the cultural context within which genetic knowledge is promulgated as well as the social processes involved in the dissemination of genetic technologies. The penetrating social impact of genetic technology and its diversified cultural manifestations should lead, for example, to a critical attitude towards moral statements that individual persons are free to chose among available genetic options and will not be directed into unwanted scenarios.

The moral debate concerning genetic technologies can be further hampered by the immediate interest in translating public fascination with new data, devices, and discoveries into practical applications. Communication of new genetic discoveries reveals a paradoxical tension between knowledge and application. On the one hand, researchers publish the results of their projects because knowledge in itself is valuable. One of the rationales for the Human Genome Project is that it will lead to gains in basic knowledge. On the other hand and at the same time, it is stipulated that genetic research has potential for medical advancement. Publication of knowledge claims, especially in the public media, seems almost always to be accompanied by expositions of the potential practical implications and the relevancy to patient care. This immediate linking of knowledge and application creates particular difficulties for moral debate. The fact that knowledge is available should not in itself dictate its application. What is necessary is prior identification of the goals that we want to accomplish in using the knowledge, a careful balancing of the benefits and harms generated through the application of knowledge, and a delineation of the norms and values that should be respected. Multiplication of technological possibilities therefore calls for a concomitant development of the moral framework guiding and regulating potential and actual application of genetic technology. In order to promote human use of new technologies, ethical reflection will be unavoidable. At the same time, such reflection is often already more or less orientated to particular applications, since these are pregiven and postulated together with the knowledge claims.

The intertwinement of knowledge and application claims also calls into question the responsibility of the human genetics community to communicate clearly and accurately about the nature and significance of genetic information. Sometimes communication is overstated, for example when it is claimed that the Human Genome Project will provide the ultimate answers to the chemical underpinnings of human existence.[2]

Moreover, such representation of genetic research can lead to particular public perceptions. It creates, for example, the impression that knowledge about many individual genes is knowledge about how the genome functions in people. It also leads to the fact, discussed by Fogle,[3] that genes are viewed by the public as entities, each of which controls one portion of the phenotype, rather than as integrated into a system.

Ethics and Genetics

The current development of genetics is a challenge, particularly to societies, to reflect upon the future evolution of human life and social existence. It is often argued that genetic information is special and that it therefore requires special ethical treatment. Genetic knowledge is not private information but necessarily implies relatives. Genetic information is also potentially valuable to third parties, such as insurance companies, employers, and prosecutors. Genetic technology can affect future generations. For these reasons, developing a framework of moral norms and regulations should involve all members of society. The purpose of the ethical debate is to develop guidelines and standards for the appropriate use of gene technology. Obtaining more information is not necessarily better unless there is a clear perception of the benefits, goals, or uses that may be approached or realized with such knowledge. The fact that genetic information is available for practical use does not imply that it is morally justified actually to use the knowledge. The morally relevant point is how to make the best possible use of the available genetic information. Various moral principles, rights, and rules have been developed to delineate what is regarded as appropriate use. The main focus of the ethics literature is precisely here: reflecting upon and evaluating the rapid evolution of genetics, ethicists try to analyze the potential effects of genetic information and to determine the conditions for justified applications of gene technology.

However, it is also possible to approach genetic issues from another perspective. While not denying that significant moral questions may arise in using and applying genetic knowledge, ethics may also raise the question of whether gene technology itself is morally neutral. A crucial concern is the moral value and meaning of genetic information. Let us assume, for the sake of argument, that the Human Genome Project will realize its claims: mapping the human genome fully and all human genes being located on the chromosomes. Diagnostic tests to identify all disease genes and predict any genetic dispositions and susceptibilities are flooding the health market. Assuming that the Genome Project has been ultimately and completely successful, we still have to concern ourselves with questions about the moral value of predictive knowledge of future human existence.

Geneticization

The development of genetics as a science is more and more associated with a growing influence of genetic knowledge and technology in particular areas of society and culture. This influence manifests itself directly through the application of genetic testing, for example in prenatal care and in various insurance arrangements, as well as indirectly through new imagery and concepts of health, disease, and disorder.[4] In the 1980s and 1990s, genetic explanations became more attractive. An analysis[5] of film, television, news reports, comic books, ads, and cartoons shows that the gene is a very powerful image in popular culture. It is considered not only as the unit of heredity but as a cultural icon, an entity crucial for understanding human identity, everyday behavior, interpersonal relations, and social problems. Nelkin and Lindee have related the growing impact of the genetic imagery in popular culture to "genetic essentialism," the belief that human beings in all their complexity are products of a molecular text.

Moreover, the expansion of the science of genetics, as well as the significance of genetics in the sociocultural context of postmodern human beings, also has repercussions for health care and medicine as well as science in general. Molecular biochemistry now has stronger claims than ever before to be the most fundamental science in medicine and the life sciences. There also is the general conviction that future genetics will drastically change medical diagnosis, treatment, and prevention. In order to identify and analyze the various cultural processes related to the biomolecular life sciences, the concept of "geneticization" has been introduced in the scholarly debate.

1. The Thesis

In 1991 Abby Lippman, a social scientist at McGill University in Montreal, introduced the concept of "geneticization" to describe the various interlocking and imperceptible mechanisms of interaction between medicine, genetics, society, and culture.[6] Lippman postulates that Western culture currently is deeply involved in a process of geneticization. This process implies a redefinition of individuals in terms of DNA codes, a new language to describe and interpret human life and behavior in a genomic vocabulary of codes, blueprints, traits, dispositions, genetic mapping, and a gentechnological approach to disease, health, and the body. Geneticization is defined as: "the ongoing process by which priority is given to differences between individuals based on their DNA codes, with most disorders, behaviors and physiological variations . . . structured as, at least in part, hereditary."[7]

Introducing the concept of "geneticization," Lippman is touching on a kind of awareness that seems to be widely shared, although it is often not well articulated. A growing number of studies nowadays are aimed at exploring our culture's fascination with genetics.[8] Genetic technology is regarded not merely as a new technology that is available for responsible use by autonomous consumers but rather as a potential transformation of human understanding and existence. For example, the opening sentences in the recent book by Barbara Katz Rothman are as follows:

> Genetics isn't just a science. It's becoming more than that. It's a way of thinking, an ideology. We're coming to see life through a "prism of heritability," a "discourse of gene action," a genetics frame. Genetics is the single best explanation, the most comprehensive theory since God. Whatever the question is, genetics is the answer.[9]

Genetic thinking is considered as a way of understanding the world; genetic practice is a way of imagining the future.

Through these more general implications, genetics has proliferated as a public issue. Despite the ubiquitous permeation of genetic thinking and despite its apparent popularity, new genetic advances are not welcomed with total acceptance or univocal acclamation. Over the last fifty years, the advancement of genetics has always been controversial. The implication of this historical lesson is drawn by José van Dijck: "The dissemination of genetic knowledge is not uniquely contingent on the advancement of science and technology, but is equally dependent on the development of images and imaginations."[10] Therefore, it is necessary to study the popular representations of the new genetics, the various ways in which the public face of genetics is shaped. The popularization of genetics is associated with contestation. The story of geneticization, therefore, is not only the story of successes and breakthroughs but also one of challenges, protests, and criticism. In her study of popular images of genetics, van Dijck distinguishes four stages in the story of geneticization: (1) the introduction of the "new biology" in the 1950s and 1960s (with new images arising, attempts to dissociate genetics from former eugenics, and disputes over "biofears" and "biofantasies"); (2) the DNA debate in the 1970s, with political controversy over biohazards (politicization of genetics, disputes over the safety of DNA research, increasing awareness of social and ethical implications), (3) the growth of the biotechnology business in the 1980s (industrialization of genetics, "biobucks" and "biomania"), (4) the initiation and implementation of the Human Genome Project (medicalization of genetics, creating new medical images for genetics, the "biophoria" of genome mapping). These stages do not represent chronological phases; they rather signify epistemological and conceptual shifts.

2. The Critique

The thesis of geneticization is apparently fruitful in generating new types of research and in directing scholarly attention to dimensions of genetic technology that are usually neglected in bioethical analyses. However, it also evokes sometimes vehement critique from geneticists and philosophers who reject it on various grounds. The critique of the thesis focuses on the following dimensions.

2.A. The Usefulness of the Concept

In response to a publication on geneticization, Niermeijer, professor of clinical genetics in Rotterdam, argues that the thesis creates misunderstandings similar to the earlier debate on medicalization.[11] The argument is twofold. First, the concept of geneticization is useless because there is already widespread public debate focusing on the social consequences of genetics. Second, the concept creates misunderstanding by suggesting that genetic technology leads to new phenomena and situations, whereas the psychosocial and cultural effects of new genetic information have been known for a long time. In reaction to this critique, it is pointed out that geneticization is not a new empirical phenomenon but a new theoretical concept that discloses particular dimensions and brings new perspectives into the debate on present-day genetics. Although there have been many types of debate and many aspects discussed up to now, the orientation to the cultural and social implications of the new genetics is rather new. Dismissal of the concept of geneticization as useless, therefore, removes those items from the agenda of public debate that it intends to bring into the discussion.[12]

2.B. The Empirical Basis

From an analysis of literature, Hedgecoe concludes that the ideas about geneticization are not based on convincing empirical evidence but rather on theory-derived polemic.[13] Close scrutiny of the thesis shows that it lacks "adequate grounding in empirical reality." For his conclusion, Hedgecoe refers to research data of Condit, showing that public perceptions of genetics are not necessarily deterministic. Hedgecoe also concludes from Condit's publications that there is no evidence that the use of genetic explanations in public discourse is more common now than in the past. This points to the fact, in Hedgecoe's opinion, that the thesis of geneticization is no more than a sweeping claim. What is needed are small-scale studies focusing on individual elements of geneticization. As a warning signal, Hedgecoe recapitulates the medicalization debate of the 1970s; he points out that overstated and inconsistent claims have been made and that the empirical basis was not sound.

This type of critique raises questions concerning the status of the geneticization thesis. Is it correct to construe the thesis primarily as an empirical claim which can then be falsified on the basis of empirical evidence? Is the thesis a sociological explanation of the facts of scientific and everyday-life reality? Or is the thesis, as I would assert, a philosophical interpretation of the self-understanding of today's human life and culture? Apart from the dispute about the status, there also is confusion about the methodology used by proponents of the thesis. Those who oppose the thesis seem to proceed from a positivistic point of view that explanatory theories should be deduced from a representative collection of empirical data, whereas geneticization seems to be a theory based on understanding the interactions of science and society. Criticisms of the geneticization thesis, which has been developed within the humanities, cultural sciences, and philosophy, seem to presuppose the priority of methodology prevailing in the natural sciences. A critique such as Hedgecoe's, therefore, is a symptom of the same phenomenon identified and criticized in the geneticization thesis, namely the uncritical predominance of mechanistic and reductionistic images; the only acceptable method of explanation and theory formation appears to be the model of the natural sciences, just as human existence is more and more explained in terms of molecular biology.

At the same time, the geneticization thesis is not an ideal construct; it is about empirical reality. The connection between reality and theory, however, seems to be different as constructed by opponents of the thesis. In philosophical discourse only a few examples will suffice to make a specific point plausible. A few well-selected examples will lead to a new interpretation of the same reality, whether or not these examples are statistically representative of the majority of cases. For instance, on a global scale, ethnic cleansing is not a widespread phenomenon. Even in locations where it occurs, it is very hard to prove convincingly its empirical reality. Nonetheless, the implications of the phenomenon for our philosophical self-understanding as human beings, for culture and politics, are enormous. It should therefore be considered a category mistake when notions and explanations from philosophical discourse are tested with instruments and methods from the empirical sciences. These category mistakes are common. For example, Michel Foucault's studies on clinical medicine have been rejected by some medical historians with the claim that they do not take into account much relevant data from the history of medicine. The same mistaken approach is used in refutations of the work of Illich on medicalization.

Finally, examples of the small-scale studies advocated by critics are already available. Processes of geneticization have been analyzed in the case of screening and counseling programs for beta-thalassemia in Cyprus.[14]

2.C. The Ambiguity of the Concept

Another critique is that the concept of geneticization is unclear.[15] On the one hand it is difficult to demarcate it from related concepts such as "genetic essentialism." On the other hand, a clear-cut definition of geneticization is missing. Descriptions of the concept used to be comprehensive, wide-ranging, complex, and therefore ambiguous. Because of the conceptual lack of clarity, geneticization will probably undergo the same fate as the discussion about medicalization; it will increasingly be regarded as unhelpful.

Indeed, it is correct that at the moment various concepts and variable descriptions of the same concept are used. There is a definite need for conceptual clarification. There is no a priori reason why such clarification is not possible. In fact, one of the newly introduced items on the agenda of bioethical research is precisely the interaction between sociocultural influences and genetic technology. Articulating and specifying the concept of geneticization will be a necessary condition for further development of this new research area. In this respect, the analogy with the medicalization debate in philosophy of medicine can prove to be more helpful than assumed by the critics.

3. *Geneticization as an Heuristic Tool in the Moral Debate*

The medicalization debate should indeed be regarded as a precursor to geneticization. Lessons from this debate should be used to develop the recent debate on the sociocultural impact of gene technology. These statements at the same time must be qualified. Medicalization as well as geneticization seem to be instantiations of more encompassing processes. Prima facie there is much similarity with Foucault's philosophy of normalization: since the early nineteenth century, medicine has created social order through its polarized distinction between "illness" and "health." The theory of medicine (classification of diseases), the human body, and society as a whole became closely interconnected. Concomitant with an epistemological shift toward the significance of knowing the interior of the body, the position and value of medical knowledge in society were elevated. Biopolitics transforms human beings into subjects. There is no escape from medical power; even the expressions of patients can be seen as an extension of medical power. Within society, modes of power are developing that admit forms of individualization while at the same time denying other forms; the same movement that empowers individuals and liberates them from some forms of oppression results in other forms of domination. This is also the Janus face of medicalization: at the same time as it provides certain benefits, it also subjects patients to certain forms of discipline.

Arney and Bergen emphasize that medicine is not simply "medicalizing." Instead of using domination and control, the field of medical power has been reformulated.[16] The locus of medical power is no longer the individual physician but is instead located in large, pervasive structures encompassing physician and patient alike. Medical power is also no longer exclusionary but has become incorporative; challenges from alternative health care, holism, bioethics, and the hospice movement are rapidly incorporated into "orthodox" medical practice. The new field of medical power, therefore, is not so much dependent on domination and control as it is on monitoring and surveillance. Technologies of monitoring and surveillance incite discourse; they make the intimacies of the patient visible, they leave visible records. Everything must be noted, recorded, and subjected to analysis.

The concept of geneticization can be further explained by relating it to the concept of medicalization. The process of medicalization can occur on different levels: (1) conceptually, when a medical vocabulary is used to define a problem; (2) institutionally, when medical professionals confer legitimacy upon a problem; (3) at the level of the doctor-patient relationship, when the actual diagnosis and treatment of a problem takes place.[17] Medicalization is also associated with several consequences: it is a mechanism of social control through the expansion of professional power over wider spheres of life, it locates the source of trouble in the individual body, it implies a particular allocation of responsibility and blame, and it produces dependency on professional and technological intervention.[18]

By analogy, the concept of geneticization can be studied on various levels: (1) conceptually, when a genetic terminology is used to define problems; (2) institutionally, when specific expertise is required to deal with problems; (3) culturally, when genetic knowledge and technology lead to changing individual and social attitudes towards reproduction, health care, and prevention and control of disease; and (4) philosophically, when genetic imagery produces particular views on human identity, interpersonal relationships, and individual responsibility. In contradistinction to medicalization, the concept of geneticization seems to be broader because it also refers to developments and differences in the interaction between genetics and medicine; there is, for example, not simply an expansion of concepts of health and disease into everyday life, but a fundamental transformation of the concepts themselves. In medicine, there is also a tendency to use a genetic model of disease explanation as well as a growing influence of genetic technologies in medical practice.[19]

Using the concept of geneticization also requires a critical analysis of theoretical developments following the introduction of the medicalization thesis.[20] In particular, the perspective that patients are not passive "docile bodies" under the control of medical power but articulate consumers and autonomous decision-makers needs to be taken seriously, because the moral requirements of nondirectiveness and respect for individual autonomy are strongly emphasized in present-day clinical genetics.

Future Society

Prima facie, it seems unavoidable that the future will bring us a society within which all potentially useful genetic information is freely available and actually applied. In principle, every member of this society will be able to foretell his individual fate from reading his genes and to adapt his personal life plan in accordance with such predictive knowledge.

In the opinion of "geneticization" authors such as Lippman,[21] this future has already partly begun. Society is involved in a process of geneticization. As an instantiation of the more encompassing process of medicalization, this process involves a redefinition of individuals in terms of DNA codes. Postmodern society is using a new genomic language to communicate about human life. Disease, health, and the body are explained in terms of molecular biology. Nelkin and Lindee, in their book, *The DNA Mystique*, examining popular sources such as television, radio talk shows, comic books, and science fiction, show how popular images "convey a striking picture of the gene as powerful, deterministic, and central to an understanding of both everyday behavior and the 'secret of life.'" [22] It seems that the cultural meaning of DNA nowadays is remarkably similar to that of the immortal soul of Christian theology. The bioinformation metaphor and cartographic metaphor, often used in the context of the Genome Project, are in fact reworkings of the mechanical metaphor that has frequently been used in the past in medical discourses on the body. These linguistic (and often also visual) representations of the body carry with them the importance of a technological approach: machinery is used to fix machinery. They represent the body as being comprised of "a multitude of tiny interchangeable parts, rendering the body amenable to objectification and technological tinkering in the interest of developing the 'perfect' human."[23]

How should this development toward a geneticized future be evaluated? It is at least important to try to identify what influences, what determines this development. It seems that this development towards a geneticized future is possible because of the consensus regarding two ideals in current moral debate: the ideal of value neutrality of clinical genetics and the ideal of individual responsibility in health matters.

1. Nondirectiveness

One of the prime tenets of genetic counseling is patient autonomy. Once genetic information is available, the basic rule is that patients or clients should be able to use the information according to their personal views. Geneticists or counselors should not seek to tell patients or clients whether they should obtain particular information or what they should do with the information if they acquire it. In other words, the goal of genetic counseling or screening is to inform patients or clients about what is possible and what their options are.[24] The leading principle of counseling and screening therefore is nondirectiveness. Accurate information should be provided to the person concerned regarding the nature of potential genetic conditions, the prognosis, possible treatments, and preventive strategies. The experts providing such information should not in any respect try to influence the decisions made by the persons who are counseled or screened.

The moral ideal underlying this practice of clinical genetics is value-neutrality. The genetic expert withholds any normative judgment regarding obtaining and applying genetic information; the aim is merely to provide information and to help the patients or clients to work through possible options. It is evident that this ideal in itself is a weak counterbalance to tendencies to make genetic tests more generally accessible. Patient values are to be decisive whenever choices have to be made on the basis of genetic information. When respect for individual autonomy is the basic norm guiding the use of genetic information, it is also reasonable to expect that predictive "combitests" will eventually be on sale in the supermarket or drugstore.[25]

2. Individual Responsibility for Health

A second determinant that may further increase the significance of genetic information is the ideal of individual responsibility for personal health. Health policy and health education, especially in times of limited budgets and reduced expenditures, increasingly appeal to the notion of "personal responsibility." If health policy defines a particular problem as undesirable, and if health education research shows the problem to be associated with a particular lifestyle, then health policy can attribute responsibility to those individuals who exhibit that lifestyle, particularly since lifestyle is supposedly the free choice of rational individuals.

Traditionally, in health care the rhetoric of responsibility is used in a specific way. In the medical model of disease, patients are usually not held responsible for the genesis and evolution of their illnesses. Diagnosing a condition as disease introduces excusability. When a person's condition is interpreted as illness, the medical judgment implies that the patient cannot be blamed for the condition and that treatment and care are appropriate and morally desirable. In this traditional model, the notion of responsibility is used with prospective force: it is equivalent to saying that we have an obligation to preserve our health. Through assigning responsibility to individuals for their future health, an attempt is made to guide and change each individual's behavior. Such practical use of the concept is different from the retrospective ascription of responsibility. The latter use implies an evaluation of what has happened. If individuals have health problems, they are held causally responsible because of their unhealthy lifestyles or risky behavior in the past. This use combines causality with culpability. Since people are the cause of their problems, they are also answerable for the consequences of their prior behavior. Retrospective use of the concept of responsibility is therefore retributive; it implies disapproval and blame.

In present-day health policy there seems to be a development towards connecting the prospective and retrospective senses of "responsibility."[26] Usually the line of argumentation is as follows. If there is an urgent need to reduce the costs of health care, and if at the same time it is scientifically argued that major expenditures are associated with certain patterns of behavior, it is tempting to create an obligation to be healthy and to introduce some system of sanction for those who do not implement such obligation. In a liberal society, individuals are normally free to do as they choose. In this respect, caring for your health is not different from other dimensions of personal life. But when individual choices turn out badly and when individuals remain uninfluenced by moral appeals of health educators, legal and financial sanctions may be thought justified.

Today, a similar argument is used concerning predictive information. It may be prudent to use genetic diagnosis to predict future disabilities, and therefore appeals to (prospective) responsibility may be justified; but this argument in practice is often linked with the argument that individuals who deliberately have not used diagnostic possibilities should be (retrospectively) responsible for adverse consequences for themselves or their offspring. When, for example, a couple decides not to use prenatal diagnosis or not to terminate pregnancy in the case of diagnosed fetal disorders, it is argued that the couple is then responsible for the suffering of the child when indeed a child with handicaps is born.[27] If suffering could have been avoided and a choice is made not to use predictive opportunities, parents should bear the consequences of their irresponsible choice; they can no longer argue that suffering has befallen them; they have themselves to blame.

This line of argumentation, if indeed taken seriously, will be a significant stimulus for individuals to obtain genetic information as much as possible, particularly when there is a threat that governments, insurance companies, and employers will work with a system of incentives and disincentives. When there is a cultural imagery that future diseases, disorders, and disabilities can be foretold by examining the individual's genome, people can no longer claim that they are victims if they have deliberately decided not to use predictive diagnosis. It has been their voluntary choice not to know and not to eliminate potential disadvantages to their health. Fate has been replaced by choice.

A Genetic Civilization Strategy?

The ideals of value-neutrality of clinical genetics and of personal responsibility for health prevailing in current bioethical debate may indeed generate a situation where the availability of genetic information in itself produces its widespread application. In this view, human beings in the next millennium will be dominated by predictive knowledge of their genome and driven by new norms in interpersonal behavior.

Such assumption is not unrealistic, given that we witnessed a similar change in normative behavior patterns at the close of the nineteenth century.[28] With the rise of new knowledge about the origin and transmission of infectious diseases, in many countries philanthropic activities were organized to civilize the public by inculcating new hygienic norms. Philanthropists launched a large-scale offensive to civilize the habits and lifestyles of the masses. As enlightened men, they coupled assistance with moralization. Norms of behavior such as cleanliness, domestic nursing, and soberness were transmitted not by repression or coercion but by the subtle means of advice, persuasion, and education. The result was the normalization of individual behavior. The new norms of a healthy, regular, and disciplined conduct passed into domestic life; the strategy succeeded in having the norms internalized. Hygienism thus produced a new behavior pattern in the general population.

Why could a similar transformation of lifestyles not occur today as a result of new genetic information? Though it is hard to forecast the future, two factors can be identified that may prevent, hinder, or at least restrict this development towards geneticization of future human existence.

The first factor is the need to make some delineation between disease and health, normality and abnormality, given the uncontrollable wealth of information that will in the end be available. In current ethical debate, the above distinctions are increasingly problematic. It is apparently difficult to make use of the traditional distinctions in determining what conditions should be screened or not. Perhaps it is even thought impossible to apply them as normative criteria guiding potential genetic screening programs. Nonetheless, the exponential growth of genetic data and resulting possibilities of detection will inevitably lead to an urgent need for selection; without selective use and meaningful criteria to make distinctions of value among the immense data available, the usefulness of data will be questionable. The multiplication of possibilities for testing will at the same time increase the necessity to reach consensus regarding those conditions and predispositions that seriously restrict the functioning of human beings within a community and those that are within the bounds of reasonable variations of human functions and structures. Of course, at the moment it is unclear how such distinctions can be made and morally justified. But the acknowledgment that it will be an extremely difficult task should not lead to the conclusion that it is impossible. Right here is a major challenge to philosophical reflection. Many moral discussions about whether or not to apply genetic knowledge seem essentially to focus on this issue in particular (for example, the debate on the development and use of human growth hormone).[29]

The second factor is the normativity of medicine. Medicine regards itself ultimately as a helping and caring profession, not merely as service institution. In such a self-conception, value-neutrality is not an appropriate position to guide medical activities. In this view, physicians adhere to professional norms that go beyond value-neutrality. Diagnosis, therapy, and prevention are guided and motivated by specific values, namely promotion of health, relief of suffering, and elimination of disease. From this value perspective, respect for individual autonomy is only an instrumental value, necessary in order to accomplish the values intrinsic to medicine as a helping and caring profession.

The norm of nondirectiveness in clinical human genetics, therefore, is inadequate from a medical point of view. It may have been prudent to introduce this norm against the background of historical misuse of genetic information. It may be desirable as a practical norm as long as genetic information is related mainly to genetic risks to offspring. But it can be argued that in the present situation, where genetic testing is more and more concerned with detecting genetic risks for the future health of the individual person who is tested, the normative attitude of clinical geneticists should shift from neutrality to prescriptivity.[30] A similar point is made by Caplan: it is likely that a shift will occur from a normative stance of value-neutrality toward "an ethic in which the promotion of genetic health and the amelioration, prevention, and correction of genetic disease are the foundation of clinical and public health practice."[31] Decisions made on the basis of genetic information should in this view aim at promoting health and alleviating disease. There is no reason to think that advocating these values in the realm of human genetics is inappropriate or unethical. Studies of the practice of clinical human genetics in fact indicate that those professionals who now offer genetic screening and testing services do not always act in conformity with their self-imposed ideal of value-neutrality.[32]

Conclusion

In postmodern society, two determinants are at work that will probably lead to a future where individual existence is to a large extent affected and permeated with predictive genetic information. First, we witness the current domination of the moral principle of respect for personal autonomy—individuals ought to choose among the potential of genetic tests those possibilities that fit their life plans. Second, society is moralizing individual responsibility in the sense that persons who do not use the opportunities to foresee and prevent future suffering have to face the consequences. Both factors give a strong push to know as much as possible about our life in the near and distant future. From this perspective, the collective destiny of human beings in Western societies will be deeply geneticized. However, there are reasons to question the prediction of further geneticization.

First, a clear opposition exists between the above determinants; the first emphasizes the interest of the individual, the second, the community interest. It is not evident which interest will prevail; it is not obvious that one interest will definitely overrule the other.

Second, autonomous individuals will not use everything available at random; they will sooner or later start to wonder what may be the meaning and relevancy of all the knowledge that is available and obtainable. Even within a fully free health market, individuals will not consume everything; they will attempt to make a distinction between appropriate and inappropriate, intelligible and unintelligible uses of genetic tests. This will instigate a public debate concerning the significance of genetic testing and genetic information, the more so since powerful parties such as insurance companies have an obvious interest in promoting testing.

Third, it is doubtful whether future medicine will depart so radically from its present-day value orientation, especially in the European setting. The autonomous request of individual patients will be a significant moral factor, but at the same time, medicine will also want to be guided by its own norms to make distinctions between disease and health, normality and abnormality. Beyond the individual demands and subjective complaints, medicine will continue to strive for a more rather than less objective determination of needs, signs, and symptoms. Apparently, a full geneticization of human existence in the future may occur only when we abandon the philosophical attempt to differentiate between "healthy" and "ill," "normal," and "abnormal."

Finally, this analysis illustrates the advantage of the concept of "geneticization:" it operates as an heuristic tool, like the concept of "medicalization" in the medical-philosophical debates of the 1970s. It discloses particular areas for philosophical scrutiny, it redirects and refocuses moral discussion. In creating and facilitating different ethical perspectives, the concept of geneticization particularly draws attention to (1) socioethical issues, and (2) an interpretative ethical methodology.

The challenge of the current development of genetics for bioethics can be explored more fully when attention is given to its social and cultural implications. The concept of geneticization can incite us to change our perspective. While not denying that significant moral questions arise in the use and application of genetic knowledge, ethics may also address the question of whether genetic information itself has any moral value and meaning. Regardless of the significance of personal autonomy, there is also the question of what the social and cultural consequences will be of new genetic knowledge. What will it mean for society and culture in general when every member of society will be able to foretell his or her individual fate from reading his or her genes and to adapt a personal life plan in accordance with such predictive knowledge? What are the implications for our notions of life and illness when disease, health, and the body are predominantly explained in terms of molecular biology? What will be the effect of a technological discourse about the human body?

The concept of geneticization is reorientating attention away from moral topics related to the current emphasis on individual autonomy, such as nondirectiveness in clinical genetics and the ideal of individual responsibility in health matters. This reorientation creates space for seriously questioning the dominant bioethical discourse, with its emphasis on individual freedom to choose. In a liberal society, it is argued, individuals are normally free to do as they choose. In this respect, caring for your health is no different from other dimensions of personal life. But when individual choices turn out badly and when individuals remain uninfluenced by moral appeals by health educators, legal and financial sanctions may be thought justified. If suffering is in practice avoidable and individuals freely decide not to use predictive opportunities, they should bear the consequences of their choices. The logic of individual choice and responsibility necessarily includes the logic of blaming the victim.

The heuristic value of the concept of geneticization is precisely here: it introduces into the bioethical debate moral issues and methods that tend to be "forgotten," neglected, or disregarded. Geneticization, in the words of van Dijck, is "a gradual expansion of loci of contestation where meanings of genetics are weighed."[33] The concept therefore informs bioethics that biomedicine and bioscience should be associated with biocriticism. A central thesis in the epistemology of the French philosopher Gaston Bachelard is that the problem of the growth of science must be formulated in terms of obstacles.[34] There is no history of science without shadows, without failures, dissent, and conflict.

Acknowledgments

This chapter was published previously as Henk A. M. J. ten Have, "Genetics and Culture," *Medicine, Health Care & Philosophy* 4:3 (2001):295–304. Republished with permission.

Notes

1. *De Telegraaf* (February 6, 1992).
2. J. D. Watson, "The Human Genome Project: Past, Present, and Future," 248 *Science* (1990):44–48.
3. T. Fogle, "Information Metaphors and the Human Genome Project," *Perspectives in Biology & Medicine* 38:4 (1995):535–47.

4. H. A. M. J ten Have, "Living with the Future: Genetic Information and Human Existence," in *The Right to Know and the Right Not to Know*, ed. R. Chadwick, M. Levitt, and D. Shickle (Aldershot, U.K.: Avebury, 1997), 87–95; also R. Hoedemaekers and H. A. M. J. ten Have, "Genetic Health and Genetic Disease," in *Genes and Morality. New Essays*, ed. V. Launis, J. Pietarinen, and J. Räikkä (Amsterdam/Atlanta: Rodopi, 1999), 121–43.

5. D. Nelkin and M. S. Lindee, *The DNA Mystique. The Gene as a Cultural Icon* (New York: W. H. Freeman, 1995).

6. A. Lippman, "Prenatal Genetic Testing and Screening: Constructing Needs and Reinforcing Inequities," *American Journal of Law & Medicine* 17 (1991):15–50.

7. A. Lippman, "Prenatal Genetic Testing and Geneticization: Mother Matters for All," *Fetal Diagnosis and Therapy* 8:supplement 1 (1993):175–88, at 178.

8. F. Koechlin and D. Ammann, *Mythos Gen* (Rieden bei Baden, Austria: Utzinger/Stemmle Verlag, 1997); B. Katz Rothman, *Genetic Maps and Human Imaginations. The Limits of Science in Understanding Who We Are* (New York/London: Norton, 1998); J. van Dijck, *Imagination. Popular Images of Genetics* (London: Macmillan Press, 1998); P. Glasner and H. Rothman, eds., *Genetic Imaginations. Ethical, Legal and Social Issues in Human Genome Research* (Aldershot, U.K.: Ashgate, 1998).

9. Glasner and Rothman, *Genetic Imaginations*, 13.

10. van Dijck, *Imagination*, 2.

11. M. F. Niermeijer, "Geneticalisering. Misleidend onbegrip door onjuiste informatie," *Medisch Contact* 53 (1998):641–42.

12. M. C. B. van Zwieten and H. A. M. J. ten Have, "Geneticalisering, een nieuw concept," *Medisch Contact* 53 (1998):398–400; M. C. B. van Zwieten and H. A. M. J. ten Have, "Geneticalisering. Naschrift," *Medisch Contact* 53 (1998):642.

13. A. Hedgecoe, "Geneticization, Medicalisation and Polemics," *Medicine, Health Care & Philosophy* 1:3 (1998):235–24.

14. R. Hoedemaekers and H. A. M. J. ten Have, "Geneticization: The Cyprus Paradigm," *Journal of Medicine and Philosophy* 23:4 (1998):274–87.

15. Hedgecoe, "Geneticization, Medicalisation and Polemics."

16. W. R. Arney and B. J. Bergen, *Medicine and the Management of Living. Taming the Last Great Beast* (Chicago: University of Chicago Press, 1984).

17. P. Conrad and J. Schneider, "Looking at Levels of Medicalization: A Comment on Strong's Critique of the Thesis of Medical Imperialism," *Social Science and Medicine* 14A (1980):75–79.

18. R. Crawford, "Healthism and the Medicalization of Everyday Life," *International Journal of Health Services* 10:3 (1980):365–88; I. Illich, "The

Medicalization of Life," *Journal of Medical Ethics* 1:1 (1975):73–77; I. K. Zola, "In the Name of Health and Illness: On Some Socio-Political Consequences of Medical Influence," *Social Science and Medicine* 9 (1975):83–87.

19. R. Hoedemaekers and H. A. M. J. ten Have, "Genetic Health and Genetic Disease."

20. S. J. Williams and M. Calnan, eds., *Modern Medicine. Lay Perspectives and Experiences* (London: UCL Press, 1996).

21. A. Lippman, "Led (Astray) by Genetic Maps: The Cartography of the Human Genome and Health Care," *Social Science and Medicine* 35:12 (1992):1469–76.

22. Nelkin and Lindee, *The DNA Mystique*, 2.

23. D. Lupton, *Medicine as Culture. Illness, Disease and the Body in Western Societies* (London: Sage Publications, 1994).

24. F. S. Collins, "Medical and Ethical Consequences of the Human Genome Project," *Journal of Clinical Ethics* 2:4 (1991):260–67.

25. G. de Wert, "De oorlog tegen kanker, de jacht op kankergenen, en de speurtocht naar de ethiek," *Tijdschrift Kanker* 18:2 (1994):41–55.

26. H. A. M. J. ten Have and M. Loughlin, "Responsibilities and Rationalities: Should the Patient Be Blamed?" *Health Care Analysis* 2:2 (1994):119–27.

27. M. T. Hilhorst, "Aangeboren en aangedane handicaps: maakt het moreel verschil?" in *Kind, ziekte en ethiek*, ed. I. D. de Beaufort and M. T. Hilhorst (Baarn: Ambo, 1993), 67–91.

28. H. A. M. J. ten Have, "Knowledge and Practice in European Medicine: The Case of Infectious Diseases," in *The Growth of Medical Knowledge*, ed. H. A. M. J. ten Have, G. K. Kimsma, and S. F. Spicker (Dordrecht/Boston/London: Kluwer Academic Publishers, 1990), 15–40.

29. T. Wilkie, *Perilous Knowledge. The Human Genome Project and Its Implications* (London/Boston: Faber and Faber, 1993).

30. G. de Wert, *Met het oog op de toekomst. Voortplantingstechnologie, erfelijkheidsonderzoek en ethiek* (Amsterdam: Thesis Publishers, 1999).

31. A. L. Caplan, *If I Were a Rich Man Could I Buy a Pancreas? And Other Essays on the Ethics of Health Care* (Bloomington and Indianapolis: Indiana University Press, 1992), 134.

32. J. Fletcher and D. Wertz, *Ethics and Applied Human Genetics: A Cross-Cultural Perspective* (Heidelberg: Springer Verlag, 1988).

33. van Dijck, *Imagination*, 29.

34. G. Bachelard, *La formation de l'esprit scientifique* (Paris: Vrin, 1938).

6

Genes and Gender: An Egalitarian Analysis

Mary Briody Mahowald, Ph.D.

AS A SCIENCE, genetics is commonly thought to be morally neutral. In its clinical and social applications, however, this is clearly not the case. The practice of eugenics has long illustrated the potential for injustice that accompanies advances in genetics. Rarely, however, has the potential for injustice in genetics been examined from the standpoint of gender differences. In this article, therefore, I wish to focus on this relatively unaddressed and worrisome potential. I will consider the meaning of gender equality as a subset of justice, apply this to a gender-related issue that arises occasionally in the clinical setting, and describe a mechanism for promoting gender equality across a spectrum of issues raised by advances in genetics. First, however, I will briefly review some of the ways in which genetics has affected men and women not only differently but disparately.[1]

EMPIRICAL GENDER DIFFERENCES IN GENETICS

Empirical differences in the impact of genetics on men and women are both biological and psychosocial. The biological differences derive not only from our different reproductive roles but also from differences in the genetic conditions themselves. Some conditions, such as X-linked diseases and breast cancer, affect only one sex or affect one sex primarily, while others, such as anencephaly, affect males and females in disproportionate numbers. Conditions such as Angelman syndrome and Prader-Willi syndrome are determined by the sex of the transmitting parent; the expression of others, such as Down syndrome and achondroplasia, is influenced by maternal or paternal age. Some genetic conditions (e.g., cystic fibrosis) render males infertile, whereas others (Turner syndrome) entail infertility in females. Pregnant women with conditions such as sickle cell anemia and cystic fibrosis are likely to experience an exacerbation of symptoms which their nonpregnant counterparts do not face.

The different reproductive roles of men and women inevitably entail greater health risks for women than for men, even when genetic conditions for which testing or intervention is undertaken are due to the male partner. Obviously, prenatal testing and interventions, whether the latter are abortive or therapeutic for the fetus, can be performed only through women's bodies. Although definitive prenatal tests can sometimes be done only if the male partner is tested as well, testing him is minimally invasive or noninvasive in comparison with the impact of common prenatal diagnostic procedures such as chorionic villus sampling and amniocentesis. Unfortunately, while genetic tests for various conditions have proliferated in recent years, there has been little progress in the development of effective therapies for treatment of these conditions; in most cases, termination of pregnancy is the only way through which potential parents may avoid the birth of affected offspring.

The psychosocial empirical differences triggered by advances in genetics relate to caregiving as well as reproduction.[2] In institutional as well as home settings, women are the predominant caregivers of those affected by specific genetic conditions, and as medical science has prolonged the lives of such individuals, women's caregiving responsibilities have been extended. Those who care for children, the disabled, and the elderly at home typically suffer a loss of income and prestige for doing so. Except for physicians, most of whom are male, the remuneration and social status of professional caregivers tend to be modest in comparison with professionals in other fields.[3]

The costs of caregiving include expenditures for therapies, medications, nursing care, hospital stays, and medical equipment, as well as increased stress, time loss, and chronic fatigue. Typically, the primary caregiver also experiences health risks, guilt, and anger. While domestic burdens are increased, activities outside the home, including remunerative employment, are restricted. Not surprisingly, these impacts are felt more by low-income and minority women than by others.

Although men and women may share caregiving responsibilities equally, some gender-based disparities are inseparable from their biological differences. For example, many women view the gestational tie to offspring as more important than the genetic tie, whereas men, whose only means of establishing a biological tie to children is through genetics, place greater emphasis on the genetic tie for both their partners and themselves.[4] Because of their more extensive biological role in reproduction, women are more likely than men to be influenced by diverse cultural expectations about having children, especially healthy children. In some cultures, women are expected to undergo whatever genetic testing and interventions are available to them—regardless of the risk—while in other cultures women are expected to avoid the use of medical technology that could expand their options.[5] In either case, social or familial pressures may compromise women's autonomy and welfare.

In a health care system that permits profound inequities, some advocates for low-income women and women of color view new means of genetic testing with suspicion.[6] Feminist critics view the latest technologies as part of a larger history of women's loss of control over birth and overuse of technologies associated with birth. Some have argued that women are pressured into accepting prenatal diagnosis by a medical system that follows the "technological imperative" of using prenatal diagnosis simply because it exists.[7] Despite difficulties of access, pressures to use prenatal diagnosis and to terminate affected pregnancies may be intensified for women who do not have the resources required to raise a disabled child, even while they are denied access to some services because of their inability to pay. Because the majority of the poor are women and their children, they are more likely than men to experience socioeconomic disadvantages associated with genetics.

Gender-based socialization and expectations seem to trigger different impacts from treatment of diseases for which genetic susceptibility tests are now available. Breast cancer, for example, is a disease that results in exacerbated impact for women because the extant treatment modalities of mastectomy and chemotherapy are disfiguring in ways that men treated similarly do not find as burdensome. Hair loss, even though temporary, is embarrassing and sometimes humiliating for women mainly because they are not expected to be bald; and breast removal, even if it is performed prophylactically, entails for many the permanent loss of their womanly appearance.

Treatment of gynecologic cancers for which genetic susceptibility tests are or will eventually be available may result in loss of the ability to conceive or bear a child. Men may also be rendered sterile through cancer treatment, but means of remediating this are less drastic, less costly, and more effective than they are for women. The fact that women can lose both their gestational and their genetic capability of having children doubles their potential losses with regard to reproduction.

In light of gender differences such as those indicated above, it hardly seems likely that advances in genetics will benefit the majority of women, at least in the short run. Rather, it is likely that these advances will increase the options of a small number of women who are already advantaged and capable of resisting social pressures to utilize or not utilize genetic services. For women who are already disadvantaged, scientific and clinical developments in genetics are more likely to exacerbate the burdens that they already experience. Whether disadvantages for one group vis-à-vis another are based on sex, race, class, or some other defining characteristic, justice demands that the disparities not only be identified but addressed in ways conducive to a more egalitarian arrangement. However, different conceptions of equality underlie different conceptions of justice. In the next section, therefore, I sketch the egalitarian perspective that underlies my analysis of gender differences in genetics.

Gender Equality as a Subset of Justice

Following Aristotle, the formal principle of justice is generally accepted as requiring equals to be treated equally and unequals unequally, or like things to be treated alike.[8] But broad disagreement prevails with regard to the material principles of justice, which involve different conceptions of equality.[9] The conception of equality that underlies my analysis is consistent with Amartya Sen's account of the term in *Inequality Reconsidered*: it focuses on human capability as differently expressed and sometimes suppressed in different individuals and groups.[10] In an egalitarian society, the suppression rather than expression of different human capabilities is to be eliminated or minimized. Differences are welcome as long as they do not advantage one group or individual over another. Equality between the sexes or races or classes does not mean that different groups are the same but that they have the same value despite their differences. To put this in chromosomal terms: XX=XY, that is, women and men are equal but not the same. Extending the equation to other chromosomal arrangements, XX=XY=XXY=XYY=XO, which means that people with chromosomal arrangements other than XX and XY also have the same value despite their differences.[11]

Gender equality hence refers to a situation in which differences between men and women are acknowledged and respected while insuring that they themselves are valued in the same degree. Because of unchangeable biological differences, women experience burdens and risks that men do not experience in reproduction and genetics. Both formally and informally, however, measures can be introduced to reduce the inequitable impact of their differences. If reasonable efforts are made in that direction, the requirements of gender equality may be met.[12]

I am not arguing here for the elimination or even for the reduction of capabilities on the part of some in order to increase those of others. Rather, I am arguing that the advantages that accrue to some because of their capabilities ought to be equally distributed or less unequally distributed to those who lack such advantages because of their lesser capabilities. Where equal distribution of advantages is not possible, there ought at least to be efforts to reduce the inequality. Morally, those who are advantaged through greater capabilities bear greater responsibilities toward those who are less advantaged. In other words, equitable distribution of advantages is a social ideal worth pursuing—despite its resonance with the Marxist maxim: from each according to ability, to each according to need.[13]

Some differences *entail* inequalities; others are merely *associated* with them. For example, women who provide gametes for reproduction experience risk and discomfort that men who provide gametes do not experience (quite the contrary, in most instances); this inequality is entailed or necessitated by one's being male or female. Once children are born, however, the unequal roles prevalently occupied by mothers or fathers are merely associated with their sex.[14] For example, the time spent with children by both parents could be the same or the ratio could be reversed. The lesser prestige and income that typically attach to caregiving in contrast with other occupations are another disparity that mainly disadvantages women; this gap could surely be reduced through changes in social attitudes and practices.

Just as differences between men and women need to be identified and examined to determine the extent, if any, to which they are associated with disadvantages for one group vis-à-vis others, the same is true for differences based on race, class, ability, or sexual orientation. Some men, after all, are disadvantaged vis-à-vis some women, and some women are disadvantaged vis-à-vis other women. In fact, the majority of us are neither wholly disadvantaged nor wholly advantaged. While I focus here on gender equality as a subset of justice, the conception of equality that underlies my analysis necessarily applies to other groups whose differences are associated with inequality. In the next section, I further limit my focus to one of the many issues that merit analysis from the standpoint of gender equality, starting with a case that illustrates this issue.

A Case of Misattributed Paternity and Gender Equality

Mary and Bill Jones, both in their early thirties, visited a genetic counselor when Mary was six weeks pregnant with their second child. They had been referred by their obstetrician for prenatal diagnosis for cystic fibrosis (CF) because their first child, Kevin, was diagnosed with CF as an infant, and died at nine years of age. If the CF mutation present in the couple were identified, the prenatal diagnosis could be made more accurately. Accordingly, tissue samples were requested from Mary and Bill. Analysis of the samples showed that Mary carried the most common mutation for CF. Bill was not found to be a carrier for that mutation, and no other CF mutation was detected in either partner.

Genetic testing for CF can detect over 90 percent of the most common mutations for the disorder.[15] While it is possible that Bill contributed a nondetectable, rare mutation to Kevin, no genetic test could verify this, and it is much more probable that he was not Kevin's biological father. If Bill was not Kevin's genetic father, the risk of CF in the fetus Mary is currently carrying is extremely slim. To provide accurate predictive data regarding the risk of CF in this pregnancy, the genetic counselor therefore faced the question of whether to disclose potentially disturbing information about undisclosed or misattributed paternity to either or both partners.

Misattributed paternity has been estimated to occur in 10 to 15 percent of the population, but this incidence is not based on published evidence. Nonetheless, parents presumed to be carriers for an autosomal recessive disorder are usually warned in advance that genetic testing may disclose unexpected information regarding paternity.[16] Either or both may then be asked whether they want to be provided with that information. Because the question itself may raise concerns on the part of either partner, asking it may likewise raise concerns for the counselor. These concerns are virtually the same as those faced by genetic counselors who have not indicated to the couple the possibility of obtaining such information. They involve confidentiality towards the woman, the man's right to know, the possibility of family disruption, and the relevance of the information to the child. These factors are relevant to a determination of whether misattributed paternity should be revealed by health professionals to individuals who come for genetic counseling.

Depending on who discloses what to whom, a finding of misattributed paternity may or may not be problematic for genetics professionals. Disclosure by the professional to the woman who is her client and disclosure by the woman to the man who had been assumed to be genetically related to their child are relatively unproblematic from the standpoint of the professional. For the woman/client, however, disclosure of this information to her partner, regardless of who discloses it, may not only be difficult but also raise fears of their consequences to her or others.

Disclosure to patients of information pertinent to their reproductive plans is a prima facie obligation of caregivers who have obtained such information, even when it was not requested. But prima facie obligations may at times be superseded by other prima facie obligations. The determination of which obligation has priority may be based on a utilitarian calculation that more harm would be done through disclosure than nondisclosure or on a deontological calculation that more compelling rights would be violated if the information were disclosed to a specific individual such as the putative father or the genetic father.

The difference between the roles and moral responsibilities of the woman-client and the genetics professional is morally significant. Assuming that the woman is morally obliged to disclose misattributed paternity to her partner, it does not follow that the genetics professional incurs the same obligation in the absence of the client's willingness to do so. Even if the woman's partner is also a client, obligations to different clients are not necessarily equally compelling.

Reasons given by geneticists for protecting a woman's confidentiality by not disclosing misattributed paternity to her partner are preservation of the family unit and the woman's right to decide what to do with the information.[17] If we assume that family disruption is a harm and family preservation is a good, the first reason may be supported by the principles of nonmaleficence or beneficence. Avoidance of potential harm to the woman if her partner is told may also be supported by the principle of nonmaleficence.[18] The second reason, however, gives priority to the woman's autonomy over the man's, assuming that the autonomy of both cannot be satisfied. Genetic counselors attribute to the woman herself the right not to have this intimate information shared with her partner.[19] Her right to confidentiality is thus associated with her right to privacy, and both are required in the context of respect for her autonomy. In other words, the male partner's right to know is subordinated to her right that he not know. That her autonomy trumps his may be defended on grounds of the other two principles, which require a weighing of the harms and benefits to all of those affected by disclosure to him.

From the standpoint of gender equality, the validity of this argument depends on the reliability of our estimates of the harms and benefits of disclosure of misattributed paternity. According to Lainie Friedman Ross, there is little evidence of harm to women or disruption to families brought on by disclosure.[20] Moreover, preservation of families is not always a good for the individuals involved. Women themselves may exercise the right that the genetic counselors want to preserve for them in these cases, that is, the right to do what they wish with the information. If the information is pertinent to the partner's own treatment or life plan, he has a right to disclosure that the woman has an obligation to respect. However, "respect" does not imply conformity. As already suggested, in complex circumstances, the woman may be morally obliged to subordinate her partner's right to know to the rights of others and her responsibilities to them. Whether or not the woman is obliged to disclose misattributed paternity to her partner or, for that matter, to the genetic father or their child, her failure to fulfill that obligation does not imply an obligation on the part of the health professional to do so.

Returning to the case with which we started, the narrative tells us nothing about the risks of disclosure to either partner, the genetic father, other family members, or the potential child. Neither does it tell us whether any of the pertinent parties would want to be given this information. To some extent, then, we are left with too much vagueness or uncertainty to support a decision by the genetic counselor about disclosure. More details might justifiably sway the decision in one direction rather than another. For example, if Mary indicated privately to the counselor that she feared for her life if such information were disclosed to Bill, or if Bill indicated privately to the counselor that he would not want to be told such information—either of these details supports nondisclosure. There may be no other family members to whom the information about Bill's carrier status is relevant. As for the genetic father, he may be deceased or it may be practically impossible to contact him; quite possibly, he would prefer not to know he had a genetic child that he might be legally called upon to support. While such specifics may not be definitively establishable, their degree of probability is morally relevant.

What kinds of details would support disclosure to Bill? Consider the following possibilities: (1) after learning that such information might be obtained through testing, Bill asked explicitly to be told; (2) Mary asked that they both be provided with whatever information was obtained through their tests; (3) both partners indicated on their visit to a counselor that their relationship would not be disrupted by such information; or (4) Bill's sister was married to a man with CF, and both were worried about having an affected child. While the details that might lead the genetic counselor to disclose or not disclose are practically limitless, an egalitarian standpoint identifies justice as the one general principle on which the details are to be weighed and a decision made. Assuming, then, that both partners are desirous of the information, if disclosure to Bill is likely to be more burdensome to Mary than to him, the information should not be disclosed to him, and if the disclosure is likely to be more burdensome to Bill, the information should be disclosed to him. Once one goes beyond the couple to consider disclosure to the genetic father, the potential child, or other family members, the rationale for disclosure needs to be yet more compelling mainly because these individuals are not clients and disclosure therefore seems intrusive or paternalistic.

In real life, as in the case described, one rarely if ever knows all of the details that are pertinent to decisions. The best we can do is to inform ourselves as fully as possible about the nuances of each situation and take account of these in formulating decisions that maximize the applicable moral values or principles. In the inevitable absence of some details, we invoke valid generalizations as the only means available by which to approximate the real situation. Here is where gender generalizations are morally relevant even though inadequate. The very fact that disclosure of misattributed paternity becomes problematic only when the one disclosing it is not the female client and when the one to whom it may be disclosed is her male partner illustrates a significant, even if unrecognized, awareness of the gender difference that governs this issue. Clearly, the genetic counselors and medical geneticists who balk at disclosure to the partner are acutely aware of the gender difference in the impact of this information.

That said, another pertinent generalization has been scarcely considered in studies of attitudes about misattributed paternity, that is, the views of female clients or potential clients. In the context of genetic counseling, they are typically the most affected group, affected even more than their male counterparts by decisions about pregnancy and child care. A recent survey of attitudes about misattributed paternity on the part of first-time genetics clients and the adult public in the United States showed important differences between their attitudes and those of medical geneticists about the issue. In general, clients and public were less concerned about privacy and individual autonomy. Three-quarters of the clients, who were mainly women, thought that the doctor should disclose misattributed paternity to men who asked about it. Most, however, thought the doctor should warn the woman first.[21] Wertz comments wryly on this contrast between the geneticists and the nongeneticists: "Perhaps these patients, as first-time visitors to genetics clinics, rather than more experienced consumers of genetics services, have not yet heard the arguments about genetic privacy that are so prevalent in bioethics circles."[22]

If women themselves are more likely to support rather than reject disclosure by doctors to their partners, policies that preempt that disclosure are paternalistic. Consistent with this view, they would probably support disclosure by other clinicians, including genetic counselors, in addition to doctors. In advocating that the women be warned, clients may simultaneously support respect for both the woman's autonomy and her partner's autonomy while minimizing potential harms to the woman. The woman may respond to the warning in whatever way she judges best. If preservation of the family unit appears threatened by the disclosure, those most affected by disclosure apparently do not see this as a worse outcome than nondisclosure.

Unless further details suggest otherwise, therefore, clients would recommend that the genetic counselor contact Mary and tell her what the genetic test reveals about Bill's paternity, mentioning that this information will be provided to Bill as well. In light of this "warning" and her anticipation of the consequences of disclosure, Mary may then decide for herself what to do. Quite possibly, Bill knows already. Regardless of whether he knows, however, decisions and policies regarding misattributed paternity may best be made by drawing on the input of those most affected. I would like to conclude, therefore, by considering a practice recently enacted by the U.S. Department of Defense to facilitate participation of those most affected in such decision-making processes. This model, I believe, suggests a means by which to reduce the inequitable impact of genetics on women as well as on people of different races, ethnicities, classes, and abilities.

Inclusion of "Consumers" as Model for Genetics

As a member of the Integration Panel for the U.S. Department of Defense Breast Cancer Research Program between 1993 until 1997, I took part in a procedural initiative to involve those most affected, that is, "consumers," in the review process. Initially, the term "consumer" referred to anyone who had survived breast cancer or who had a family member with the disease; later its meaning was limited to survivors. Not surprisingly, this innovation evoked a great deal of skepticism on the part of investigators, who felt that the participation of consumers would compromise the integrity of scientific review, impede its efficiency, and increase its cost.

While we agreed as a panel to include consumers on the scientific and technical review panels in the 1995 funding cycle, our plan for the implementation of this decision included two important components. First, we would prepare both the consumers and the scientists who would serve on these panels through meetings designed to inform them of the background and rationale for inclusion and to address their concerns about it. Second, we planned to assess the impact of consumer inclusion in scientific and technical review through a formal evaluation process.[23]

The evaluation of consumer inclusion on the scientific and technical review panels was undertaken through questionnaires distributed before and after the panels were convened to panel chairs, scientific reviewers, executive secretaries, and consumers. Although the prepanel data showed an overall openness on the part of the scientists to consumer participation, many of them doubted that the consumers would be able to engage meaningfully in discussions of the scientific merit of proposals. Some worried that unpreparedness and emotionality on the part of consumers would slow the process and dilute its scientific rigor. Others felt that consumers would be inclined to support immediate clinical relevance at the expense of basic science, and many indicated that the validity of scoring might be jeopardized.

According to the postpanel questionnaires, none of the above concerns materialized in the scientists' experience of consumer participation on the panels. The scientists found the consumers hardworking, intelligent, reasonable, open, articulate, and well prepared. In addition, all of the panels finished before their scheduled time, and consumer judgments generally coincided with those of the scientists. Although it is impossible to tell whether the consumer input influenced the scientists' decisions, or vice versa, it is clear that members of both groups found the experience a positive one. A majority indicated that they felt the practice of inclusion should continue.

Although the costs of having consumers and scientists as participants parallel each other, additional costs were incurred by the preparation of potential panelists, chairs, executive secretaries, and staff for inclusion of consumers. These costs were considered reasonable by the reviewers as well as by government personnel. Had adequate orientation of participants to this novel approach not been achieved, fears about compromising the quality of the review process might have been fulfilled. In time, as the rationale for inclusion of consumers is more widely disseminated and its practice spreads, the costs of preparation are likely to diminish. From an egalitarian perspective, the cost incurred may be regarded as an essential expenditure.

The inclusion of consumers in assessment of research proposals for breast cancer provides a model for genetics. As genomic research continues, possibilities for genetic testing not only for susceptibility to breast cancer but for other complex or multifactorial diseases and traits will proliferate. Advances in gene therapy will also increase. Decisions and policies about applications of these developments may or may not promote social equality. To insure that they are as equitable as possible, the input of those who are affected or at risk is indispensable. The rationale for this input is that those affected are better able than others to educate unaffected decision-makers about the impact of alternatives being considered. For their input to be effective, however, the unaffected group members must recognize that their own knowledge base or expertise inevitably involves limitations that can be reduced by listening to those whose experience and perspective is different from theirs, particularly to those most affected by the decisions to be made.

If equality in genetics is a social goal, participation by consumers of genetic services in decisions about the distribution of those services is indispensable. Currently and in the foreseeable future, women are the principal consumers of genetic services on their own behalf as well as on behalf of others. In many cases, women also predominate in other consumer groups that are significantly affected by advances in genetics—the poor, the disabled, and ethnic minorities. Because gender equality is inseparable from other forms of equality, it is more than a subset of justice. Its pursuit through consumer involvement in decision-making is a means of promoting justice for everyone.

Notes

1. With considerable revision and additions, material in this article is drawn from various chapters of my book, *Genes, Women, Equality* (New York: Oxford University Press, 2000).

2. Although the enumeration is not exhaustive, biological and psychological gender differences related to genetics are well documented in Mary B. Mahowald, Dana Levinson, Christine Cassel, et al., "The New Genetics and Women," *Milbank Quarterly* 74:2 (1996):239–83.

3. See Mary Briody Mahowald, *Women and Children in Health Care: An Unequal Majority* (New York: Oxford University Press, 1996), 25.

4. Amy J. Ravin, Mary B. Mahowald, and Carol B. Stocking, "Genes or Gestation? Attitudes of Women and Men about Biologic Ties to Children," *Journal of Women's Health* 6:6 (1997):639–47.

5. For example, Chinese Americans tend to embrace new technology with enthusiasm, whereas African Americans and Latino Americans tend to mistrust it. See Nancy L. Fisher, ed., *Cultural and Ethnic Diversity: A Guide for Genetics Professionals* (Baltimore, Md.: Johns Hopkins University Press, 1996).

6. Laurie Nsiah-Jefferson and Elaine J. Hall, "Reproductive Technology: Perspectives and Implications for Low-Income Women and Women of Color," in *Healing Technology: Feminist Perspectives*, ed. K. S. Ratcliff, M. M. Ferree, G. O. Mellow, et al. (Ann Arbor, Mich.: University of Michigan Press, 1989), 93–117.

7. E.g., Barbara Katz Rothman, *Recreating Motherhood: Ideology and Technology in a Patriarchal Society* (New York: W. W. Norton, 1989); and Rothman, *The Tentative Pregnancy: Prenatal Diagnosis and the Future of Motherhood* (New York: W. W. Norton, 1993).

8. See, generally, Aristotle's *Nichomachean Ethics* V, 3–5. Many authors cite this principle before delineating various material principles or arguing for their own.

9. E.g., Tom L. Beauchamp and James F. Childress list six principles, each designating a different criterion for distribution to individuals: equal shares, need, effort, contribution, merit, and free-market exchange. *Principles of Biomedical Ethics* (New York: Oxford University Press, 2001), 228.

10. Amartya Sen, *Inequality Reexamined* (Cambridge, Mass.: Harvard University Press, 1992).

11. XXY is called Klinefelter syndrome and XO is called Turner's syndrome, both of which are associated with infertility. XYY has been alleged but not proved to be associated with greater aggressiveness.

12. Admittedly, the criteria for determining "reasonable efforts" are problematic, as problematic as criteria for determining the "reasonable person" (or "reasonable man") standard in law and bioethics.

13. Robert C. Tucker, ed., *The Marx-Engels Reader* (New York: W. W. Norton, 1972), 388.

14. An exception to this point is the woman's capacity for lactation. While others, including fathers, are also able to feed infants, only women can nurse them.

15. E.g., Genzyme Genetics offers screening for 87 CF mutations on its Web site at www.genzymegenetics.com. The DNA Diagnostic Program at

Michigan State University stipulates the following detection rates on its Web site: 97 percent for Ashkenazi Jews, 90 percent for Caucasians, 69 percent for African Americans, and 57 percent for Hispanic populations. See www.phd.msu.edu/DNA/cf_fact.html.

16. Theoretically, misattributed *maternity* can occur as well, but it is extremely improbable because of the tie between gestation and genetics. While the genetic contribution of the father is provided in private, that of the mother is typically visible to others through her pregnant body.

17. Dorothy C. Wertz, John C. Fletcher, and John J. Mulvihill, "Medical Geneticists Confront Ethical Dilemmas: Cross-Cultural Comparisons among 18 Nations," *American Journal of Human Genetics* 46 (1990):1201.

18. Wertz, Fletcher, and Mulvihill link preservation of the family unit with protection of the child and mother, interpreting both as based on the principle of nonmaleficence, Wertz, Fletcher, and Mulvihill, "Medical Geneticists Confront Ethical Dilemmas," 1203.

19. Deborah F. Pencarinha, Nora K. Bell, Janice G. Edwards, and Robert G. Best, "Ethical Issues in Genetic Counseling: A Comparison of M.S. Counselor and Medical Geneticist Perspectives," *Journal of Genetic Counseling* 1:1 (1992):23.

20. "The dominant theme in the family counseling literature," Lainie Friedman Ross writes, "is that the woman's behavior has no causal influence on the man's decision to batter, but rather, that the problem lies exclusively in the man" (126). However, even if the woman's welfare were threatened by disclosure to her partner, Ross would oppose a policy of nondisclosure on grounds that "most men do not and would not abuse their partners" (127). Ross, "Disclosing Misattributed Paternity," *Bioethics* 10 (1996):114–30.

21. Dorothy C. Wertz, "Society and the Not-So-New Genetics: What Are We Afraid Of? Some Predictions from a Social Scientist," *Journal of Contemporary Health Law and Policy* 13 (1997):314.

22. Wertz, "Society and the Not-So-New Genetics," 314.

23. Yvonne Andejeski, Erica S. Breslau, Elizabeth Hart, Ngina Lythcott, Linda Alexander, Irene Rich, Isabelle Bisceglio, Helene S. Smith, and Fran M. Visco, "Benefits and Drawbacks of Including Consumer Reviewers in the Scientific Merit Review of Breast Cancer Research," *Journal of Women's Health & Gender-Based Medicine* 11 (2002):119–36.

7

Using Human Dignity to Constrain Genetic Research and Development: When It Works and When It Does Not

Jan C. Heller, Ph.D.

WESTERN PHILOSOPHERS and theologians have long argued that there exists a quality or property peculiar to humans that ought to constrain what is done to and by humans. That quality or property is often discussed as human dignity, and humans are said to possess it either, according to Western religious traditions, by virtue of having been created in the image of a personal God or, according to common secular thinking, by virtue of being persons. In this sense, then, human dignity functions as an important deontic constraint on actions we undertake that affect humans, and this includes (it is commonly assumed) those actions that are made possible as a result of genetic research and development.

However, I will argue that the research and development of a certain type of genetic application utterly undermines this important constraint, namely, any application that can or might be used to *create* humans. We can get an intuitive sense of how human dignity is undermined by such applications if we first consider those applications where dignity remains an important constraint, that is, those that are developed for use *on* humans. To do this, it will also be helpful to distinguish at least four classes of human persons.

FOUR CLASSES OF HUMANS

For example, probably none of us would think it ethically justified to enter an adult (with decision-making capacity) into an experimental gene therapy protocol without first securing that person's informed consent. And we would seek this consent not only because we believe this person is likely to be the most qualified to evaluate the possible risks and benefits of participating in the protocol, but also (and perhaps primarily) as a way of respecting this person's dignity. Similarly, many of us would raise ethical questions about a procedure being considered for a person who had just died that would, without the *prior* informed consent of that person, use his or her gametes to produce a child. Whatever we might think about the use of genetic applications to produce such a child, we are not likely to think it ethically justified to proceed in this particular case without the deceased person's consent, and this again is because we respect his or her dignity. Finally, many of us would find it ethically troubling to develop genetic engineering applications that might adversely affect future persons. That is, we might worry about developing applications that would signif.cantly advantage one group of future persons over another whose genetic constitution is not similarly engineered. This worry could spring from a concern for the justice of such deliberately engineered inequities, which in turn could be related to the dignity that these future persons will enjoy when they come into existence as persons.

There is no need to list additional examples. It is obvious that we use the concept of human dignity to constrain the research and development of genetic applications with respect to at least three classes of human beings: living humans who are persons, dead humans who were persons, and future humans who will be persons.[1] This said, there is a fourth class of humans whose dignity cannot be used to constrain the research, development, or use of certain genetic applications; indeed, they cannot be said to have dignity. This fourth class of humans are those whose future existence depends on the genetic applications in question. They are distinguished from the other future persons mentioned above in that these other future persons will exist whether or not these genetic applications are developed. As we shall see, this dependency renders the humans created by genetic applications particularly vulnerable to the interests of the agents who created them.

The Problem of Contingent Future Persons

Genetic applications that are used to create particular human beings undermine dignity as a moral constraint with respect to those who are actually created by them because they involve agents in a more general problem that was first defined only in 1967.[2] Since then, this perplexing problem has been discussed in the literature under a number of provocative names. Gregory S. Kavka discusses it as the "futurity problem" and as "the paradox of future individuals."[3] Thomas Schwartz discusses it as "the problem of the disappearing beneficiaries."[4] R. M. Hare discusses it as a problem of "possible people."[5] Derek Parfit discusses it as "the non-identity problem."[6] David Heyd discusses it as a "genesis problem."[7] Finally, I call it the "problem of contingent future persons."[8] This problem arises when we are considering whether to bring a potential or, as I typically say, a contingent future person into existence.

The future human in question is potential or contingent in the sense that its future existence and thus its identity depend on the choices or actions of the relevant agents in question. Derek Parfit claims that this situation can arise because personal identity is time-dependent. By this, Parfit means that the identities of all people necessarily depend on when they were conceived, that is, on when a particular egg and a particular sperm cell successfully join to form a new human embryo.[9] This fact is not usually morally problematic, but it could be under conditions like those that accompany the use of the relevant genetic applications. Indeed, there is only one condition that must be satisfied for this problem to arise: the action in question must affect the timing of the conception of a future person so that any change in that timing results in a *different* child actually coming into existence. We can get an intuitive sense of this problem by considering another example.

Suppose we are clinicians who are counseling a couple who know they are both carriers of the same recessive gene for a very serious genetic disease, say Tay Sachs, an invariably fatal disease that kills its youthful victims after several years of increasingly intense suffering. (If they are like most couples in this situation, they will have already had such a child, which is how they would have learned that they are carriers.) If they want to have a child who is genetically their own, they currently have two options. Either they can conceive the child naturally, and run a 1 in 4 chance of it being born with Tay Sachs, or they can use a procedure called the preimplantation diagnosis of genetic disease, which will permit them to select an embryo created *in vitro* that is free of both recessive disease genes. They can then transfer that embryo to the woman's uterus and, assuming it implants and develops normally, have a child who is free of the disease. (Of course, they are likely to use more than one embryo in such cases, but I avoid this complication here.) There are many ethical questions raised by such applications, but the one I want to focus on is whether human dignity can be attributed in any coherent sense to the future child who might come into existence by using this particular application.

The answer to this question is: No. The couple's goal is to have a healthy child, either one who is completely free of the disease gene or one who is merely a carrier of one recessive gene. However, before the child is actually conceived *in vitro*, it is merely a potential child, that is, one whose future existence is contingent on the action made possible by the genetic application in question. What is ethically significant about the status of this potential child is that it may never be born, and this again because any change in the timing of its conception will result in a *different* child actually being born. (This different child would be born because it is likely that a different egg and almost certain that a different sperm cell would join *in vitro* if the timing of the procedure is changed). And hence, because a different child would be born under such conditions, it makes no sense to claim that this child has had its dignity either violated or honored by not being born earlier. For if agents had not delayed its conception, this child would not exist. Said differently, there is no earlier "state of being" during which this child can be said to have had dignity that can be either violated or honored by being brought into existence. Conversely, and more controversially, if the same couple under the same conditions decides *not* to delay conception, either naturally or *in vitro*, and then tragically has a second child who is affected (either by chance, if conceived naturally, or because a mistake was made in interpreting the test results on the embryo conceived *in vitro*), it also makes no sense to claim that this child's dignity has somehow been either violated or honored by being born with Tay Sachs. Again, if its parents had delayed this child's conception, a different child would have been born and this one would not exist.[10]

Two Approaches to Value

Now, our concerns for the dignity of human persons routinely arise in moral arguments that are based in an approach to value that is oriented toward the interests of particular persons, whether present or future. An approach to value is concerned with the very nature of value, that is, with the most basic or fundamental conditions under which value can be attributed to *any* object. The approach to value that concerns us here (of two that are possible) is discussed in the literature as a "person-oriented" or "person-affecting" approach. As a condition of attributing value to an object, a person-affecting approach requires that a person exist for whom the object can be valuable. This may seem like an obvious requirement, but the alternative approach to value does not require the prior existence of persons. It is typically discussed as an "impersonal" approach and it is not concerned with the effects of our choices or actions on particular persons; indeed, the existence of value in this approach does not depend in any way on the prior existence of persons.[11] In any case, when a person-affecting approach to value is utilized by agents in moral arguments to make choices about future persons, those persons must exist in the future *independent* of the agents' choices or action if value is to be attributed to them. This is, of course, exactly what is at issue with potential future persons: their existence is utterly dependent on or contingent on the agents' choices or actions, and hence they may or may not exist in the future.

Hence the problem of contingent future persons undermines what may be referred to as person-affecting deontological moral arguments. These are moral arguments that need not refer to the harms or benefits that may be visited on future persons as a result of our actions but may refer instead to something about such persons as persons that purportedly requires them to be brought into existence in certain ways or under certain circumstances—for instance, that they possess or will possess human dignity which confers on them certain rights or that creates certain duties or obligations which are owed to them. However, even when we can agree that contingent future persons will possess human dignity when they come into existence, they do not possess it *when the choice is being made that will bring them into existence*. For this reason, then, the moral philosopher David Heyd argues that the use of deontological arguments to make choices that bring potential or contingent future persons into existence results in a form of moral reasoning that is "viciously circular" and thus utterly self-defeating.[12] Such arguments become circular because the problem of contingent future persons undermines our ability to make any *direct* moral reference to the potential child who might come into existence as a result of the choices or actions we are trying to evaluate. (I do not discuss it here, but this same inability undermines any person-oriented consequentialist moral argument that attempts to attribute harm or benefit to the potential or contingent future persons who might come into existence as a result of such choices.)

Some Implications

We now turn to consider some implications of this problem for a number of moral arguments, both secular and religious, that have been made in the literature about the research and development of genetic applications. First, consider French Anderson's much-debated categorization of genetic applications.[13] He outlines four categories of genetic intervention that he believes are both descriptively and ethically distinguishable, and then argues that only the first should be morally permitted at this time. The four categories are somatic-cell gene therapy, germ-line gene therapy, enhancement genetic engineering, and eugenic genetic engineering. Somatic-cell gene therapy is defined as a medical intervention that modifies the genetic material of living cells. Germ-line gene therapy uses techniques similar to gene therapy, but is distinguished from it by the facts that the "patient" has had the genetic cause of his or her disease removed and that this "fix" can be passed to succeeding generations. Enhancement genetic engineering is aimed at "improving" what are perceived to be "normal" human traits, whereas eugenic genetic engineering is aimed at modifying complex traits such as intelligence.

By now it is probably well known that many commentators do not believe that Anderson's distinctions can be sustained. Some argue that the impermissible enhancement applications (which are already being done in any case; e.g., human growth hormone) are hard to distinguish from permissible somatic-cell gene therapies. They also argue that some eugenic engineering applications may be hard to distinguish from enhancement or other germ-line interventions (e.g., genetically engineering children for culturally influenced aesthetic reasons). Hence conceptually the categories are not very useful and morally they set up Anderson for slippery-slope problems.

Based on the above discussion, however, I suggest that it might be more helpful to distinguish between genetic applications that are used or intended for use *on* humans and those that are used or intended to be used to *create* humans.[14] This categorization is helpful because it signals to agents that the applications which are developed for use on humans may be constrained in the same way other new biomedical technologies are constrained, namely, by concerns for safety, efficacy, cost, and access. Further, these constraints can be justified by reference to the human dignity of those on whom the application is used. But these same constraints, insofar as they can be justified by reference to human dignity, are undermined for those applications that are used to create human beings. For, because of the problem of contingent future persons, there can be no reference to the future person who may be brought into existence through such applications. This fact removes a *very* significant constraint on the research, development, and use of these applications. Said differently, there may be no sustainable reason not to develop Anderson's impermissible applications.

Another area where the problem arises is in a consideration of the research and development of somatic-cell nuclear transfer cloning techniques in humans. I have written about this elsewhere,[15] so I mention it only in passing here. A number of constraints on human cloning have been suggested that attempt to reference the dignity of the cloned child who might exist as a result of cloning. The problem of contingent future persons undermines these constraints as well. Consider, for example, what one Roman Catholic commentator said about cloning in a letter to the National Bioethics Advisory Committee. He argues that human cloning should be regarded as a violation of human dignity because it would "jeopardize the personal and unique identity of the clone (or clones) as well as the person whose genome was duplicated."[16] However, assuming the "parent" of the clone gave his or her consent, it is not clear how having a cloned "twin" would jeopardize his or her identity and it is also not clear how having a cloned "twin" would violate the parent's dignity.

But more problematic, I think, is the claim that cloning somehow violates the human dignity of the child who is born as a clone. If you follow the cloning literature, you will know that no Roman Catholic moral theologian has claimed that the child itself would not be fully human for having been born a clone. Nevertheless, the problem of contingent future persons implies that, whatever *future* dignity a cloned child might possess, appeals to dignity *prior to its conception* cannot be used as a basis for arguing that such a child should not have been born or, at least, born in this way. It is an incoherent argument that cannot be sustained.[17]

Some Possible Responses

So how can we respond to the problem of contingent future persons, especially in view of the fact that many of us will want to continue using human dignity as a constraint on the research and development of genetic applications that are or will be used to create humans? If we want to make a coherent appeal to dignity, we have one option. We cannot appeal to the dignity of the potential or contingent future persons who might be brought into existence with those applications; they have no dignity to which we can appeal. But we can appeal to the dignity of those persons, living or future, *who will be affected* by our choice to bring or not to bring potential or contingent future persons into existence. This option requires a move to consequentialism, and elsewhere I refer to it as an *indirect* response to the problem. It is indirect because the agents in question cannot include contingent future persons directly in their moral considerations; rather, they can only consider the effects of their possible existence on others who, without question, either currently possess or will possess human dignity. This indirect response will do much of the moral work we need done, but not all of it.

Consider Anderson's categories again. An indirect response to this problem implies that whatever constraints we develop on human germ-line applications or on the engineering of future children, we will not be able to reference the children who might be brought into existence as a result of the applications. However, if the existence of such children were adversely to affect other persons, such as their parents or siblings, or even human society generally, honoring the dignity of *these* persons could coherently constrain the research and development of the genetic applications that might be used to bring the potential persons into existence. Similarly, if other living or future persons are adversely affected by the birth of a cloned child, this could count as a reason not to clone a particular child or, perhaps, any children. However, if clinicians and the parents of engineered or cloned children did not have their dignity violated by using such applications, and if we could be reasonably sure that human society in general would not be adversely affected, there may be no moral reason not to pursue such applications in an indirect, person-orientated, deontological approach to value. There would, in other words, be little to constrain such agents.

Notes

1. I avoid here the controversial discussion of when personhood may justifiably be attributed to developing humans. The problem of contingent future persons, discussed below, is not dependent on this determination.

2. Jan Narveson, "Utilitarianism and New Generations," *Mind* 76 (1967):62–72.

3. Gregory S. Kavka, "The Futurity Problem," in *Obligations to Future Generations*, ed. R. I. Sikora and Brian Barry (Philadelphia: Temple University Press, 1978), 180–203; Gregory S. Kavka, "The Paradox of Future Individuals," *Philosophy and Public Affairs* 11 (1982):93–112.

4. Thomas Schwartz, "Obligations to Posterity," in *Obligations to Future Generations*, ed. R. I. Sikora and Brian Barry (Philadelphia: Temple University Press, 1978), 3–13.

5. R. M. Hare, "Possible People," *Bioethics* 2 (1988):279–93.

6. Derek Parfit, *Reasons and Persons* (Oxford: Clarendon Press, 1984), 351–79.

7. David Heyd, *Genethics: Moral Issues in the Creation of People* (Berkeley, Calif.: University of California Press, 1992).

8. Jan Christian Heller, *Human Genome Research and the Challenge of Contingent Future Persons* (Omaha, Nebr.: Creighton University Press, 1996).

9. Parfit, *Reasons and Persons*, 351–55. This is not, of course, a claim that personal identity can be reduced to the genetic makeup of a person, which is (largely) determined at conception.

10. Parfit argues in this case that the logic of the problem is constrained by the claim that the child's life must be at least minimally "worth living." Not everyone is convinced that this logic can be sustained, however, and even Parfit seems to think the claim is weak. At one point he says: "I also will call certain lives 'worth living.' This description can be ignored by those who believe that there could not be such lives." Parfit, *Reasons and Persons*, 358.

11. For a more complete discussion of these two approaches, see Derek Parfit, *Reasons and Persons*; David Heyd, *Genethics*; and Jan Christian Heller, *Human Genome Research and the Challenge of Contingent Future Persons*.

12. David Heyd, *Genetics*, 39–64.

13. See W. French Anderson, "Human Gene Therapy: Scientific and Ethical Considerations," *Journal of Medicine and Philosophy* 10 (1985):275–91.

14. Of course, many will do both, because they will use the genetic material of living humans, whose informed consent will be needed, to create new humans. But I believe the distinction is valid nevertheless.

15. Jan C. Heller, "Religious Perspectives on Human Cloning: Revisiting Safety as a Moral Constraint," *Valparaiso University Law Review* 32:2 (1998):661–78.

16. Quoted in National Bioethics Advisory Commission, *Cloning Human Beings* (Rockville, Md.: National Bioethics Advisory Commission, 1997), 50.

17. The problem arises for similar appeals in at least two of the Ethical and Religious Directives for Catholic Health Care Services. See, for example, directives 40 and 42 in United States Conference of Catholic Bishops, *Ethical and Religious Directives for Catholic Health Care Services*, 4th ed. (Washington, D.C.: United States Conference of Catholic Bishops, 2001). I discuss these in Heller, *Human Genome Research and the Challenge of Contingent Future Persons*, 158–59.

8

Engineering Our Grace: An Old Idea and New Genetic Technologies

Brent Waters, Ph.D.

IN THE FIRST EDITION of *The Foundations of Bioethics*, H. Tristram Engelhardt contends that we are on the verge of reshaping human nature.[1] With recent advances in biomedical research, what were once perceived as intractable physiological and behavioral boundaries are now becoming subject to our willful control. In Engelhardt's words: "In the future our ability to constrain and manipulate human nature to follow the goals set by persons will increase. As we develop the capacities to engage in genetic engineering not only of somatic cells but of the human germ line, we will be able to shape and fashion our human nature in the image and likeness of goals chosen by persons."[2] In short, we will increasingly be able to construct rather than accept our destiny as a species.[3]

Why this sudden impulse to reshape human nature? According to Engelhardt, there are two principal factors. First, we are simply procuring the knowledge and technological capability to do so. Second, given the Enlightenment's failure to establish a universal morality, there are no compelling reasons why we ought not to do so. In the absence of any given landmarks guiding our quest for meaning and value, we are "left to our own devices" in constructing our destiny.[4] We must turn to ourselves rather than to God to save us from an undesirable fate. The principal means of imposing our will upon the future is our offspring, whom we will create in an image and likeness of our choosing. We are entering an age when we will engineer our grace through a "baptism" of technological rather than supernatural means.[5]

Engelhardt is half-right. Recent scientific discoveries are inviting us to ponder if human nature is more malleable than once assumed. A combination of genetic and social manipulation may yet make the future subject to the human will instead of to capricious natural processes. Yet he is wrong to imply that the failed Enlightenment project spawned our desire to control our destiny. The notion of engineering our grace, particularly through offspring, is an old idea that keeps reemerging in different guises.

The purpose of this essay is to examine some illustrative manifestations of this old idea, especially in light of the development of new genetic technologies, and then ponder if we are witnessing the emergence of a renewed heresy and, if so, what this may portend for the future.

The Old Idea

Around 420 A.D. Augustine argued that only sin and not grace is transmitted from parents to offspring. This is the case because all humans carry Adam's sin in their bodies.[6] Through procreation a child is given the "carnal birth" of her parents but cannot receive their spiritual rebirth.[7] What is born of the "flesh" is a child of the world, and a child of God is born only through a baptism of water and the Spirit. This is why even godly parents may give birth to unregenerate offspring, and why the children of notorious sinners may be among the elect.[8] The only hope for healing our sin is through baptism in Christ's resurrection, for it is not flesh or blood but water and the Spirit that will inherit the Kingdom of God.[9] We are to place our hope solely in God's grace and not in the progressive perfection of the human body through generations of offspring.[10] Or, more tersely, there is no salvation through procreation. It is our second, rather than first, birth that should concern us.

In making this argument, Augustine was refuting a notion that was evidently prevalent enough to spark his ire, namely, that if Adam's original sin was transmitted from parent to child, then so too was Christ's grace transferred among the elect. The elect carried in their bodies something akin to a "godly seed" that was passed on to offspring. Yet this implied that humans could engineer their grace over time through the progressive breeding of the most spiritually fit. The fruition of God's Kingdom would be, after all, a matter of flesh and blood. Through this original infusion and subsequent transmission of grace across generations, humans could eventually achieve their own perfection. Moreover, this implied that humankind was the architect of its own destiny and consequently that the object of its hope was a projection of itself.

One of the most obvious ways of projecting ourselves into the future is through offspring, so children become the practical objects of hope. Rome provided a splendid example, for the empire could perpetuate and perhaps eventually perfect itself only through its children. Augustus, for instance, enacted laws penalizing childless couples or men who did not marry by a certain age, while rewarding married couples producing at least five legitimate children. Additional legislation discouraged and in some cases prohibited marriage among individuals of differing social strata, in the belief this would help preserve if not improve the physical and mental vigor of the aristocracy.[11] In effect, human bodies in general and children in particular were public resources for shaping the empire's destiny.[12]

According to Augustine, however, it is God alone who determines our destiny. Since the Kingdom of God will displace all earthly empires, we can never place our hope in children because they merely magnify the futility of a passing age. Since Christians are citizens of a heavenly rather than an earthly realm, they have little interest in helping temporal regimes achieve their quantitative and qualitative population objectives. This is one reason why early Christians had little interest in procreation, for since in Christ the child who seals the world's destiny had already been born, there was no urgency to produce any more offspring.[13] If God is the true object of hope, Augustine contended, then grace is a divine gift given to every generation instead of a possession imparted from one to the next.

Although the concept of "godly seed" did not receive much attention from subsequent theologians, Augustine did not quash the idea entirely. It reappeared with the Puritans, albeit in a different guise. If the elect were to form a holy nation, then procreation instead of evangelism offered a more reliable method for achieving this goal. God's Kingdom on earth was not to be comprised of converts but consisted of a covenanted people sharing a common lineage and destiny. The elect were stewards of a "godly seed" through which God bestowed both judgment and grace upon his creation. Moreover, this "godly seed" would bear good fruit when family, church, and state worked together, inculcating and reinforcing a series of moral codes, religious beliefs, and social principles. Thus the voluminous literature produced by the Puritans on household, ecclesiastical, and social ordering.[14]

Richard Baxter provides a representative figure of seventeenth-century Puritanism. According to Baxter, the family is both God's instrument for governing the world and a community sanctified by God to achieve "the public and universal good that he has ordained."[15] Parents carry an especially heavy burden in fulfilling these expectations because children "bear the image and nature" of their parents, born rebellious sinners who must be disciplined and brought to a saving knowledge of Christ.[16] Indeed, Baxter penned the sobering admonition: "Understand and lament the corrupted and miserable state of your children, which they have derived from you, and thankfully accept the offers of a Saviour for yourselves and them."[17] The best way for parents to exercise their heavy duty is to instill in their children a life of holiness and to direct them toward a useful calling for church or state.[18]

Baxter's account of parental duty, however, contains a theological dilemma. If *all* children share a "natural corruption,"[19] then they are born sinners saved only by God's mercy and grace. Since saints are not born, procreation has no inherent redemptive significance.[20] Yet Baxter blunts his argument by asserting: "A holy and well-governed family, doth tend to make a holy posterity, and so to propagate the fear of God from generation to generation."[21] Children of "godly seed"[22] seemingly do not suffer as extensively the ill effects of original sin as those begotten by "ungodly parents."[23] The children of saints are more likely to be among the elect.

With Augustine, Baxter presumes that only sin and not grace can be transmitted from parent to offspring. Yet Baxter also admits that children born of successive generations of the elect are more easily prepared by their parents to come to a saving knowledge of Christ. If all children are born sinners, how can it be that the offspring of saints are more receptive and responsive to parental instruction and discipline? The only explanation he can entertain is that despite a universal corruption of sin, these children must nonetheless contain some innate orientation toward a godly life that is not present in other children. And what better way to explain this phenomenon than that they must be the products of "godly seed."

Baxter, however, does not attempt to resolve this quandary. That task was undertaken two centuries later by Horace Bushnell. Bushnell's thesis is that rather than preparing children for conversion, parents should instead ensure that a child *"grow[s] up a Christian, and never know[s] himself as being otherwise."*[24] In short, Christian parents give birth to Christian children because they share "something like a law of organic connection as regards character," so the "faith of one will be propagated in the other."[25] More succinctly, the "Christian parent has, in his character, a germ which has power, presumptively, to produce its like in his children."[26] Since an acquired sanctification may be passed on to offspring, children must be reared in pious households in order that the progressive and cumulative effects of Christian virtue are reinforced and magnified over time. Moreover, the ability to pass on acquired as well as natural characteristics means that the organic unity of parent and child mitigates the transmission of original sin, so that procreation is reoriented as an instrument of grace rather than a "vehicle of depravity."[27]

The principal implication Bushnell draws from his concept of the organic unity of parent and child is that God's Kingdom will be established on earth through a complementary relationship between procreation and child-rearing. He contends that God created the first humans that "he might have a godly seed," and established laws of propagation that "piety itself shall finally over-populate the world."[28] It is through the quantitative and qualitative growth over time of a people stemming from this godly seed that humanity shall find its salvation. Appealing to the selective breeding of livestock, Bushnell asserts that since acquired traits are passed on and amplified to subsequent generations, "civilization is, in great part, an inbred civility."[29] Likewise, a Christian lineage produces an "inbred piety."[30] Through "a kind of ante-natal and post-natal nurture *combined*, the new-born generations will be started into Christian piety, and the world itself over-populated and taken possession of by a truly sanctified stock."[31] Procreation and child-rearing are the principal instruments of humanity's salvation and perfection, because God's grace must first permeate every aspect of our fallen nature. If heredity is the means of transmitting sin, then it must also be a means of receiving grace. Hence the world's hope and destiny lie in the propagation and nurture of godly seed through which God "will reclaim and resanctify the great principle of reproductive order and life."[32] It is, after all, sanctified flesh and blood that will inherit in the Kingdom of Heaven.

Regarding the theme of genetics, we may view this notion of "godly seed" as a curious relic having little relevancy for the issues we are facing. Unlike Augustine, we tend to believe that "sin," whatever it may be, is more a result of one's actions and circumstances than a condition one is born into. Unlike Baxter, we have more sophisticated explanations for why some children respond better than others do when exposed to good "parenting" skills. Unlike Bushnell, we read Darwin instead of Lamarck and know that learned behavior can no more be passed on to our offspring than can the physical outcomes of vigorous bodybuilding. Moreover, we have learned our lesson from the ill-informed eugenics programs of the late nineteenth and early twentieth centuries and will not repeat them in the twenty-first. The renewed interest in genetics is not driven by some religious zeal envisioning the perfection of human beings and society. Rather, it is motivated by a desire to use our growing knowledge of genetics to improve the quality of life for many individuals through the development of various preventive and therapeutic treatments.

This sober and chastened optimism may prove to be true. Yet it would not be prudent to simply ignore or dismiss this old idea of "godly seed." It is part of our cultural heritage that has helped shape our current perceptions of human nature, of what constitute deviations from this standard, and how they may be corrected. In addition, a largely inarticulate belief in the progressive and cumulative improvement of the human species through managed breeding and efficient parenting may still be motivated by more a religious zeal, albeit in secular guise, than we realize or care to admit. At the very least it may prove interesting to view some recent research in genetics and its implications for biomedical applications in light of this old idea.

The New Technologies

If the daily deluge of headlines is any indication, recent discoveries in genetics have enabled us to take our first, tentative steps into a new realm of biomedicine. A few representative news items over recent months will suffice as landmarks of the terrain we are entering.

Item 1 In late June 1999, a dramatic improvement in preimplantation genetic diagnosis (PGD) for Cystic Fibrosis (CF) was announced. The technique allows embryos created through *in vitro* fertilization (IVF) to be tested for the deleterious mutation. Unaffected embryos can then be chosen for implantation in the womb. Although PGD was first developed earlier in the nineties shortly after the discovery of the defective gene causing CF, it can identify the mutation in only 60 percent of the cases. Given this limitation, many couples still faced the bleak dilemma of either giving birth to an affected child or terminating the pregnancy. The newly developed technique, however, is 99.5 percent reliable. What the news report characterized as the "genetic game of chance" will be largely replaced with a means for ensuring that only "unaffected embryos . . . will grow into children." Not only will many children be spared unnecessary suffering; the announcement stressed that couples could now avoid the "painful decision" of either rearing a CF child or aborting a fetus.[33]

Item 2 The University of Minnesota School of Medicine reported in the fall of 1999 that its researchers had cured a rare liver disease in rats by repairing, rather than replacing, a defective gene. A molecule is injected into an affected cell, which, according to the principal investigator, "tricks the cell into thinking that there is a defect in its DNA sequence for the particular gene and by tricking the cell, the cell basically repairs (what it perceives as) its own defect." The technique holds great promise for treating not only liver disease in humans but also a range of other diseases, such as hemophilia and sickle cell anemia. It is expected that a clinical trial application will be submitted to the Food and Drug Administration early next year, to be conducted, interestingly enough, among the Pennsylvania Amish population.[34]

Item 3 In the fall of 1999, scientists at Princeton University announced the development of a new strain of mice called "Doogie" mice. These genetically modified rodents are smarter than their normal counterparts. A Doogie can learn quickly, is more attentive, and has a superior memory. In short, the technique keeps the mouse brain young for a longer period of time. The research not only has important implications for treating dementia in humans but it may also enable us to enhance intelligence.[35] There is, however, an important caveat. Earlier this year researchers discovered, much to their surprise, that when "genetically identical mice" were placed in "nearly identical situations" they failed to act in identical ways but responded differently to various laboratory technicians.[36] Evidently Doogies will remain dependent upon conducive learning environments and the pedagogical skills of their teachers to achieve their full potential.

Item 4 In the summer of 1999, a group of scientists announced they had discovered the genes that may unlock the mystery of immortality. Their research with fruit flies disclosed genes with a capacity to repair themselves indefinitely. The hope is that "doctors will ultimately be able to offer anti-ageing cell repair pills or injections for human patients." The scientists predicted that people born today could live well into the twenty-second century, "reaching ages of 140 or more."[37] It did not take long for commentators to make more immodest claims. One pundit, for instance, asserted that: "Aging could become fully reversible," with a "normal" life span reaching "at least" 150 or 225. Moreover, when combined with recent advances in stem-cell research, we will not live as "frail, feeble" senior citizens but will remain "young, active, and vigorous." The only thing preventing us from achieving greatly enhanced longevity is outmoded "religious beliefs" restricting more aggressive research and development.[38]

Competing Destinies

These recent headlining stories are noted to illustrate the great breadth and rapid pace of genetic research being conducted and its potential for new breakthroughs in health care. In each case there is little with which we may quarrel. What is wrong with developing techniques to prevent passing on deleterious genetic traits (such as CF), repair defective genes, cure dementia, or increase longevity? In each case a strong argument can be mustered that research is being conducted in order to achieve a good end.

It is, rather, when we stand back to observe the larger picture of the cumulative effect of these research projects (in tandem with the hundreds of other headlines appearing during the same brief time period) that we may ask a more troubling question: Is there an underlying principle or belief driving these multifaceted research projects toward a common end? I want to explore the thesis that there is such an underlying principle or belief, and that it is not unlike a notion that through a godly seed humans may engineer their grace or some state approximating it.

Like Augustine, do we not believe that children inherit from their parents a genetic legacy that is less than desirable? Deleterious and undesirable traits occur with sufficient regularity to justify investing vast sums of time and money to correct or improve upon what is perceived to be an expanding circle of biological and behavioral deficiencies. If we do not believe there is something "wrong" with the human reproductive process, why would we try to "trick" genes into doing something we judge to be good rather than bad? Do we not also share with Augustine the belief that young adulthood represents the most desirable stage of human development, as witnessed by his speculation that in the resurrection we will each be thirty years old and at the height of our physical and intellectual capabilities?[39] The principal difference between Augustine and us is that we believe this state can be achieved in the earthly rather than heavenly city.

With Baxter, do we not believe, however, that children are not equally deficient? Some children suffer more than others the ill effects of a less-than-perfect human biology. As our knowledge of human genetics grows, we will be able to prevent more and more diseases and disabilities. Preimplantation screening for CF, for instance, is but a first step in preventing a growing list of conditions that are thought to have some genetic basis or factor. Will these interventions not promote the passing on of a more contemporary and subtle version of "godly seed"?

Not unlike Bushnell, are we not also coming to believe that if physical and behavioral characteristics cannot be passed on to offspring, they can, at least, be added or engineered? Moreover, our offspring offer the best hope of improving the human species, for the good effects of managed breeding *and* child-rearing will eventually overcome our biological deficiencies. Along with Bushnell, we are sophisticated enough to recognize that it is not a question of either nature *or* nurture, but both. Yet are we not also coming to believe that our nurture would be more effective if we had a better nature to work with? Imagine how much more efficient our social-engineering schemes of the home and classroom would be if our efforts were directed toward the human equivalent of "Doogies."

I have tried to draw these connections between recent developments in genetic and biomedical technologies and earlier notions of original sin and godly seed to suggest that current controversies surrounding these developments reflect more than disputes over the status of embryos or procedures ensuring fair access. Rather, they mark the latest round of an old theological debate on destiny and grace. On the one hand, there is the belief that the future is largely what we make of it, so we must engineer our grace in order to construct our destiny. On the other hand, there is the conviction that our destiny is a gift to be inherited, so that grace is bestowed upon us by the giver of the gift. I do not have the time to develop adequately the theological frameworks of these contending accounts or tease out their contemporary moral implications in their various secular guises. Instead, I want to pose three considerations regarding the outcome of the present round of this old debate.

First, if predictions regarding the medical applications of current genetic research prove true, then a wondrous range of preventive and therapeutic treatments will be developed. In addition, we will be able to enhance selected physical and behavioral traits. A side effect of this progress, however, will be a diminished capacity to endure or tolerate suffering. As we come to believe that more and more diseases, disabilities, and physical and behavioral limitations can be prevented or treated, the category of "unnecessary" suffering will enlarge. Why should someone endure physical pain or emotional distress if it is within our power to prevent or alleviate the genetic factors contributing to the unwanted condition? Why keep company with the suffering, for they serve to remind us of our failure to correct our biological deficiencies? In practical terms, children born with severe congenital defects may become increasingly stigmatized because their plight is preventable and thereby unnecessary. In addition, their parents may be held accountable, if not morally or criminally culpable as some writers urge,[40] for failing to prevent such unnecessary suffering. In short, if we come to believe that we are the masters of our destiny and engineers of our grace, then suffering also becomes increasingly inexplicable.

I do not mean to imply that suffering is inherently good. We should not allow people to suffer or to inflict suffering on others to help build their character. Rather, there is a more tacit admission at stake that, given the gift and grace of community, there are times when enduring suffering for our own or the sake of others is explicable. It is for the sake of these larger values and relationships that we may be called to suffer for or with others. Within this larger context, the purpose of medicine or health care is to help us live a life we believe to be good, and right, and true. In pursuit of such a life, suffering is not something we seek, but neither is it to be avoided at all costs. When medicine or health care becomes fixated on avoiding suffering, it not only corrupts itself as a moral art and community,[41] but also has a corrupting influence on these larger human relationships, such as the one between parent and child.

This leads to the second consideration: applying techniques designed either to prevent undesirable traits or to enhance desirable ones may exacerbate a growing perception of children as artifacts of parental will. Rapid advances in reproductive technology have already altered our perception of how conception, gestation, and child-rearing can and should be pursued. Through the use of donated gametes and surrogates, these are now discrete tasks that can be managed or manipulated in order to achieve certain reproductive goals. John Robertson, for instance, likens parents to "commissioners" who enter into contracts with "collaborators" in order to obtain a child.[42] As a number of writers have noted, we are moving away from "procreation" to that of "reproduction" or "baby-making," replete with the manufacturing imagery these terms imply.[43]

The advent of various genetic and biomedical technologies will enable us to exert greater quality control over the outcomes of our reproductive projects. This capability will magnify a growing perception that parents have a right not to be burdened by the prolonged care of a seriously ill or disabled child. Indeed, they may come to believe they are entitled to desirable or even "perfect" offspring. Presumably, the development of genetic technologies will enable parents to exercise these so-called "rights." Moreover, it is becoming difficult to object to parents' asserting greater quality control over their reproductive projects. As some commentators eagerly remind us, there is little difference between providing our children with genetic advantages as opposed to social advantages. What is the difference, for example, between sending children to a prestigious private school or rigorous summer camp and enhancing the requisite genetic traits for intelligence and strength, when in either case the goal is to give them a good start in life by making them smarter and stronger?[44] Designer babies do have an appeal in a society that prides itself on planning, management, and outcomes assessment.

I realize that the rejoinder to this uncharitable portrayal is that future generations will live longer, healthier, and more productive lives. What better gift could we give our offspring? Although I grant that future generations may be healthier and smarter, I wonder if these qualities will be the benefits of a gift or the dividends of an investment. If it is the latter, then the cost will be to place a heavy burden on our offspring, for they must simultaneously be the object of our hope and the means of our self-fulfillment. In the fashioning of children in an image and likeness of what we want them to become, they also personify the goal for what we are willing the future to be. Yet this goal is a projection and magnification of ourselves, so it is through our children that our dreams and aspirations are fulfilled. Our children are not so much like us as they are artifacts created by and for us. This is one reason why Augustine argued so strenuously against the notion that the birth of any child—save one—can ever be a proper object of hope. The future is not simply an extended and improved version of the present. We live instead in the hope of a destiny that awaits us. Hence children are properly to be regarded as gifts entrusted to the care of their parents, for together they share the gift of a common destiny and grace rather than the task of constructing both.

This leads to the third consideration: our growing capability to manipulate genetic characteristics may reinforce a perception of both our individual lives and corporate life as projects to be undertaken instead of gifts to be safeguarded. "Project" is an apt image for our age. We really do believe that life is largely what we make of it. We create our lifestyles, we sculpt our bodies, and we make babies. We are contemplating the development of genetic technologies in a social context that may regard them more as tools in fashioning our individual lives and corporate life than as medical treatments. And if we choose to admit children into our life construction zones, an ability to manipulate selected genetic traits will help us obtain happy and fulfilling outcomes of our parental projects. If quality-control techniques become widely deployed, it will be difficult to dismiss a prevalent notion of parental proprietorship over offspring.

If life is a gift, however, we cannot own our children because our lives are not our own.[45] We are creatures who belong to God, and we belong with other creatures created in God's image and likeness. Our lives, as well as the lives of others, are not our property but are, in Karl Barth's words, "held in trust."[46] The metaphor of "life as gift" should not be pressed too far, given its inherent limitations, but it does offer countervailing imagery to the equally distorting metaphor of "life as project."

One of the more interesting things about a gift is its sheer givenness and gratuitousness.[47] A true gift cannot be begged, borrowed, or bought. Pestering someone to buy us something we want or purchasing an item for ourselves is not the same thing as having a gift bestowed upon or entrusted to us. Moreover, a gift carries with it an implied purpose for how it should be used, as well as something of the character and expectations of the giver. If we are given a painting, for instance, we may assume its purpose is to be hung on a wall rather than used to start a fire in the fireplace. In addition, the giver implies some expectation of how an otherwise identical gift is to be used. For example, it is one thing to receive a gun from a skilled hunter and another to receive it from a notorious bank robber.

If we tend to perceive children more as gifts entrusted to our care than as outcomes of our parental projects, we confront a different set of concerns in developing and applying various genetic technologies. If children are gifts, does this not imply a given and gratuitous structure to the parent-child relationship that is not subject to our willful manipulation? And if children are not properly a means of one's self-fulfillment, then does this not suggest an unconditional quality to the parent-child relationship that is not premised upon offspring possessing desirable traits? This is simply another way of asking, as we enter this so-called revolutionary age of genetics and biomedicine: Are we giving adequate attention to determining if there is a difference between using these new technologies to attend to legitimate health care needs and using them to make desirable children? And if so, can we sort out the practical implications of this difference?

My concern is that much of the moral inquiry accompanying the development of genetic technologies has fixated on questions of safety, efficacy, and access, so we may fail to take into account a matrix of subtle cultural factors and expectations in which these technologies will be introduced. For example, we are making available a widening range of quality-control techniques at the very time when we are growing increasingly uncertain about what it means to be a parent or child. This is witnessed by the awkward vocabulary we are concocting to describe our present circumstances. We now engage in "parenting," suggesting that parents are adults who do something to children rather than sharing with them a unique bond and unconditional relationship. With the genetic technologies we are contemplating, we will be able to do quite a lot to the bodies, minds, and souls of children. As we exert greater qualitative control over the outcomes of our reproductive projects, will we come to see offspring simply as programmed young people "childing" in response to their "parenting"? I do not think this is a fanciful or alarmist specter. We already talk casually about investing in our children's future by obtaining the best education, activities, and gadgets money can buy. Why not add genetic endowment to the list?

It needs to be emphasized that the issue at stake is not the prospect of new genetic and biomedical technologies per se. Individually, many of the technologies envisioned promote more effective and humane medical treatments. It is the cumulative effect of these developments, however, that should concern us. As each new piece of technology is placed to form a larger pattern, the moral landscape in which it is perceived and assessed will also be altered. If the new genetic technologies envisioned prove safe and effective, our perceptions and expectations regarding longevity, health, autonomy, familial relationships, community, and the ordering of larger human associations will change accordingly. Although these new technologies may not result in either our best dreams or our worst nightmares, we should at least acknowledge that they, like any technology, will transform the course of our moral deliberation on what constitutes a good life and how it should be pursued. As George Grant has written: "The coming to be of technology has required changes in what we think is good, what we think good is, how we conceive sanity and madness, justice and injustice, rationality and irrationality, beauty and ugliness."[48]

This is why our first steps into a new terrain of genetic technologies should be prompting a fundamental theological debate over destiny and grace. We need to be aware of the beliefs and hopes that are motivating us to discover the genetic basis of longer, healthier, and more productive lives. And this debate needs to be initiated in a manner ensuring that the subsequent reshaping of the moral landscape in which these technologies will be used is prompted more by deliberate conviction than by the sheer force of technological capability and efficiency. In short, it is one thing to ponder the prospect of new genetic technologies as instruments enabling our stewardship of the gift of life, and quite another to see them as tools for mastering our fate and engineering our grace.

Conclusion

In the second edition of Engelhardt's *The Foundations of Bioethics*, allusions to a baptism of genetic engineering have disappeared. In its place are more modest observations about a pluralist society in which various individuals and communities subscribe to multiple accounts of the good life.[49] The growing ability to manipulate selected genetic traits will assist us in attaining these good lives; we will turn to what he characterizes as "Dr. Feelgood" to aid us in pursuing "numerous visions of wholeness, health, and human fulfillment."[50] This more benign characterization, however, should nonetheless trouble us, for we are turning to Dr. Feelgood at the very time when we are growing increasingly uncertain about what the good is that we aspire to feel. If we do not or cannot engage in a prior and ongoing deliberation on a common good that binds us together—about what it means to be a good parent, a good child, a good citizen, a good society—then we may become, to invoke some old theological terms, more adept at engineering our sin than our grace.

Notes

1. H. Tristram Engelhardt, *The Foundations of Bioethics* (Oxford and New York: Oxford University Press, 1996). See also H. Tristram Engelhardt, "Persons and Humans: Refashioning Ourselves in a Better Image and Likeness," *Zygon*, 19:3 (1984):281–95.
2. Engelhardt, *The Foundations of Bioethics*, 377.
3. See Ibid., 380.
4. See Ibid., 375. Also see Alasdair MacIntyre, *After Virtue: A Study in Moral Theory* (Notre Dame: University of Notre Dame Press, 1984), 36–78, and Stephen Toulmin, *Cosmopolis: The Hidden Agenda of Modernity* (Chicago: University of Chicago Press, 1990).

5. See Engelhardt, *The Foundations of Bioethics*, 383–84.
6. See Augustine, *On Marriage and Concupiscence*, in Philip Schaff, ed., *A Select Library of the Nicene and Post-Nicene Fathers of the Christian Church*, Vol. 5 (Edinburgh: T&T Clark, n.d.), I/7, 266.
7. See Augustine, *On Marriage and Concupiscence*, I/37, 278.
8. See Ibid., I/21, 272–73.
9. See Ibid., I/38, 278–79.
10. See Ibid., I/20, 271–72. See also Romans 8:23–25.
11. See Eva Maria Lassen, "The Roman Family: Ideal and Metaphor," in *Constructing Early Christian Families: Family as Social Reality and Metaphor*, ed. Halvor Moxnes (London and New York: Routledge, 1997), 103–20; and Beryl Rawson, ed., *The Family in Ancient Rome: New Perspectives* (London: Routledge, 1992).
12. See Peter Brown, *The Body and Society: Men, Women, and Sexual Renunciation in Early Christianity* (New York: Columbia University Press, 1988), 5–32.
13. See Augustine, *On Marriage and Concupiscence* I/14, 269; see also Paul Ramsey, "Human Sexuality in the History of Redemption," *Journal of Religious Ethics*, 16:1 (1988):56–86.
14. For critical overviews of this literature, especially in regard to the role of the household and child-rearing, see Christopher Hill, *Society and Puritanism in Pre-Revolutionary England* (London: Secker and Warburg, 1995); Edmund S. Morgan, *The Puritan Family: Religion and Domestic Relations in Seventeenth-Century New England* (Westport, Conn.: Greenwood Press, 1966); John Morgan, *Godly Learning: Puritan Attitudes towards Reason, Learning, and Education* (Cambridge, U.K.: Cambridge University Press, 1986); and Levin Ludwig Shucking, *The Puritan Family: A Social Study from the Literary Sources* (London: Routledge and Kegan Paul, 1969).
15. See *A Christian Directory. Part II: Christian Economics (or Family Duties)* in Vol. 4 of *Practical Works* (London: James Duncan, 1830), 57. Ibid., 55.
16. Ibid., 118.
17. Ibid., 175–76.
18. Ibid., 189.
19. Ibid., 99.
20. Ibid., 110.
21. Ibid., 99.
22. Ibid., 30.
23. Ibid., 109.
24. Horace Bushnell, *Christian Nurture* (New Haven, Conn.: Yale University Press, 1947), 4 (emphasis in original).

25. Ibid., 18.
26. Ibid., 30.
27. Ibid., 91–94.
28. Ibid., 165.
29. Ibid., 171–72.
30. Ibid., 172–73.
31. Ibid., 173 (emphasis added).
32. Ibid., 183–84.
33. "Gene Test Cuts Out Cystic Fibrosis," BBC News Online Network (June 28, 1999).
34. See Penny Stern, "Gene Repair in Rats Raises Hope for Genetic Diseases," Reuters online News Service (September 10, 1999).
35. See BBC News Online Service, "Genetic Engineering Boosts Intelligence," (September 1, 1999); and Rick Weiss, "Mighty Smart Mice," *Washington Post* Online News Service (September 2, 1999).
36. See CNN Interactive, "Mice Study Shows Genes Not Always Destiny," Reuters (June 4, 1999).
37. See Lois Rogers and Steve Farrer, "'Immortal' Genes Found by Science," *Sunday Times* (4 July 1999), 12.
38. See Ben Bova, "Living to 200 Within Our Reach," *USA Today* Online Opinion (June 6, 1999).
39. See Augustine, *Concerning the City of God against the Pagans* (London: Penguin Books, 1984), XXII/15, 1055–56.
40. See, e.g., John Harris, *Clones, Genes, and Immortality: Ethics and the Genetic Revolution* (Oxford and New York: Oxford University Press, 1998).
41. See Stanley Hauerwas, *Suffering Presence: Theological Reflections on Medicine, the Mentally Handicapped, and the Church* (Notre Dame, Ind.: University of Notre Dame Press, 1986), 23–83.
42. See John Robertson, *Children of Choice: Freedom and the New Reproductive Technologies* (Princeton, N.J.: Princeton University Press, 1994), 119–45.
43. See, e.g., Gilbert C. Meilaender, *Body, Soul, and Bioethics* (Notre Dame, Ind., and London: University of Notre Dame Press, 1995), 61–88; Oliver O'Donovan, *Begotten or Made?* (Oxford: Clarendon Press, 1984), 31–48; and Paul Ramsey, *Fabricated Man: The Ethics of Genetic Control* (New Haven, Conn., and London: Yale University Press, 1970), 32–60. Cf. Ted Peters, *For the Love of Children: Genetic Technology and the Future of the Family* (Louisville, Ky.: Westminster John Knox Press, 1996), 33–57.
44. See, e.g., Harris, *Clones, Genes, and Immortality*, 171–95; and Robertson, *Children of Choice*, 165–67.

45. See John Calvin, *Institutes of the Christian Religion*, 2 vols. (Grand Rapids: Eerdmans, 1975), III/7.1, 7–8.

46. Karl Barth, *Church Dogmatics* (Edinburgh, U.K.: T&T Clark, 1961), III/4(55), 327.

47. See Stephen H. Webb, *The Gifting God: A Trinitarian Ethics of Excess* (New York and Oxford: Oxford University Press, 1996).

48. George Grant, *Technology and Justice* (Notre Dame, Ind.: University of Notre Dame Press, 1986), 32.

49. Engelhardt, *The Foundations of Bioethics*, 411–13.

50. Ibid., 416.

9

The Impact of Genetics and Genomics on Views of Aging

David Schlessinger, Ph.D.

IT IS MY PLEASURE to review with you some examples of the increasing impact of genome approaches and modern human genetics on thinking about aging. The Human Genome Project and its goals have become commonplace in our thinking, and we all expect that in a very few years we will have in front of us the complete sequence of the Human Genome from nucleotide 1 to 3 billion, with an accompanying catalog of all the human genes and auxiliary information about the genome sequence and gene contents of a number of other organisms. My message here is that initial applications of genome studies are modifying the way in which we think about aging—and in an unexpected direction. Traditionally, aging has been thought of as a separable phase of life after development has ceased. As discussed here, aging can rather be seen as a process with links to much earlier and, ironically, even embryological events.

OLD AND NEW IDEAS OF THE GENETICS OF YOUTH AND AGE

Before turning to some examples of how genome studies are leading us in this direction, it may be useful to comment on well-worn ideas about the genetics of aging. Deeply embedded in Western thought, such ideas date back to the period of high culture in Athens. As codified by Aristotle and his contemporaries in the fourth century B.C., the life of an individual consisted of a period of development, or "becoming," culminating in the fulfillment of the form of a species. As we see it in classic Greek sculpture, this brief moment, "acme," occurred at about the age of eighteen; and it was all downhill from there. Such views of human life have had remarkable staying power, persisting to the present day. One recent formulation[1] has it that "first we ripen and then we rot," or as Samuel Beckett put it for the theater: "We grow old. We lose our hair, our teeth, our ideals."

Some may agree with such a formulation, but many of us spend our careers actively fighting it. Aristotle himself might have been influenced in his thinking by the example of his student, Alexander the Great, who at the age of eighteen was in fact preparing to embark on his campaign to conquer the world. We all admire the verve and energy of the young, but productive life has scarcely begun at eighteen. At the opposite end of the scale, no one has to be reminded of the deficits and problems that accumulate as part of aging; but productive life continues, often with extraordinary creativity and all the more impressive in the face of the problems of aging.

One of many persuasive examples is provided by the lifelong brilliance of Giuseppe Verdi. Verdi had written nothing by the age of eighteen but then began to produce increasingly strong work. In his fortieth year he wrote *Il Trovatore*, *La Traviata*, and *Rigoletto*. Moving forward from his "maturity" to his "old age," we find that Verdi's last opera, *Falstaff*, premiered when he was eighty. Yet it is also his most complex treatment of life, achieving a view of the human comedy that eliminates all traces of earlier bombastic melodrama and ranges easily, with Shakespeare, from slapstick to wry irony. He had, in the way that is permitted only to the old, achieved balance based on accumulated experience and skill. We might also note that had he not written *Falstaff*, Verdi's last opera would have been *Otello*, often considered to reach the highest level of tragic intensity in all of his creative work. He was only seventy-four when he wrote that one. Again, experience provided the groundwork for increasing mastery in later life.

This theme of the interaction of early and later development in human life is paralleled in recent studies of the genetics of aging.

The Conundrum of the Genetics of Aging

For almost a century, geneticists have had a difficult time attacking the genetics of aging and aging-associated diseases. The reason is clear. By definition, if we wish to study the genetics of a trait such as longevity, it is easy to score: obviously, individuals will either have reached the age of ninety or not. But even if we identify and collect DNA for study from considerable numbers of such individuals, DNA from their parents is unavailable; and their children cannot be scored for achievement of comparable old age for another thirty years! Denied the DNA from successive generations that permits the study of inheritance in families, most of the approaches of traditional genetics are stymied.

The genetics community has responded to the challenge of the genetic analysis of aging in several ways. One productive approach for those groups who continue to concentrate on humans and other mammalian species is to study a condition or disease in a single generation by looking at many sibling pairs. In a very large family, for example, all the brothers and cousins who have a particular inherited disease will share the same modified bit of DNA, and this can be detected and localized by extensive comparisons of all of the DNA of the affected individuals. Such studies can also be done efficiently in certain populations in the world (e.g., Sardinia, French Canada). In those special populations, a small initial number of "founders" have given rise to a large number of descendants (1.5 million in Sardinia, 4 million in French Canada). Such populations are something like a single very large family, making all sibling pairs partially related. There is only a small number of founder groups, but they are increasingly valuable for the study of genetic diseases and aging-related traits. In such studies, sibling pairs can be investigated to reveal genes involved in osteoporosis, Alzheimer's disease, even longevity itself.

Another approach to circumvent the problem of doing genetics of diseases of the old is to look for early-onset forms of inherited conditions associated with aging. If a condition begins early enough in life, it becomes possible to score and study DNA from several generations of individuals. An example is the study of early-onset Alzheimer's disease, for which the brilliant work of a number of groups has produced our knowledge of presenilins, proteins that give clues about the pathophysiology of the disease. In a similar way, the important work of George Martin and his collaborators[2] has found the protein that is mutated to produce premature aging in Werner's syndrome.

Links of Aging and Embryology: Setting Limits for Later Life

Our group has been working on another instance of an early-onset condition, one that provides us with a demonstration of the use of genetic and genomic mapping approaches. The condition is premature ovarian failure (POF). In contrast to the recorded universal average age of menopause of about fifty for several centuries in many cultures, about 2 percent of all women either never go through menarche or undergo early menopause in their late twenties or early thirties. POF is a problem of increasing severity, as many women postpone childbearing to a later age and then may find that early menopause precludes reproduction. Such premature ovarian failure is a condition rather than a disease. As such, it can arise from a number of environmental factors such as radiation therapy or certain drugs. But in an appreciable number of affected women, the causation is genetic.

A single autosomal locus has been implicated in a number of instances. In several families, the condition can be traced by standard genetic mapping to a region of chromosome 3. The same region was found to be deleted in other women with POF; and in at least five instances, affected women had a translocation, with about a third of chromosome 3 exchanged with a segment of another chromosome. In such cases, it is assumed that the translocation interrupts a gene on chromosome 3, and that in its intact form, the broken gene ordinarily contributes to reproductive life span.

A collaborating group headed by Dr. Giuseppe Pilia[3] has started from such a translocation patient and used modern gene-finding techniques to identify the causative gene. In such positional cloning exercises, large pieces of cloned DNA are assembled across the region, and a hunt is done progressively to show which one crosses the breakpoint—that is, contains sequences from chromosome 3 that fall on both fragments of chromosome 3. This work is very much in progress, and the breakpoint has now been localized in a clone of DNA that is being sequenced to determine the identity and character of the gene involved in ovarian function.

There is another reason why I have focused on the instance of POF. It is based on the course of egg and follicle formation and utilization during a woman's lifetime. Early in fetal life, a very large number of eggs are formed. Most are then discarded, but a small number are protected in follicles and are then preserved until progressive ovulation begins at puberty. The critical point for this discussion is that at the moment of birth, a female baby has all the follicles she will ever have; and it is the number of these follicles that determines when menopause will occur. In women with POF, from the information we have, egg and follicle formation begins normally, but the rate of attrition in the fetus is too rapid. Consequently, the number of follicles that remain at birth are insufficient to sustain a full, normal reproductive life span. So we are discussing a phenomenon, menopause, that is clearly age-associated; but if we want to understand its basis and the mechanism of POF, we have to investigate processes of follicle formation and dynamics in the fetus.

A similar line of thought applies to a great many age-associated phenomena. It has led us, rather than thinking of aging as a compartmentalized phase of life occurring subsequent to development, to think that aging has genetic determinants as an integrated part of human development, with a profound dependence on processes that are initiated *in utero*. A number of links can be made in which one can speak about the primacy of embryology as it affects aging. Menopause and POF offer instances in which embryological events determine ground rules of aging. But as we discuss briefly below, embryological events also set the balance of immortal and mortal cells; set the course of tissue development in such a way that reserves are determined; and set up some mechanisms of "rejuvenation" that are retained in a quiescent form but can be reactivated in aging tissues and organs.

Links between Aging and Embryology: Immortal versus Mortal Cells

One of the most systematic studies of relevant processes in embryology is the work of my colleague Dr. Minoru Ko. Dr. Ko uses the mouse as a model system and has been carrying out systematic studies of the global repertoire of genes that are expressed at each stage in embryonic and fetal development.[4] The studies include an approach to a question that is fundamental to many theories of aging: the difference between totipotent, "immortal" cells which can grow and divide indefinitely, and differentiated "mortal" cells which can undergo only a limited number of cell divisions before exhausting their replicative potential. In Ko's work he turns to the stage in embryonic life 3.5 days after the initiation of fetal development. At that time, the bulk of the embryonic cells (the inner cell mass) is still composed of immortal embryonic stem cells; but the first mortal cells have assembled around the growing embryonic ball as a monolayer of trophectoderm, destined to become the placenta and auxiliary tissue.

This is another case where standard genome approaches can be of use. Ko has prepared libraries of the transcribed genes active in the immortal and first mortal cells of the embryo and is now comparing the genes selectively turned on or off during the immortal-to-mortal transition. This can be done either by one-by-one sequencing of individual transcripts or with the use of microarrays (chips) bearing sets of all the genes active in the 3.5-day embryo. In either case, candidate genes are detected which can then be tested for their capacity to regulate or carry out critical steps in the process. In fact, the line of work started by Dr. L. Hayflick[5] has in recent years produced the discovery of some genes that are selectively turned on or off during the transition of immortal to mortal cells, including telomerase and several marker genes. The genomic approach promises to provide us with an increasingly complete tally of all the genes differentially expressed during the transition.

Links between Aging and Embryology: Tissue Formation and Regeneration

As for the general development of tissues, comparable genomic approaches are defining the complement of genes responsible for the development and organization of all tissues and organs. This is of course the substance of modern embryology or, as we can now call it, developmental genomics, and occupies the full attention of thousands of laboratories interested in one or another physiological system. One should immediately disclaim any preposterous notion that the decline of function or complete loss of cells is in any way a "reversal" of the way in which those cells are formed during fetal life; but knowledge of the way, for example, in which glomeruli are formed can help to prolong their function—or even to regenerate them—in patients otherwise headed for end-stage renal failure.

As a typical example in which our laboratory is involved, one can consider the case of ectodermal dysplasias. There are at present about 175 clinical entities distinguished by dysplasias affecting some combination of skin, teeth, hair, sweat glands, and nails. In one instance, the X-linked form of anhidrotic ectodermal dysplasia, the affected individuals have an early block in the formation of skin appendages, and have sparse hair and sweat glands as well as rudimentary teeth. Thus this is largely a cosmetic condition, but a considerable number of individuals are very interested in the formation or replacement of hair.

The gene mutated to cause the hereditary condition of anhidrotic ectodermal dysplasia (EDA) was isolated two years ago by my collaborators Dr. Juha Kere and Anand Srivastava.[6] Once again, this was an example of the now-standard gene-finding by positional cloning. The work started with two individuals with translocations marking the position on the X chromosome where a putative gene involved in skin development was interrupted. Starting from clones covering the region, generated by our group during the construction of a map of the X chromosome, the gene was isolated. Work is now ongoing to analyze the way in which the protein encoded by the EDA gene participates in the differentiation of skin appendages based on epithelial-mesenchymal signaling. Hints have been obtained from the fact that the protein is bound to the cell membrane of cells in which it is produced but juts out into the environment with a long tethered tail that includes a collagen segment capable of interacting with other cell surfaces. In this case, model systems in ongoing work include cells that round up when the EDA gene is expressed within them, and Tabby mice, a well-known mouse strain which also has fine hair, few sweat glands, and poor teeth and proves to be lacking in the mouse gene equivalent to human EDA. These models make it possible to begin the search for proteins that interact with the EDA protein, a first step in functional studies.

Links between Aging and Embryology: Recapitulation in Aging Tissues

I will not provide detailed examples but should also mention that in a number of aging tissues, including heart and muscle, "fetal" proteins and groups of interacting proteins are seen that are not found in those tissues during earlier adult life. Because such studies are just getting under way, the significance of such findings is unclear; but one teleological notion is that the response of tissues to stress during aging may include two phases: a quick first-line defense that is mounted in immediate "panic," and a slower partial recapitulation of developmental processes that is aimed at recovering flexibility and strength.

Conclusion: An Adage at the Molecular Level

We have thus touched on four aspects of the "primacy of embryology." In addition to setting ground rules (for reproductive life span, for example), embryological events determine the transitions of immortal to mortal cells (and the levels of remaining immortal stem cells); development of specific cells, tissues, and organs; and possible backup processes that may be called into play to sustain or "rejuvenate" aging tissues. Stepping back from the immediate limits of our topic, we can easily recognize that the formulation we have been discussing is really a traditional one. We can see it exemplified in the well-known painting by Ghirlandaio (1470) in the Louvre of a grandfather and his grandchild.[7] The complexity of the interactions of the child and the grandfather is transparent, and we are reminded of the venerable adage that the child is father to the man (and to the grandfather). Our message in this article is that using modern genomics and human genetics, the process by which the child determines the father can now be described in analytical molecular terms. We will not be able to restore the innocence of the child to his grandfather; nor will the child be able to avoid the accumulation of age-related deficits as he gains the wisdom of his grandfather. But we can hope to use the understanding of process to optimize function throughout the human life span.

Notes

1. G. M. Martin and I. S. Mian, "Aging: New Mice for Old Questions," *Nature* 390 (1997):18–19.
2. M. D. Gray, J. C. Shen, A. S. Kamath-Loeb, A. Blank, B. L. Sopher, G. M. Martin, J. Oshima, and L. A. Loeb, "The Werner Syndrome Protein Is a DNA Helicase," *Nature* 17 (1997):100–103.

3. L. Crisponi, M. Deiana, A. Loi, F. Chiappe, M. Uda, P. Amati, L. Bisceglia, L. Zelante, R. Nagaraja, S. Porcu, M. S. Ristaldi, R. Marzella, M. Rocchi, M. Nicolino, A. Lienhardt-Roussie, A. Nivelon, A. Verloes, D. Schlessinger, P. Gasparini, D. Bonneau, A. Cao, and G. Pilia, "The Putative Forkhead Transcription Factor FOXL2 Is Mutated in Blepharophimosis/Ptosis/Epicanthus Inversus Syndrome," *Nature Genetics* 27 (2001):159–66.

4. M. S. Ko, "Embryogenomics: Developmental Biology Meets Genomics," *Trends in Biotechnology* 19 (2001):511–18.

5. L. Hayflick, "The Illusion of Cell Immortality," *British Journal of Cancer* 83:7 (2000):841–46; M. H. Linskens, C. B. Harley, M. D. West, J. Campisi, and L. Hayflick, "Replicative Senescence and Cell Death," *Science* 267 (1995):17.

6. J. Kere, A. K. Srivastava, O. Montonen, J. Zonana, N. Thomas, B. Ferguson, F. Munoz, D. Morgan, A. Clarke, P. Baybayan, E. Y. Chen, S. Ezer, U. Saarialho-Kere, A. de la Chapelle, and D. Schlessinger, "X-Linked Anhidrotic (Hypohidrotic) Ectodermal Dysplasia Is Caused by Mutation in a Novel Transmembrane Protein," *Nature Genetics* 13 (1996):409–16.

7. http://www.ibiblio.org/wm/paint/auth/ghirlandaio/

10

Genetic Labels and Long-Term-Care Policies: A Winning or Losing Proposition for Patients?

Ruth Purtilo, Ph.D.

FOR MOST of the twentieth century, people battling the effects of persistent debilitating conditions (e.g., cystic fibrosis, Huntington's disease, Lou Gehrig's disease, sickle cell anemia, muscular dystrophy, and diabetes) have been classified by health policy-makers and practitioners as having a "chronic disease" or "chronic condition." A chronic disease is characterized by a pathology that may manifest itself in many different systems but results in persistent or permanent functional debilitation and, in most instances, death. The primary response from U.S. policy-makers has been to develop "long-term care" services, the majority of which are health care interventions but also include a network of social supports such as home care, job retraining, and transportation. Today a subtle shift has begun to take place in the way many chronic conditions are viewed in health care practice and policy: as more and more information about the genetic component of chronic diseases becomes available, numerous conditions are now discussed solely as *genetic diseases.*

In the medical research arena, the search for DNA-based prognostic tests and treatment must be heralded as a harbinger of hope for people whose pathology is found to be located in their genetic make-up. Approximately 3 percent of all children are born with a severe disorder presumed to be genetic in origin. Juengst predicts: "Among the most immediate . . . consequences of human genome research will be the localization and identification of medically interesting genes . . . and their mutations and the subsequent development of DNA-based tests to detect specific alleles in individuals."[1] A recent article in the

Philadelphia Inquirer declares: "Scientists take giant steps towards new tests." What tests? Trials of injection of Vector I, the gene that makes cystic fibrosis transmembrane regulator. It reads like a detective story and probably correctly conveys the anticipation of the 30,000 or more affected people in the United States alone who now catch a glimpse of the possibility that this disease's power can be brought down through a genetic understanding of their condition.[2] However, generally speaking, there is a long shadow between an effective DNA-based test for the presence of a mutation and subsequent therapeutic advances made on the basis of this new knowledge. Given what we know (and do not know), a rush to redefinition of chronic conditions to make them "genetic" for purposes of health policy initiatives should raise serious ethical questions.

What are the trade-offs of these changes for already affected individuals (as well as for the increasing number of people who suffer from chronic disabling conditions that have no known genetic basis)? What additional burdens might be unintended side effects of such a change in emphasis? Will health professionals experience these changes as facilitating them in—or hindering them from—practicing in a manner that honors the patient as a person? Finally, will the change in emphasis encourage a deeper understanding of, human identification with, and more humane policies towards such people?

The goal of this paper is to encourage further discussion on these broader ethical issues through an analysis of the use of labels in this situation, recommending resistance to changes that would demean the affected person and weaken the ethical foundation of the patient's relationships within the health care setting.

The Ethical Use of Labels

Potter describes a label as "denoting the assignment of an individual to a category according to properties or qualities a person possesses in common with other people in the category." He proposes that three criteria create a moral justification of classification by label:

1. The label effectively identifies the need for a specific type of assistance in the labeled person/group;

2. The label is temporary and "erasable." It does not continue to be a mechanism for classification once the need for placing a person or group into the category has been removed;

3. The label confirms the treatment of people as people.[3]

How do the labels of "chronic disease" and "genetic disease" stand up against the three criteria?

1. A Label Identifies the Need for a Specific Type of Assistance in the Labeled Person/Group

1.A. Chronic Disease

The effectiveness of the present label, "chronic disease/condition," has been debated over the last two decades on the basis of whether this label accurately conveys the needs of the affected person. In the balance of the debate, the shortcomings of the label to meet this criterion outweigh the justifications. This conclusion rests on the type of response the label has generated in regard to summoning appropriate societal responses, especially in the services termed long-term care. The shortcomings of long-term care are hauntingly well illustrated. For example, in the September 23, 1999, issue of the *New England Journal of Medicine*, a study reported the shortcomings of policies geared to services for noncancer chronic disease patients. Investigators report that among the 988 patients in their study, 41 percent of the noncancer patients reported having substantial unmet need for help with their daily needs over time—23 percent had problems in being left unattended and 20 percent relied on out-of-pocket paid caregivers. The differences are most pronounced in patients who follow a roller-coaster pattern of exacerbation and remission of symptoms over several years.[4]

One problem is that U.S. health policy-makers have never been able to come to an agreed-upon list of the health care services required over a long period of time for people with chronic conditions (i.e., "long-term care"). Therefore, in the United States, long-term care itself has become a political battlefield. Some identify it with the *sites* where people with persistent symptoms go, some emphasize how long-term care *interlocks with related services* such as rehabilitation and hospice or geriatric care; others rely on the *months or years of required interventions* to ameliorate or remove symptoms; and still others orient long-term care around the *range of services* needed. Long-term-care policies are often directed to people who have slowly developing, life-threatening diseases leading to eventual death (e.g., amyotrophic lateral sclerosis, cystic fibrosis, muscular dystrophy, AIDS) and who almost invariably require acute, chronic, rehabilitative home care and often hospice care as well. But another large body of policy beams attention on people suffering from nonacute and usually nonfatal conditions (stroke, asthma, myocardial insult, traumatic head injury). Mental retardation, chronic mental illness, and learning disabilities may or may not be included in long-term-care policy. In an era when the needs of the very old are too often played off against those of the very young, it is not surprising to find advocates who claim a stake on resources for one over the other. Some policies focus on people with special symptoms (e.g., chronic pain of arthritis) or sensory loss (e.g., Hansen's disease). This fragmentation and politicization of long-term-care services does not increase confidence that the chronic disease label upon which the services are based will ever completely generate beneficial policies and practices in response to people with disturbing debilitating symptoms.

A second problem is that people labeled with chronic disease are often a source of frustration to cure-oriented health professionals. In a now-classic study by Groves, he describes four characteristics of patients whose presence creates frustration and even rage in health professionals. Two of the four characterize people with persistent symptoms that "won't" respond to the physician's best efforts at cure or amelioration.[5] This, too, does not bode well for people classified as having a chronic disease.

1.B. Genetic Disease

Does a label of genetic disease fare better? The genetic disease label does not have as long a history of application as chronic disease. Consequently it is more difficult to assess whether it could meet the challenge of producing effective policies and practices for the wide range of diseases that have debilitating and persistent symptoms traditionally labeled as "chronic."

A genetic disease label does have one advantage in that it calls attention to a group's need for special services: it is "interesting" to the physician! The identification of genetic factors in a patient's history and physical examination is increasingly being judged significant *clinical* information. In a *Journal of the American Medical Association* editorial entitled "Family History and Genetic Risk Factors, Forward to the Future," the author notes: "in the evaluation of new patients . . . until recently much of the focus of medical genetics has been on rare disorders. However, during the past decade, increasing attention has been directed to genes as part of the causal nexus of common disorders."[6] Therefore if the only ethical criterion of labeling was to call special attention to the person's symptoms, it would be beneficial to change the label from "chronic" to "genetic." Unfortunately, there is more to it than this.

2. The Label Is Temporary and Erasable

2.A. Chronic Disease

The label of chronic disease will adhere to the person if visible symptoms (weakness, lack of coordination, memory loss, disfigurement) are present. Conditions that do not include physical signs (e.g., pain, sensory loss) may seem less convincing. Therefore the label is temporary and erasable.

But is it *appropriately* temporary? Not necessarily. As the above observation suggests, the label has often relied too heavily on what the naked eye can see to assure the timeliness of its application and removals. The result is that this label is an unreliable marker of real need and may not follow the course of the person's disease. In fact, nonvisible symptoms are likely to lead to allegations that the person is a malinger, a hypochondriac, or is in other ways seeking "secondary gain."[7] The label may disappear at a time that exacerbates a person's problems rather than ameliorating his or her distressing symptoms.

2.B. Genetic Disease

There is sound cause for pessimism about a labeled group's ability to shuck off a genetic disease label at such time as it no longer is helpful to the person. Whenever the notion of "genetic" is applied to a disease, the label stands at high risk of becoming *indelible,* lying deep within the person's makeup. At least that is the conviction of the public. As chair of the recent Governors' Commission on Human Genetic Technologies in Nebraska, I heard hours of testimony from Nebraskans regarding the already experienced (or feared future) misuse of the label of genetic disease/condition in their—or their children's—classroom, in the workplace, and when they sought medical attention. The mention of their children is also a reminder that since genes are a family affair, "to know that an individual is genetically disposed toward a specific disease is an indicator that members of this individual's family may also be saddled with the genetic disease label."[8] Genetic labels have the danger of spreading across a kinship and following throughout future generations unchecked, through either facts or overly deterministic interpretations of genetic health risks.[9]

In short, neither the chronic disease nor genetic disease label appears to meet perfectly the criterion that it will be applied to a person only as long as his/her benefits will be assured or enhanced.

3. *The Label Should Confirm the Treatment of People as People*

The idea that a person must be treated as a person points to a deeper frame of reference drawn from humanistic and religious teachings and expressed in ethical guidelines regarding the health professional and patient relationship. Inherent in those guidelines are many cautions against attitudes or actions that reduce the patient or health professional to anything other than a person with an inherent dignity that must be honored. How do the two labels measure up in that regard?

3.A. Chronic Disease

Does this label serve to honor human dignity? Its basic intent is to do so. But concerns from the independent-living and patient-empowerment communities point out that the label generates many *indignities*, among them the social expectation that an individual with chronic disease be grateful for his/her services and, more fundamentally, that he/she remain "dependent" in order to be eligible for care, thus bearing the stigma of dependency in Western societies that themselves value independence and self-determination. The Americans with Disabilities Act, designed to prevent discrimination against people with symptoms that could prevent full social participation, is a sign that legal measures are required to maintain even basic conditions respecting the dignity for such people in the larger society. A fragmented set of interventions in an imperfectly realized long-term-care network of services adds to the burden of the label for the person and loved ones.

3.B. Genetic Disease

Does a label of genetic disease correct for these indignities? Yes and no. Critics of long-term care point out the remarkable progress being made in the science of genomics (marking the evolution from an understanding of single genes to an understanding of the actions of multiple genes and their control of biological systems) and predict that "identifying human genetic variations will eventually allow clinicians to sub-classify diseases and adapt therapies to the individual patients."[10] Labels that promoted individualized approaches combined with a continual move to prevention and amelioration of disturbing symptoms would be impressive expressions of respect from the society at large.

What, then, stands to be lost to the cause of human dignity by categorizing people with persistent debilitating symptoms as having a genetic disease? Insight into problems that might emerge can be found by examining ways such a shift would affect the interdisciplinary health care teams (IHCTs) presently providing health and social services to this group. Why? Short of the ultimate goals of genomics to prevent or ameliorate all debilitating symptoms throughout everyone's lifetime, the best solution the health care system has found to date to minister to the complex symptoms of people with chronic diseases is through the interventions of IHCTs.

LABELS, PEOPLE, AND INTERDISCIPLINARY HEALTH CARE TEAMS

IHCTs emerged in the 1950s and 1960s.[11] Among the several forces leading to that mechanism for the delivery of health care, one was the recognition of the necessity for ways to address the needs of people who required comprehensive and coordinated health care interventions over a long period of time. Today the idea of IHCTs is taken almost completely for granted as the customary way to offer a range of services over time in a more comprehensive and complete fashion than could be achieved "solo." One might conclude, then, that the IHCT is a "perfect fit" as a health care system mechanism for offering services consistent with the moral ideal of respect for people. In general this is so, though not all teams will be equally successful in achieving that ideal because today "the team" is not a homogenous phenomenon in terms of the basic functions different teams might serve. As a new millennium approaches, some IHCTs serve the *moral* ends of health care directly and others, *instrumental* ends. Each is designed to provide high-quality professional services, though those designed to serve a moral end are more relevant for assuring that such services occur.[12]

What are the moral and instrumental ends that characterize different teams? Some IHCTs give their professional attention directly to the good of the whole patient. Because the patient's well-being *qua* person, and not just as a bundle of symptoms, has traditionally been viewed as the appropriate moral goal of health care, teams engaged in such activity can accurately be characterized as teams that serve moral ends. These include teams that frame a total treatment plan sensitive to patient preferences and lifestyle and the larger social, psychological, and spiritual needs of people (e.g., geriatric care teams, hospice teams, rehabilitation teams, discharge teams). In other words, coordinated and complete care over time is the job of such a team.

Other IHCTs are characterized by functions that directly serve instrumental ends in the total endeavor of health care, so called because the good of the whole patient is not the *direct* goal of the intervention(s). Instead, the goal is to use technical skills as instruments to that (ultimate) end. This, of course, in no way suggests that they are immoral or amoral, rather they are to the ultimate moral ideal of health care what a small but highly sophisticated supporting stage light is to the successful illumination of a stage.

In short, teams providing instrumental functions may play an essential role in the overall treatment plan for a person requiring ongoing services. However, a team focused on the moral ends of health care will carry the burden of more fully addressing the patient's life situation as that situation is shaped by goals, fears, hopes, and joys. It follows that a chronic disease label is likely to incur a claim on services from an IHCT with a moral function.

Will a genetic disease label be consistent with requiring the services of a moral or instrumental team? The technology of genetic approaches for use in the diagnosis and treatment of human disease promises to modify radically the patient's relationship with physicians and others in the direction of supporting the activities of an instrumental IHCT. In most cases the patient will never need to encounter the physician (or other health professional) face to face. It will be sufficient to send in a hair sample, fingernail, or other body part, have it analyzed, and be presented electronically with a treatment option. The most important team will be made up of those who conduct the genetic analysis. This is an efficient and technologically effective means of carrying out such a service and is completely consistent with a well-working instrumental IHCT.

Is this a problem? Well, there is the nagging reminder that so far in the history of health care, the well-being of people has seldom been reducible to technological "fixes." Until all human suffering from chronic diseases and conditions is removed from the human fabric, nothing should create a barrier to the type of treatment approach that takes the "whole person" into account. At present, the most effective means we have developed for the provision of such services is the moral IHCT, with its emphasis on prevention, healing, rehabilitation, and palliation. All of these are known means of honoring the patient's dignity as the quest for new genetic treatments is pursued.

Labels and Justice

An overarching goal of seeking the ethically justified approach to labeling is to assure that people with debilitating symptoms will receive what is due them.

Both chronic disease and genetic disease labels carry some stigmatizing characteristics and as such have the potential for further marginalizing the people who bear the label. In some cases the substitution of a genetic label for a chronic disease label may intentionally or unintentionally exclude people who have been eligible for services on the basis of the original (chronic disease) label. Since not all chronic diseases are genetically linked, society must continue working toward preventing this type of injustice. The excitement about genetic prevention, treatment, and cures must not be allowed unnecessarily to deflect attention away from long-term care for everyone who can benefit, whether or not there is a genetic component to their condition.

Finally, vigilance must be exercised so that the identification of genetic factors does not serve as a springboard for further discrimination and suffering among already marginalized groups. We must be reminded of the excitement that followed the identification of the sickle cell trait and the large-scale screening that ensued. Sickle cell trait appears primarily in African Americans. As a full-blown disease it is a painful condition manifesting itself in infancy and continuing throughout a shortened life span. Now the bad news. The screening to identify affected individuals and therefore (it was implied) to offer them assistance was not accompanied on a large scale by policies mandating treatment programs that would attend to their basic medical symptoms.

Instead, in many instances, African Americans identified through the screening process have suffered discrimination in insurance, employment, and, yes, in the health care environment because of records carrying stigmatizing information regarding their genetic status as having sickle cell disease or even as carriers of the sickle cell trait. (Sometimes carriers are reported as having the disease!) And once the disease is full-blown, it has often been difficult for patients to access traditional long-term-care services designed for people who have been diagnosed with more "traditionally" conceived "chronic diseases." The sickle cell screening experience highlights that shifts in category of disease may be made selectively. For instance, critics are quick to point out that people labeled in a racial/ethnic group who have suffered previous social wrongs can be simply further stigmatized by a genetic label. This must not be allowed to happen again.

A Concluding Comment

The promise of genetic and genomic information has led to understandable enthusiasm among sufferers from debilitating diseases, health professionals, and, in some instances, society itself. At the same time, an analysis of the use of two less-than-perfect labels, chronic disease and genetic disease, highlights that a simple shift of category does not necessarily uphold the high ethical standards that have guided the health professions throughout the ages. Prudence and its sister, patience, may serve individuals and the common good better than a rush to the fountain of new discovery.

Notes

1. Eric Juengst, "The Ethics of Prediction: Genetic Risk and the Physician-Patient Relationship," in *Health Care Ethics: Critical Issues for the 21st Century*, ed. John F. Monagle and David Thomasma (Gaithersburg, Md.: Aspen Publishers, 1998), 212–23.

2. Donald Drake, "Editorial," *Philadelphia Inquirer* (Monday, July 24, 1995), A1–A2.

3. Ralph Potter, "Labeling the Mentally Retarded. The Just Allocation of Therapy," in *Ethics in Medicine*, ed. S. J. Reiser, et al. (Cambridge, Mass.: MIT Press, 1977), 623–30.

4. Ezekial Emanuel, "Assistance from the Family Members, Friends, Paid Care Givers, and Volunteers in the Care of Terminally Ill Patients," *New England Journal of Medicine* 341:13 (1999):956–63.

5. J. Groves, "Taking Care of the Hateful Patient," *New England Journal of Medicine* 298:4 (1978):883–87.

6. Roger Rosenberg, "Family History and Genetic Risk Factors, Forward to the Future," *Journal of the American Medical Association* 278:15 (1997):1284–85.

7. Ruth Purtilo and A. Haddad, "Challenges to Patients," in *Health Professional and Patient Interactions* (Philadelphia, Penn.: W.B. Saunders Company, 1996), 117–35.

8. Goldworth Amnon, "Informed Consent in the Genetic Age," *Cambridge Quarterly of Healthcare Ethics* 8 (1999):393–400.

9. Juengst, "The Ethics of Prediction," 222.

10. Francis Collins, "The Shattuck Lecture—Medical and Societal Consequences of the Human Genome Project," *New England Journal of Medicine* 341:1 (1999):30–37.

11. Theodore Brown, "An Historical View of Health Care Teams," in *Responsibility in Health Care*, ed. George Agich (Dordrecht, Netherlands: D. Reidel Publishers, 1982), 3–21.

12. Ruth Purtilo, "Interdisciplinary Health Care Teams and Health Care Reform," *Journal of Law, Medicine and Ethics* 22:2 (1994):121–26.

11

The Challenge of Advances in Genetic Medicine for the Biomedical and Organizational Ethics of Health Care Delivery

Dennis Brodeur, Ph.D.

INTRODUCTION

This paper addresses the ethical issues that health systems will face in response to the promise of new genetic developments over the next decade. Some of the issues, while specific to health care systems per se, also reflect a wider ethical concern about the Human Genome Project, genetic technology in general, and genetic interventions in particular. The "promise" of further genetic information, new insights into the genetic relationship of many diseases, and new genetic interventions may aggravate an already ethically troubled health care system.

On the horizon, many scientists, medical experts, and others engaged in the Human Genome Project promise an entire mapping of the human genome with information then more readily available about the causes of disease from a genetic perspective, or the relationship between genetics, environment, and lifestyle in preventing and curing human illness. The Human Genome Project alone will not provide all of the necessary information to create major breakthroughs in medical care, but it is a first step in a series of steps that will help to understand more fully the nature of disease and illness and the future shape of health care delivery.

HOPES AND PROMISES

Despite difficult ethical concerns, some people hope that the genetic future may bring successful therapeutic germ-line cell interventions to prevent some genetic disease. This remains to be seen. Some people think somatic-cell interventions are more likely, and in fact experimental protocols already exist. Individuals who are the subjects of a particular genetic disorder might receive therapeutic interventions so as not to experience certain diseases in their lifetime or to correct existing genetic-based illness.

Multifactorial disease poses a more difficult issue. Understanding the relationship between genetic variation in an individual or group of individuals and the response to environmental and lifestyle factors is thought to be important to understanding the causes of some diseases. While there is still great uncertainty about the relationships between the genetic makeup of the human person and many diseases, research continues in the hope that the results of this research will help to prevent or cure disease and extend life span.

There are numerous ethical issues that can be raised by the very activity of genetic research. Some of these issues have been described and partially resolved by the Ethical, Legal, and Social Implications Program at the National Center for Human Genome Research. Additional ethical studies have been undertaken and are under way, but these are not the subjects of this ethical analysis. However, some of the findings of the above groups may highlight or further the understanding of ethical concerns in health systems delivery.

Ethical Issues and the Health Care Delivery System Today

Health care delivery systems are variously described. For the purposes of this paper, health care delivery systems refer to those combinations of institutions, providers, and other places where health services are delivered with the purpose of preventing, ameliorating, or curing the effects of disease and illness. The connection between providers, such as physicians and institutions, may be through credentialing, ownership, employment, or any other common affiliated structures. There are biomedical, professional, and organizational ethical concerns in any kind of health care setting. These can and most likely will be affected by future advances in genetic information.

The Human Genome Project is variously evaluated from an ethical, scientific, technological, or genetic breakthrough point of view.

Often an evaluation of this scientific project will combine more than one of these perspectives. The genetic insights or information will be embedded in the entire health care delivery system for the benefit of present and future persons. To the extent that today's health care delivery system is fraught with serious ethical problems, genetic advances can resolve or aggravate already existing problems. The suspicion, however, is that advances in genetics will not cause major reform in the health care delivery system sufficient to overcome the kinds of ethical concerns that already exist.

The ethical problems of today's health care system are biomedical, organizational, and professional. A short list might include at least the following.

The health care system is filled with unconnected personal encounters. This means that individuals meet with particular physicians or other health professionals who are not necessarily interacting one with another for the well-being of the patient in the system. This problem is seen over and over again, and is frequently blamed on managed care, but it existed long before managed care was on the horizon. How to create a seamless network for care so that people move from one provider of care to another to ensure the right care at the right time remains a critical medical and ethical issue.

There are more mistakes in health care than anyone is willing to admit. These include unnecessary cesarean sections, constant problems with medication errors, a lack of appropriate referral to other health care resources, a general denial of alternative or complementary medicine, unnecessary falls, unnecessary restraints, and a variety of waits and delays, whether in the physician's office, the ambulatory-care center, the emergency room, the acute-care hospital, or the long-term-care facility. This results in poor-quality care and adverse outcomes, and such incidents are hardly "beneficent" interactions.

There remains less than full use of present scientific knowledge, particularly in acute-care medicine. Because many therapeutic interventions have not been "fully" tested or allow for reasonable questions, people deny, refute, refuse to use presently available data to create clinical breakthroughs to improve patient care. This is seen over and over again in the Institute for Healthcare Improvement's clinical collaboratives.

There is a strong love affair with technology and the technical. Health care organizations share society's belief that technology and technological interventions are the promise for tomorrow's better health care, yet evidence for that does not always exist. This raises serious issues about ethical principles of beneficence, malfeasance in some instances, justice, and the use of resources.

There is a constant moral assertion by the professions and the institutions about "care" followed by unfulfilled promises in terms of both delivering care and following through on care. In part this may be due to managed care, as often claimed, but it is also due to some physician unresponsiveness to patient needs and to a breach of care in the continuum of care as it presently exists.

There is a naïve belief that the market will resolve all of our health care problems, even thought the market does not seem to be doing that and has not for the last ten years. As a result, health care systems' responses to patient, family, and community needs are skewed.

There are serious ethical questions about individualized choices in the social reality of health care and the social constraints of medicine in general. We are unwilling to recognize any limits in medical care and believe that physician advocacy for ultimate patient care in all circumstances is an absolute moral imperative even though it carries significant social consequences.

There is little or no connection in the arena of health care at the physician, hospital, or long-term-care level with the general commitments of public health. Even though we are deeply aware that public health has probably contributed greatly to longevity and to the general health advance of the health of the population, there is little connection. The ethical problems that are created by this lack of connection remain unaddressed. We cure people and send them back to the same sources of disease. This is waste of resources that raises numerous questions of justice.

There are multiple ethical questions in the care of people with chronic disease. Chronic diseases are the largest reason for care that any health care institution faces and yet there is no continuum of care in addressing these issues. Even with genetic interventions in the future, chronic disease is unlikely to go away, and if its treatment patterns are not improved, the ethical and medical problematic issues for patients will remain.

There remains confusion in the present health care system about enhancement issues versus care and cure issues. As a result, some institutions are very willing to forego care to a population who could be cared for or cured, instead engaging in lucrative enhancement fields such as plastic surgery. This diminishment of services for care and cure to cover those issues of personal enhancement is already problematic and likely to become more problematic in the future.

Our health care institutions seek to ameliorate, cure, and care for patients with disease, but the notion of preventive medicine is still a sideline issue. In part, this may be due to the market propensities not to cover such services and therefore for physicians and health care institutions not to provide such services. This lack of preventive care is likely to be aggravated in the future.

There are numerous issues of justice and access to health care services that continue to develop within our present health care system. The number of people without adequate access to health care services continues to grow at unacceptable rates. These issues are not likely to be solved by new advances in genetic technology.

But There Are Some Good Things

Although the ethical issues mentioned above are major concerns within the present health care system, there are also many good things within the system. For example, there have been many advances in technology that have seriously reduced the effects of the top killers in health care—AIDS, cardiac disease, and cancer.

There are better drugs available then ever before. Some of these may be defined as lifestyle enhancement drugs, as is the case with Viagra. But other drugs do a better job at curing or ameliorating disease without serious side effects. There is some continuing focus on quality, although the earlier insistence on continuous quality improvement in clinical services by many physicians and hospitals has been forgone as a fad. Nonetheless, a focus on quality outcomes remains key. The question will be whether people are willing to focus on the processes necessary to achieve quality outcomes for patients. There is more patient information available through the physician's office, the hospital's information department, and the Internet. This information furthers the goals of the principles of informed consent and patient choice. It can help to reduce some physician paternalism and can help to arrive at more informed decisions for treatment and care.

The health care system has also contributed—even if not predominantly—to an increased life expectancy for the population as a whole. Unfortunately, some studies would indicate that this increase is differentiated on the basis of race and income. These are the issues of justice and access that remain to be addressed. Finally, there has been the eradication of some diseases, such as smallpox and for the most part polio, although the eradication of disease has not been full-blown, and many people may have a false expectation that the continued developments in genetic technology will solve what today's medicine cannot.

There are other positive comments one could make that would reflect positively on the moral and ethical commitments of medicine to patients, families, and the communities served. Nonetheless, the ethical problems that do exist create massive difficulties in the system. The additional ethical question confronting health care is whether the advances of genetic medicine will in fact change, solve, or further aggravate these ethical concerns.

The Ethical Issues for Health Systems in a New Genetic Era

The first ethical issue in the genetic era is not unlike one of the issues that already confront health systems. This is the definition of disease, illness, health, and health care in general. If one were to adopt the pervasive World Health Organization definition—"a state of complete physical, mental, and social well being"—one would have to question what the genetic implications of such a definition would be. Does "health" occur only in a context in which one has the maximum genetic potentiality? How does one deal with the relationship between genetic factors and environmental factors in multifactorial disease? Should one be allowed to enhance one's genetic code to minimize the possibilities not only of real health problems but also of possible health problems sometime in the future? In addition, should one be allowed to manipulate one's genetic code in order to minimize or to reduce the risk of genetic and environmental or lifestyle diseases?

Alternatively, one could adopt, as many have, the definition of health used by Norman Daniels to reflect opportunities for normal species function. However, in the advent of genetic information and the possibility of genetic manipulation, does normal species function change? Even if one were not inclined to be optimal on the definition of health and inclined to be minimal on the definition of disease and illness, ethical difficulties of definition would continue. Could society decide, based upon generalized human genetic information, that there is an optimum genetic code that anyone would wish to have? Would interventions to achieve this optimum genetic code then be identified as a therapeutic intervention regardless of whether the disease was certain or simply probable at some time in the future? Would human need correlate in some way with needed health care services?

From a health systems perspective, the definitions of health, disease, illness, and the ability of the health system to intervene effectively are already problematic. A lack of clear definitions leads systems at times to expend resources to provide services that may be humanly desired but do not necessarily correspond to general definitions of disease or illness. Since most health systems have limited resources to respond to illness and disease, decisions must be made about what services will or will not be provided, the level of services that will be provided, and the relationship that the system may have with other facilities to provide care that a health system cannot provide. Advanced genetic information will only complicate this issue and probably strain the decision-making processes within health systems for how to determine what services ought to be provided, when, to whom, and where. This raises questions of justice, stewardship of resources, and the organization's commitment to the common good.

A second ethical issue arises in the debate over whether there exists any inherently moral commitment on the part of health care professions and/or institutions to care for persons or whether health care relationships are fundamentally contractual. At issue here is a shift in the paradigm used to understand the nature of disease and illness— from an infectious disease model to a more molecular model. Health care systems and the profession were shaped by the infectious disease model of care in this century. This affected everything from the organization of the profession of medicine itself, with its various departments, to the way acute-care beds were laid out in a given institution.

A molecular view of medicine may require a very different arrangement, possibly with less attention to acute-care beds, and may decide where health services should exist in the community.

In an infectious disease model, great moral weight is placed upon communities having at least some hospital presence. Simply witness the uproar in small and rural communities when hospitals close, even when they have been unused. If a molecular view of medicine results in more attention to prevention and diagnosing disease in a more "probabilistic" setting, and if interventions are judged best at the somatic-cell level, our current concept of hospitals and health care could change dramatically. The ethical concerns of the stewardship of available resources and the switch of capital from present infectious-disease medicine to molecular understandings of disease will have an impact on the system and hence on the people who rely on that system. The ethical analysis here can cut either way. Less reliance upon and presence of acute-care beds could reflect a more just delivery system and a better use of resources. Preventive strategies could be enhanced. The avoidance of disease could be promoted — a good for both the individual and the community.

Alternatively, molecular medicine might be accessed only by the well-off, creating further injustices, less access, and worse health care status for the least well-off. When infectious disease was the paradigm, professionals often responded to a sick patient because of the randomness of disease and the vulnerability of the person. One would not or should not abandon one's vulnerable brother or sister in the community. In a molecular model, especially if both genetic information and environment/lifestyle issues are the source of an illness, one's professional response for some diseases could be curtailed and victim-blaming could be more prevalent than it already is.

A third ethical problem lies in the relationship between how health care delivery is reconfigured and its relationship to public health functions or activities. As noted above, the connection between public health and acute-care delivery systems is not as strong as it should be. In part, this is due to the fact that most patients who come into the health care delivery system are seen in an isolated setting. The patient has a particular problem; the physician or other health professional interacts with that patient; informed consent, autonomy, other issues are addressed in the context of the therapeutic encounter; the patient gets better or does not; the intervention ends; and generally the patient goes home. Unless there is widespread public fear about a particular disease, such as tuberculosis or hepatitis A, few medical interactions really engage public health activities. In fact, we are sometimes all too willing to return "cured" patients to sick environments.

The promise of genetics in the future, especially as it is related to multifactorial issues in disease, may or may not enhance the relationship between health care services and public health in general. A community perspective is not likely to be achieved only by understanding the genetic makeup of human persons. The organizational values and commitments of tomorrow could lead institutions to focus on public health philosophy and initiatives to complement and promote health care services made available by new genetic information. Systemic organizational change to address root causes of disease at both the molecular and the environmental/lifestyle level could better promote the common good and the goals of medicine. But the fascination with new genetic information, in the absence of other changes, continues a technological drive that may do more to harm than to heal the human community.

One difficulty is in helping people to understand common genetic polymorphisms, as they are known, and the role that they may play in illness or the susceptibility to certain diseases, especially when these occur in connection with other genes and/or environmental conditions. (The concept of informed consent in this regard will pose additional ethical challenges.) Another difficulty for the delivery system arises as the health care system moves from looking at the good of individuals to the good of groups who may be at risk for the possibility of certain future illnesses. How will surveillance and testing be done in a way that adequately protects individual freedoms while promoting the common good in the molecular view of medicine? Will the common good require all persons of a particular group to be genetically tested in order to provide information that may enhance their ability to change the environment in which they live, to change lifestyle, or possibly both? Will epidemiological studies in public health be used against certain persons or groups of people?

Yet another ethical issue arises from an assumption that health care is inherently a social issue, despite the fact that health care encounters tend to be atomistic and autonomy tends to be the reigning ethical principle. The ethical question—very difficult to answer—is how an increase in information about our genetic makeup will affect us as social beings. Will, for example, those with less genetic variation or those who are "endowed with better genes" be valued more because they are less likely to be predisposed to certain illnesses, less likely to cost health care dollars in the workplace, or less likely to use health care resources? How will society determine what the new health care priorities should be? Will greater genetic information aggravate the seeming inability of society to set those priorities today?

Many health care institutions grapple with the fact that there are limited resources. This is clearly seen in the choice of which clinical services to provide. Sometimes the most needed services cannot be provided because of costs, the availability of professional personnel, or competing demands. Since it is highly unlikely that future genetic information, even for multifactorial disease, will eradicate all diseases, including cancer and heart disease, gene therapy will become one more competing service.

Organizational ethical commitments around fairness and acting justly should ensure that these choices are public, not done at the expense or exclusion of the poor, able to be changed across time and in accord with need, responsive to the community served, and that those who make the decisions remain accountable. Ethical issues persist with the difficulties involved in the technological imperative in which medicine is so steeped. One could argue that the Human Genome Project and other forthcoming genetic tests and treatments are simply the next step in our fundamental belief that science will find an answer to all human disease and add significant numbers of years to people's life span. From an organizational and biomedical ethic, it is clear that the system is already unable to handle the sometimes-false promises of technology. Witness the difficulty of accepting that there are simply no curable interventions for certain diseases or to what lengths some people will go to attempt to cure any disease. We often fail to care and heal when technology fails us. When genetic promises fail, organizations will be confronted with the same ethical problems and may not fare any better in resolving them.

As noted above, the ethical issues that health systems will face in the relationship between genetic information and informed consent will likely loom large. Many believe, for reasons beyond new genetic knowledge, that the future of health care is in the provision of information rather than care per se. Facts and knowledge will be key drivers for health and health services. Care decisions will shift more clearly to patients and their families and away from the "leftover" experiences of paternalism both in the medical profession and in the health care delivery system. This assertion may not be true. Will genetic information, especially when it needs to be considered in light of environmental or lifestyle activities, be clearly understood and aid people in life choices? Will the notion of predispositions to disease be clear and helpful or will the information be ambiguous, lead people to false conclusions, or tempt people to believe they are ill when they are not symptomatic for any disease? Can information be communicated in such a way that people's consent to genetic testing, for example, is genuine and real?

The health care system is already unable to ensure true consent from a moral perspective. Unless massive change takes place within the system itself, the promise of genetic information will not ensure this and indeed will make our institutional encounters more difficult. The present health care system exhibits numerous ethical issues in light of the problems of the haves and have-nots. These display themselves in access to health care services, in the availability and portability of health care insurance, and in the treatment and care of the poor, including where health care delivery entities are located. These ethical issues are likely to be aggravated in the face of new genetic information. Most of the therapeutic interventions that might take place will be in experimental form first. Who qualifies for them is a first ethical issue. Should those who are usually denied health insurance because of disease be included in experimentation if health insurance coverage for these services is later denied? Do we risk using the vulnerable to gain a good for the wealthy? Traditional ethical concerns about human experimentation and the use of people's tissue or body fluids, including future use of these, have also raised numerous questions in terms of informed consent. These questions remain in a reformed system. Additionally, a new, "more valued" group of persons may be created, raising issues around the cost of insurance, hence insurance coverage, and by implication access to health care services. The information may be valuable and seen as an advance in itself but may not be perceived as quite as valuable from a societal view.

Ethical issues about access may also be raised as people are penalized for lifestyle-related concerns, especially when genetic information had been provided to them. With the knowledge of one's genetic makeup and with some explanation of the relationship between environment, lifestyle, and possible disease, access could become more difficult for those who do not accept a genetic intervention, change lifestyles, or both.

Conclusion

The issues are not all new and they do not have to be seen as draconian. The fact is that an imperfect health care delivery system that finds it difficult to realize the "good" in the context of organizational values and biomedical ethical questions will have a difficult time incorporating the best of scientific advances, therapeutic improvements, and new scientific knowledge. In the face of new genetic information, it is sometimes hard to grapple with the questions of limits. As society grapples with the ethical implications of the limits to life, wherever that might be, and therefore the limits of medical treatment, it must also grapple with the limits of the organizations within which medicine, health care, and these social values take place.

12

The Challenge of Physician Education in Genetics

Carol Bayley, Ph.D.

NOT TOO MANY YEARS AGO, appropriate opioid use for pain relief was constrained by a profound lack of clinical knowledge. Clinicians and patients alike feared addiction. Surgical patients who had a history of drug addiction were considered dangerous candidates for pain relief. Patients were evaluated and treated for pain based on the physician's or the nurse's estimate of how much pain the patient should have rather than on the patient's self-report. Clinicians were afraid of dispensing a lethal dose of narcotics for someone with extreme pain.

For a while, these things—fear of addiction, fear of a lethal dose—masqueraded as "ethical issues in pain management." With education in new understandings about pain and the way opioids are metabolized, including the relative difficulty of addicting a person in pain and the low risk of killing a person by the appropriate administration of opioids for analgesia, the real ethical issue emerged: the routine undertreatment of pain in clinical settings where relief of suffering should have been paramount.[1] And why was this the case? Because pain management was a subject that physicians knew little about.

There is a parallel in this situation with the current state of physician education in genetics. For a variety of reasons that I will address below, physicians as a clinical group are undereducated about new developments in clinical genetics, and their patients are therefore at risk of not getting what they need. The purpose of this paper is to outline the challenges facing those charged with the education of physicians and to describe the changes in attitude, knowledge, and skills that physicians need to serve their patients well in an age of genetic medicine.

ATTITUDES

The first hurdle to overcome in preparing the ground for physician education in genetics is one of attitude. As an ethics educator in a large multihospital system, I have heard a variety of responses to inquiries about the state of physician education in genetics. Four attitudinal postures toward inquiry about the state of physician education in genetics are common.

The first common response is: "It's interesting, of course, but it is off in the future. I need to keep up with today's clinical developments today, and read my science fiction on the weekends." What this response does not account for is the fact that there are already clinical tests for more than 600 human genetic disorders, and more tests are being developed every month. The Human Genome Project is progressing ahead of schedule. Pharmaceutical companies have great financial motivation to develop useful tests. These two factors have converged to produce genetic test after genetic test in quick succession.

A second common response to genetics is: "Genetic diseases are rare. I don't meet that problem in my practice." This response is no doubt prompted by the fact that early progress in genetic testing *was* made in single-gene disorders that *are* relatively rare. Huntington's disease, for example, as devastating as it is, affects a small minority of the population. However, this response ignores the explosive advances in our understanding of genetic factors in disease risk. Genetic tests are now being developed to predict an individual's risk of common diseases such as breast cancer, ovarian cancer, colon cancer, Alzheimer's disease, cystic fibrosis, sickle cell disease, and diabetes. The number of people who worry that they may contract one of these diseases is many times greater than the number who will contract a genetically linked form of one of them—and the "worried well" are the ones who will show up in a physician's office asking for the genetic test.

Breast cancer is a good illustration. Much attention has been given by breast cancer activists to the statistic that a woman has a one-in-nine lifetime risk of contracting breast cancer. "Risk factors" such as a mother or a sister with breast cancer or older age at childbearing are cited. This means that, even though a small proportion of breast cancers are of the type for which genetic testing might identify a predisposing risk, women who are concerned about their health, particularly those whose mother or sister had breast cancer or who were older when their first child was born, may ask about genetic testing from their primary care provider. The same is true of Alzheimer's disease, which currently affects about four million people in the United States.[2] The proportion of individuals for whom testing is appropriate may be small, but the proportion of worriers is large.

A third common response to genetic testing is: "It's a lab test, just like other lab tests." There is a grain of truth in this response. Genetic tests are, for the most part, ordinary blood assays.[3] Like any blood test, they are relatively simple to perform. However, as with other laboratory tests, physicians must understand a test well enough to decide on its appropriateness for any given patient. Results come back to the physician, who must then interpret them for the patient and suggest a course of treatment if appropriate. As I will describe, understanding a genetic test and its limitations and exploring its appropriateness with a patient are two major challenges for physicians. However, even within the wide universe of lab tests, genetic tests are unique in their implications for current and future generations of a family.

Under ordinary circumstances, a lab test performed on a patient results in knowledge about that patient for that patient's treatment. Most genetic tests, however, are designed to detect inherited mutations. That means, for example, that if a woman fears, on the basis of her family history, that she may have inherited a mutation that increases her risk of breast cancer, at least one affected relative may need to be tested first. That means that not only is the test often performed on others, but the knowledge gained from the test applies to family members beyond the patient. Again, this is very different from an ordinary lab test.

Finally, a typical response to genetics from physicians is: "We're just primary care. We don't do that stuff here." This response is indicative of the highly specialized venues in which genetic testing was first performed and the kinds of diseases that were initially detected. But once again, several factors in medicine are converging that emphasize the inadequacy of this response. First, as mentioned above, genetic tests are being developed for common diseases that hospitalize patients. Second, specialized genetic knowledge is now becoming more generally utilizable. For example, as we learn more about the kind of cancer associated with BRCA1 mutations, we inadvertently learn more about all breast cancers. Third, primary care is becoming more and more the focus of the patient's encounter with the medical system. As Alan Guttmacher has pointed out, in the next several years it will become as impossible to practice in primary care without thinking genetically as it is now to practice in primary care without understanding infectious diseases.[4] These factors together make it imperative that primary care physicians understand the promises and limitations of genetic testing in order to deliver quality care to their patients. Gene therapy treatments, by most estimates, are a long way off and will likely be within the purview of specialists. By contrast, diagnostic testing, which is already here, will fall within the purview of primary care. In some diseases, such as hemachromatosis, early diagnosis is itself lifesaving simply by indicating the need for conventional treatment.[5]

Knowledge

The explosion of knowledge in medical genetics is recent; this would be a serious problem even if the pace of new knowledge had grown at a relatively constant rate over the last ten years or so. It has not. New understanding of the genetic components of ordinary and rare diseases is in fact exhibiting a mushroom effect—it seems to double overnight. This trend will intensify as the sequencing of the human genome is completed and tests are developed for more and more disorders with genetic components. Furthermore, the roles of previously identified genes in diseases are becoming clearer.

Physicians in active practice vary in their knowledge of many aspects of medicine, and genetics is no different. A 1991 genetic knowledge survey of nongenetic nonacademic physicians in several specialties showed that the participants were able to respond correctly to less than three quarters of the questions deemed important by other nongenetic, nonacademic physicians who helped develop the survey.[6] The same survey showed that the more recently a physician had graduated from medical school, the better he or she would score. Still, a 1995 survey of 125 four-year medical schools in the United States showed that only little over half of them (68) required genetics courses.[7] Medical schools are improving in their curricular offerings of genetics, but there are many physicians in current practice who graduated from medical school when genetics was not well understood or who practice in specialties other than pediatrics, internal medicine, or obstetrics, where physicians are more likely to have been exposed to patients with heritable conditions.

The knowledge necessary for physicians to meet the needs of their patients adequately falls into two main areas. One is the scientific foundation of medical knowledge. Physicians need to understand patterns of heritability, types of genetic mutations, and the range of manifestations of a genetic disorder. A general understanding of the science on which the processes for genetic testing are based is an important fundamental.

The other important component of physician knowledge has to do with the appropriate application of scientific knowledge in a clinical context. This involves the critical assessment of science in the service of quality medical care. For instance, the clinical validity of a test is an important factor in a patient's decision about whether to undergo genetic testing. Physicians must also recognize that important patient choices and changes in family relationships may spring from a positive or negative test result, and must be competent to discuss the potential consequences prior to the patient's decision about testing. A physician must understand the differences between positive, negative, and uninformative test results. If they themselves do not possess and transmit this knowledge, their patients will be unlikely to give informed consent for the test, and an important standard of care will not be achieved. Furthermore, some patients who are *not* at elevated risk of a particular disease may mistakenly think they are. In addition, understanding the circumstances under which genetic testing is appropriate is just as important as being able to interpret a test result correctly.

There have not been many studies of physician knowledge of genetics, but one will serve as an illustration. In 1994 a genetic test became commercially available for familial adenomatous polyposis (FAP), an autosomal dominant disorder that causes colorectal cancer if prophylactic colectomy is not performed. The test can identify the mutation in affected members of 80 percent of families with FAP, and when the particular mutation is known in a family, the test can differentiate affected from unaffected family members with virtually 100 percent accuracy. The appropriate use of the test can confirm a diagnosis of FAP, justify surveillance with regular colonoscopy, and aid in the surgical management of the disease. In addition, negative test results in families with an identified mutation can exclude some family members from invasive (and expensive) surveillance.

In March of 1997, Giardello and colleagues published research on a sample of 177 patients from 125 families who underwent testing in 1995. The research assessed whether physicians had recognized indications for the test and ordered it appropriately; whether informed consent was obtained for the test; whether genetic counseling had been offered; and whether test results were accurately interpreted. What did the researchers find?

> A high number (83%) of those tested had valid indications for the test, which is not surprising since the tests were ordered by specialists in the disease. Still, 17% of tests were ordered for "unconventional indications" and when so ordered, the test positive rate was low. More worrisome is the finding that, although genetic counseling before the test and informed consent to the test are considered essential, neither was done in about 80% of cases. A fifth of the presymptomatic patients had the test before the precise mutation was identified in an affected family member. Finally, one third of physicians would have misinformed their patients about uninformative test results: if a mutation was not detected, the patients would have been told that they were mutation-free, rather than informed that they must continue surveillance. Since surveillance and early intervention are the keys to saving lives in persons with FAP, this mistake could have been deadly.[8]

This example illustrates the importance of up-to-date clinical knowledge in meeting the challenges of patient care in an age of genetic medicine. Real harm can come to patients if their physician's knowledge is not adequate. Relaxed surveillance in a patient whose test result has been misinterpreted can result in a cancer that otherwise would have been watched for, caught, and perhaps cured. Other genetic tests would illustrate other lessons: a mistaken interpretation of a particular gene sequence in a person thought to be at increased risk for familial breast cancer could result in an unnecessary mastectomy or oophorectomy.

SKILLS

In addition to improving physicians' knowledge and understanding of genetics in both science and clinical application, another challenge to be met is the improvement in certain clinical skills. With overall increase in knowledge of the kinds of tests available, indications for tests, and the clinical ramifications of test results, physicians' skill in applying this knowledge in their practices will also improve. Along with these cognitive skills, however, physicians must develop others.

The first I will call the skill of nondirectiveness. In training and in traditional practice, physicians learn to be directive. Their role is to assess and diagnose, using clinically appropriate tools, and to present their findings to the patient with a suggested course of action. While it is true that informed consent to treatment requires the patient's ultimate authorization and agreement, physicians can be accused of abandoning their patients if they leave them without a clear recommendation for treatment. Directiveness is the order of the day in traditional medicine.

It is easy to imagine how this approach might carry over into clinical conversations between physician and patient about presymptomatic testing or prenatal diagnosis of genetic defects, and yet this transfer would not serve patients well and could undermine patient autonomy. In these situations, the physician's role is to explain and communicate—risk probabilities, options, treatment possibilities, degrees of confidence about this or that genetic test. Perhaps no other endeavor of medicine better illustrates that the expertise of a physician is *not* in making value-based decisions for patients, since every patient's values may be different. A decision about whether to undergo genetic testing for oneself or one's fetus is rife with value decisions, particularly in the absence of symptoms in the one being tested. It is essentially different from the clinical presentation physicians are accustomed to dealing with, in which a sick person asks for advice about what medicines to take or surgery to undergo or even lifestyle change to make in order to feel better.

A second skill to be developed by physicians concerns sensitivity to ethical issues. Genetic testing and diagnosis present a whole array of opportunities for examination of ethical issues. I will mention just a few.

Confidentiality, long a hallmark of a physician's relationship with a patient, is already undergoing challenges from new electronic forms of charting, storing, and retrieving medical information. This challenge will intensify with the development of enormous amounts of genetic information that will be readily available on a "smart chip" or other new technology. But confidentiality undergoes stress in another dimension as well in genetics. It is the nature of genetic assessment that mutations found in families can then be identified in individuals. Testing an individual often means testing family members as well. A genetic test may reveal information not only about the patient but also about the patient's family members, both those who have agreed to be tested and, to some extent, those who have not. How physicians can simultaneously honor confidentiality for the patient and the family and protect the autonomy of patient and family members is an ethical challenge.

Another ethical issue deserving critical attention of physicians who promise to "do no harm" is the possibility of actually causing harm by the test itself. As mentioned above, unless a physician is aware of the risks, he or she cannot communicate them to gain informed consent. But a physician is not necessarily aware of the policies of HMOs or life insurance companies or of employers who may discriminate on the basis of a genetic test, and therefore has an ethical duty to become educated about them. There is also the possibility of psychological stress induced by knowledge that the test will reveal. Physicians need to be aware of this risk of harm in advising patients about any genetic test. The reigning paradigm of knowledge in the medical profession is that knowledge itself is neutral. This paradigm deserves critical scrutiny in the light of genetic developments and the inability of our social schemes to keep pace with them.

There are other ethical issues surrounding the testing of certain kinds of patients or testing for certain kinds of disorders. For example, tests for late-onset disorders or those with no effective therapy may be problematic for some patients. For children, they are even more ethically problematic. A number of professionals have concluded that in the absence of medical management issues, children should not be tested because they cannot give informed consent, and further, such testing deprives them of the right to make their own decision about testing when they reach adulthood.

Genetic testing, together with the required genetic counseling and informed consent, is a practice option that a physician might reasonably decide whether or not to offer. However, every physician must gain sufficient knowledge to recognize the conditions under which a genetic test might be considered and to refer patients and their family members for appropriate evaluation and care.

Conclusion

The challenges of physician education in attitude, knowledge, and skills require determination and creativity on the part of both the physicians and those who wish to provide educational opportunities. Ordinary educational wisdom is important here if we are serious about filling in the gaps.

One piece of educational wisdom, which holds for kindergartners and physicians alike, is that different people learn in different ways. Some physicians will learn best in conference settings, like the conference on Genetic Medicine and the Practicing Physician held by the American Medical Association in New Orleans in March of 1998. Others will learn best by reading, either randomly or in some directed way. Some physicians will learn by taking advantage of Internet- or Web-based opportunities, particularly those through academic institutions such as Stanford; University of California, San Francisco; the Mayo Clinic; and others. Interactive sites are becoming more available to the individual user, both for physician education and for patient educational support. Finally, some physicians will always learn best in conversation with collegial experts who share their particular interests or experience. In one region of the health system I work for, a medical geneticist has set up a monthly telephone conference call during which issues in breast cancer genetics are discussed. Participants register for the call and are sent articles and illustrative slides in advance. The call lasts an hour and is structured around a particular topic, such as the assessment of various predictive models or the usefulness of tests for a particular kind of breast cancer.

The other bit of wisdom many educators know is that skills and knowledge are improved in different ways. Knowledge may be more amenable to increase by cognitive means such as reading and the Internet. Skills are less likely to be improved in any substantial way without actual practice. The skill of being less directive in the genetic aspects of one's practice and skills in dealing with ethical issues are more apt to improve in educational settings that are deliberately structured to address both the cognitive and the affective dimensions of the clinician. Only if these are integrated will the fundamental reorientation of physicians to the new challenges of genetic successfully be accomplished.

Notes

1. In a large study published in 1992, a stunning 81 percent of physicians and nurses reported that the "most frequent form of narcotic abuse" in their hospital was the undertreatment of pain. See Solomon, et al., "Decisions Near the End of Life," *American Journal of Public Health* 83:1 (1993).

2. Somewhere around 5 percent of Alzheimer's disease is familial, which means about 200,000 persons are currently affected with the familial AD. Even if only half of these persons wished to be tested, and each of those had only one relative who wanted to know his or her risk, the number is still large.

3. The number of genetic assays is increasing all the time.

4. "Genetic Medicine in the Next Five Years: An Interview with Dr. Alan Guttmacher of the Human Genome Project," *Health Progress* 80:5 (1999):43–45.

5. In fact, hemachromatosis is not only effectively treated by conventional means (regular phlebotomy), it is also common enough in the population to make genetic screening a viable possibility.

6. Neil A. Holtzman and Michael S. Watson, *Promoting Safe and Effective Genetic Testing in the United States: Final Report of the Task Force on Genetic Testing* (Washington D.C.: National Institutes of Health, 1997), 63.

7. Holtzman and Watson, *Promoting Safe and Effective Genetic Testing*, 65.

8. F. M. Giardiello, J. D. Brensinger, G. M. Petersen, M. C. Luce, L. M. Hylind, J. A. Bacon, S. V. Booker, R. D. Parker, and S. R. Hamilton, "The Use and Interpretation of Commercial APC Gene Testing for Familial Adenomatous Polyposis," *New England Journal of Medicine* 336:12 (1997):823–27.

13

Pharmacogenetics, Genetic Screening, and Health Care

Ruth Chadwick, LL.B., Ph.D.

Pharmacogenetics

The publication of the Human Genome Project results has increased predictions of a paradigm shift in medicine,[1] and genetic screening and testing are at the heart of the debate. Over the past few years much work has been done on developing criteria for the implementation of population genetic screening, including the seriousness of the condition screened for, the reliability and predictive power of the test, and the possibilities for effective intervention or scope for action in the light of a positive result. It has been argued, however, for example by John Bell in the *British Medical Journal*, that the "development of drugs along genetic guidelines will be a major force driving the implementation of screening by healthcare providers."[2] The term used to describe the use of genetics to show how variations in patients' DNA may diminish or increase the effects of a drug or render it harmless is "pharmacogenetics."

There are predictions that pharmacogenetics might lead to a new understanding of disease.[3] Whereas common diseases are currently defined by their clinical appearance, it will become possible to subdivide heterogeneous diseases into discrete conditions, in other words, to change our perception of what the condition is for which a treatment is sought.[4] As genetic variants are identified that are associated with drug response, there is likely to be a move towards widespread testing before prescribing—in fact it may come to be considered unethical not to carry out such tests.[5] The type of testing involved, however, is different from testing for single-gene disorders: it will involve testing for single nucleotide polymorphisms (SNPs), and hence the transferability of guidelines developed for other kinds of testing cannot be assumed.[6]

The first criterion frequently referred to in discussions of genetic screening is whether the condition sought is an important health problem—or whether it is "serious" (e.g., Nuffield Council on Bioethics).[7] There has been considerable discussion over what counts as serious, but despite the difficulties over a precise definition, there is a widespread consensus on particular examples of conditions that are life threatening, including some of the cancers and the hemoglobinopathies such as thalassemia.

In the case of screening related to pharmacogenetics, however, the condition sought is susceptibility to drug toxicity—in other words, a manufactured or iatrogenic condition. Does this count as an "important health problem" or "serious condition"? It is estimated that adverse drug reactions account for more than 2 million hospitalizations and 100,000 deaths per annum in the United States.[8] We cannot use these figures, however, to justify any given screening program unless what is sought is a predisposition to find all the drugs implicated in these figures toxic. What is envisaged is screening for risk factors for toxicity for particular drugs, for example, women who would be likely to suffer from blood clots from birth control pills or who would be at risk of adverse side effects from drugs such as tamoxifen in breast cancer provision or treatment.

The second criterion to be discussed concerns what can be done in the light of a positive result. Where a genetic diagnosis of an existing or presymptomatic condition or a prediction of a late-onset condition or predisposition is sought, what might be at issue is the availability of treatment. In the case of pharmacogenetics, however, this criterion again has problematic applicability. What is being tested for is the potential toxicity of the treatment itself, so it is difficult to use availability of treatment as a criterion of screening since the screening is being carried out to establish the extent to which *this* treatment is an "available" treatment.

What may be envisaged, however, is not populationwide screening but individual *testing*. Pharmacogenetics has been said to have the potential to individualize prescribing. This potential for predicting individual susceptibility to responsiveness to drugs has major implications not only for therapy but also for participation in clinical trials and research. As regards *therapy*, one of the principal benefits, it is suggested, is that more genetically informed prescribing will reduce the rates of morbidity and mortality due to iatrogenic disease. It is important to bear in mind the negative-positive distinction in relation to pharmacogenetics: its use to avoid adverse events, on the one hand, and to facilitate better-targeted treatment, on the other. It has been estimated that about one in fifteen hospital admissions is due to adverse drug reactions.[9] Pharmacogenetics could affect a prescribing decision for a given patient in at least three different ways: (1) adjustment of dosage of drug A; (2) a choice between prescribing drug A or drug B; (3) drug A or nothing (where there is no alternative treatment available).

The "positive" applications of pharmacogenetics, however, need not be confined to treatment. For example, concern about aging, combined with a move towards predictive testing and prevention and with developments in pharmaceuticals, could conceivably lead to the whole population being construed as a market for aging-prevention drug products *throughout* life. Currently individuals may be encouraged to take antioxidant vitamin pills, for example; in the future they may be taking new pharmaceuticals to prevent specific age-related diseases such as Alzheimer's and, if there is a mechanism related to an underlying aging process, antiaging drugs which would be relevant to every single person in the population.[10]

Clinical trials in this area may have features that distinguish them from traditional clinical trials: (1) they are likely to involve storage of DNA samples as responses to drugs are tracked over time; and (2) the nature of the risks and benefits to which the participants may be liable are of a different kind, such as the possible (mis)use of genetic information on the one hand and genetically informed prescribing on the other. The potential impact on research, however, has other aspects, including the extent to which it will be possible for clinical trials to become more targeted towards specific groups. These potential developments in therapy and research give rise to questions in bioethics of three kinds: (1) substantive ethical issues; (2) professional ethics; and (3) challenges to existing ethical frameworks.

Substantive Ethical Issues

As already indicated, some of the literature on this topic has described developments in pharmacogenetics as facilitating "personal pills,"[11] the suggestion being that awareness of genetic variation between individuals will facilitate prescribing in accordance with the specific needs of the individual, hence arguably in accordance with a principle that health care resources should be allocated according to need at the point of delivery. The possibilities of this with regard to monitoring of appropriate dosage as compared with choice of medication need to be considered. The situation where the choice is between drug A and *no* medication gives rise to the ethical problem of (perceived) abandonment. How pharmacogenetics will affect patient perception is important.

A major feature of the debate about the introduction of other genetic screening and testing program has been the question of right to know versus the right not to know, supported by competing interpretations of concepts such as autonomy and solidarity.[12] It has been argued that there might be a right not to know genetic information about, for example, one's future health status. But it might appear that the same considerations would not apply in relation to susceptibility to drug toxicity — surely it could only be beneficial to have information enabling one to avoid the side effects of drugs? A right to know one's genetic status vis-à-vis susceptibility to drug toxicity might be supported by an autonomy-based argument, where autonomy is interpreted in terms of self-determination — facilitating the choice of the individual in relation to treatment. Let us imagine that in the event of multiplex testing, however, it were possible to test at the same time for predisposition to a disease and for susceptibility to toxicity for the standard treatment. Then the question arises as to whether having this information is a benefit or a burden, because this is analogous to the situation where there is *no* treatment available. In such a case the argument for a right not to know comes into play.

What other reasons might ground a right not to know about susceptibility to drug toxicity? One possibility would be the converse of a placebo effect. The knowledge that one has a higher risk of toxicity might in itself increase that risk. Furthermore, genetic susceptibility to drug toxicity may have insurance implications in the way that genetic predisposition to health problems might — people who are slow because of their genotype to clear drugs from their bodies or to convert them to nontoxic form may be identified as belonging to a higher insurance risk category.[13]

Connected with this problem is the issue of quality control in a situation where hundreds of thousands of tests are carried out annually. External quality assessment schemes (EQAs) of genetic tests in Europe have demonstrated a low but significant error rate in cystic fibrosis testing,[14] and the number of laboratory tests carried out annually as pharmacogenetic testing comes onstream is set to increase dramatically. Mistakes may arise not only through technical error but also through clerical error or sample mix-up.[15]

Apart from the possibility of error, there are problems with uncritically accepting that an identification of genetic risk factors will determine or assist in determining the appropriate treatment for a particular patient. Other factors such as food intake, general state of health, and age may account for someone's response to a drug;[16] drug efficacy and toxicity may be considered as multifactorial traits that involve some genetic component(s) in much the same way complex diseases do. Beyond the issues for individuals, if we return to the "patient stratification" possibility, whereby patients could be classified according to genetic risk factors as they are presently classified by other risk factors such as high blood pressure,[17] in this connection the possible implications for particular population groups should be considered in the light of possible differences between ethnic groups as regards, for example, slow or rapid rates of metabolizing a drug.

Thus patient stratification could have discriminatory implications. The Council on Ethical and Judicial Affairs of the American Medical Association, in an article on "Multiplex Genetic Testing" in the *Hastings Center Report* (1998), argued that: "ethnic heritage may contribute to particular concerns, it is clinically relevant and should be considered. Offering multiplex tests that are bundled according to race or ethnicity, however, serves to categorise patients rather than to address their distinct needs. . . . The profession can ill afford the perception that science is being used to bring attention to the genetic flaws present in lines of inheritance."[18] If genetic susceptibility were correlated with some other characteristic such as ethnicity, it might lead in effect to a presumption of effective treatment for that condition for that particular group, although there might be considerable variation within the group. Indeed, there is some support for the view that the significance of ethnic variation in drug response might have been overstated.[19]

In clinical trials, the extent to which research in pharmacogenetics raises ethical questions that are distinctive needs to be addressed, for example, the implications for informed consent, feedback of information, privacy issues. Is it the case, for example, that pharmacogenetic information is *less* sensitive than other kinds of genetic information, which might lead us to believe there is a need to think differently about ethical issues?[20] We have to examine what interests need to be protected for research participants in this field, and how.

Professional Ethics

Questions for professional ethics arise when considering how pharmacogenetics will affect health care delivery. Different modes of delivery will raise different ethical questions, and countries may differ in how they integrate pharmacogenetics into health care. If genetic testing becomes a standard accompaniment of prescribing, there are questions about how this will be carried out. If doctors carry out pharmacogenetic testing at the time of prescription, then this will affect the doctor-patient relationship, at the very least in terms of workload, raising questions about the extent to which counseling, for example, should be required. On the other hand, what may be envisaged is that there will be a central database containing patient genotype information, which will be accessed at the time of prescription. If the latter is the case, then quality-control issues, mentioned above, become particularly important to prevent errors being perpetuated over time.

The person who accesses this database, however, need not be the doctor—it may be, for example, the pharmacist. There may be an expanding role here for pharmacists if, for example, doctors *prescribe* generically and pharmacists *dispense* according to genotype. There is a need, however, to think through the ethical implications for doctors and pharmacists arising out of these possible changes to their roles. The last scenario may be more appropriate in certain applications of pharmacogenetics, for example, when the choice is between drug A and drug B. There will also be a need for education and training in the ethical implications. What form this training should take will depend on how the ethical issues should be addressed, and this also needs to be considered.

CHALLENGES TO EXISTING ETHICAL FRAMEWORKS

In addition to the implications for practice, it is important to consider ethical frameworks themselves. It is not possible to think about the ethics of new health technologies without also attending to research on ethics itself. In the last few years, developments in the life sciences and biotechnology have not only challenged deeply held intuitions but have also tested the adequacy of ethical theories and principles. For example, when the first "test-tube baby" was born, people were unsure how to think about it. Did the arguments for a right to reproduce apply both to a right against involuntary sterilization and to a right to artificial means of reproduction? The debate about the possibility of human reproductive cloning has now added a new dimension to this question. The boundaries of our very concepts are also challenged by new developments: the concept of "embryo," for example, in the context of the stem-cell debate.

Developments in genetics also have the potential to change the way we look at things and to challenge the boundaries of our concepts. In the case of pharmacogenetics, the implications for concepts of disease have already been mentioned, but the impact is wider than that: the ethical frameworks we use may need to be revised. It cannot be assumed that principles of bioethics are immune to revision: there may be a "value impact."

It has already been mentioned that there is a view that there is a lesser privacy interest in pharmacogenetics. In the relatively short history of bioethics, a consistent theme, however, has been the desirability of patient access to information, as found, for example, in the doctrine of informed consent. The advent of predictive testing, however, has led to arguments for a right not to know, discussed above, in the light of arguments that information is not necessarily empowering: there are circumstances in which it can be a burden rather than a help. Developments in genetics have led to rethinking, then, the meaning of autonomy, the extent and limits of the duty of confidentiality, and the right to know. There is a growing body of opinion that it is not sufficient to continue with the traditional principles of biomedical ethics and simply seek to apply them in the new context, and there is specific concern about the transferability of existing guidelines to pharmacogenetics: "It is . . . incumbent that medical guidelines for mendelian- or susceptibility-gene testing do not extend automatically to discussions of other types of genetically based profiles in pharmacogenetics. Clear language and differentiation of respective ethical, legal and societal issues are required. . . ."[21]

It is important to distinguish different things that might be at stake in value impact. I want to distinguish three levels. First, there is the relationship between developments in science or technology and social attitudes. An example of this can be found in work on how developments in genetics might influence attitudes towards disability. In the context of pharmacogenetics, there are questions about whether people are likely to become less tolerant of adverse reactions.

Second, there are questions about the ways in which emerging technologies test the adequacy of existing ethical theories. An example here would be the suggestion that traditional theories of justice are inadequate because they depend on assumptions that construct who counts as disabled, who should be considered for the purposes of just distribution, and what there is to be distributed.[22] Third, there is the complex question of the relation between the above two levels—the complex interplay between theory and social values. What is being suggested is not a simple linear relationship.

For present purposes I shall focus on the second question and on how the doctrine of informed consent might be held to be particularly problematic in the light of the new genetics, in particular pharmacogenetics. The suggestion that informed consent is more complex in genetics might be made for a variety of reasons, some of which arise out of the special nature of the risks and benefits in pharmacogenetics, while others relate to the genetics arena more generally: the degree of public awareness of genetics; the sensitivity of the "information"; doubts about whether it is possible to be informed in this field; and the identification of the subject who gives consent.

In the context of pharmacogenetics, it has been argued that the information concerned is not as sensitive as information about single-gene disorders or susceptibility to common disease. Let us put this point on one side for the time being. The concern about the possibility of being "informed" in this area, however, is highly relevant. This relates to the idea that it is simply not possible to be genuinely informed of all the risks and benefits in this area. The thought behind this concern is that no one can be adequately informed, because it is not possible to foresee the range of uses to which genetic information, even pharmacogenetic information, about someone might be put. This point applies particularly to long-term storage of samples for pharmacogenetic research. The legal and ethical regulation of this area is still developing, so individuals are likely to be making choices in an uncertain situation. "The pace of progress in biomedical research poses . . . a serious threat to the future of healthcare development. The healthcare scene would need a regulatory framework in place with clear rules to ensure equitable and appropriate care for all, in particular for those individuals 'at risk' through genetic predisposition."[23]

In this situation there are at least two possible responses. The first is to take the view that much more effort should be put into the informing process. The second is to suggest that we need to think again about medical ethics, to examine the "value impact" of genetics on doctrines such as informed consent and ask to what extent they can do the work that we need them to do in the age of molecular genetics as we think about the implications of pharmacogenetics for health care delivery. Onora O'Neill[24], in her address to the American Society of Bioethics and Humanities in 1999, argued that we should do the latter, that we need to rethink our views on informed consent and indeed on confidentiality, revisiting what it takes to respect persons in the new context, which will include revisiting concepts of human agency. O'Neill was in this address particularly concerned with the accumulation of genetic information in large databases and the use of increasingly sophisticated systems for handling them.

This is a good example of the value impact of genetics. The implications for informed consent in pharmacogenetics need to be thought through both in the research and clinical contexts. In research, the issues of long-term storage of samples are already giving rise to concerns about the feasibility of an informed-consent model. Special regard may be needed to the possibility of less tolerance among patients of adverse reactions to pharmaceutical products. If it is true that pharmacogenetic data are not as sensitive as disease-related data, it may yet be necessary to explain why this is the case. As far as health care delivery is concerned, much will depend on who is doing the testing, the roles of different health care professionals involved, who is informing, and who controls the data. A model of applied ethics which relies on "applying" models developed in one context to another context without modification is unsatisfactory. New paradigms in medicine require new thinking in ethics.

Notes

1. K. Schmidt, "Just for You," *New Scientist* 160:2160 (1998):32–36.
2. J. Bell, "The New Genetics: The New Genetics in Clinical Practice," *British Medical Journal* 316:7131 (1998):618–20.
3. Ibid.
4. A. D. Roses, "Pharmacogenetics and Future Drug Development and Delivery," *Lancet* 355 (2000):1358–61.
5. C. R. Wolf, G. Smith, and R. L. Smith, "Science, Medicine, and the Future: Pharmacogenetics," *British Medical Journal* 320 (2000):987–90.
6. A. D. Roses, "Pharmacogenetics and the Practice of Medicine," *Nature* 405 (2000):857–65.
7. Nuffield Council on Bioethics, *Genetic Screening: Ethical Issues* (London: Nuffield Council on Bioethics, 1993).
8. K. Schmidt, "Just for You"; G. Stix, "Personal Pills," *Scientific American* 279:4 (1998):10–11.
9. K. Schmidt, "Just for You"; G. Stix, "Personal Pills"; Wolf et al., 2000.
10. I owe this point to David Schlessinger.
11. A. Persidis, "The Business of Pharmacogenomics," *Nature Biotechnology* 16 (1998):209–10.
12. R. Chadwick, M. Levitt, and D. Shickle, eds., *The Right to Know and the Right Not to Know* (Aldershot, U.K.: Avebury, 1997).
13. K. Schmidt, "Just for You."
14. E. Dequeker, et al., "Quality Control in Molecular Genetic Testing," *Nature Reviews Genetics* 2:9 (2001):717–23.
15. Dequeker, E. et al., "Quality Control in Molecular Genetic Testing."
16. See G. Stix, "Personal Pills"; and R. Chadwick and M.A. Levitt, "When Drug Treatment in the Elderly Is Not Cost-Effective: An Ethical Dilemma in the Environment of Healthcare Rationing," *Drugs and Aging* 7:6 (1995):416–19.
17. R. Chadwick, "Criteria for Genetic Screening: The Impact of Pharmaceutical Research," *Monash Bioethics Review* 18 (1999):22–26; Wolf et al., 2000.
18. Council on Ethical and Judicial Affairs, American Medical Association, "Multiplex Genetic Testing," *Hastings Center Report* 28:4 (1998):15–21.

19. J. Hodgson and A. Marshall, "Pharmacogenomics: Will the Regulators Approve?" *Nature Biotechnology* 16 (1998):243–46.

20. A. D. Roses, "Pharmacogenetics and Future Drug Development and Delivery"; Roses, "Pharmacogenetics and the Practice of Medicine."

21. A. D. Roses, "Pharmacogenetics and the Practice of Medicine."

22. A. Buchanan, D.W. Brock, N. Daniels, and D. Wikler, *From Chance to Choice: Genetics and Justice* (Cambridge, U.K.: Cambridge University Press, 2000).

23. O. Mendoza, "Patient-Centred Healthcare," *Nature Biotechnology* 17:supplement (1999):BV15.

24. Onora O'Neill, "Informed Consent and Genetic Information," *Studies in History and Philosophy of Biology and Biomedical Sciences* 32:4 (2001):689–704.

14

Ethical Aspects of Prenatal Screening and Diagnosis

Norman M. Ford, Ph.D.

PREVALENCE OF FETAL CONGENITAL MALFORMATIONS

The risk of a live-born child with Down syndrome increases with maternal age: 1 in 1,500 at 20 years of age, to 1 in 1,350 at 25 years, 1 in 900 at 30 years, 1 in 380 at 35 years, 1 in 85 at 41 years, 1 in 50 at 43 years, and 1 in 28 at 45 years.[1] The incidence at conception is higher, but more than 60 percent miscarry and about 20 percent are stillborn.[2] The frequency of trisomy 21 is higher earlier in the pregnancy, when prenatal diagnostic tests are done.[3] From 8 to 19 weeks, the background rate of miscarriage is about 6.8 percent, usually due to congenital anomalies.[4] Most chromosomal and other congenital anomalies, not being inheritable, arise unpredictably at, or after, conception, and affected offspring are born to women who are not in recognizable risk groups.[5]

PREGNANT WOMEN'S ANXIETY

Judith Searle has analyzed why pregnant women feel at risk and fear the worst outcome for their unborn children.[6] Her research was on routine screening tests for blood infection, diabetes, and ultrasound (U/S) and screening tests for Down syndrome. She found that women's perception of risks of fetal abnormalities is higher than objective epidemiological risks. Of 367 participants questioned, 242 (65.9 percent) reported that they felt anxious sometimes or a lot about having an abnormal baby. She found that 53.4 percent of participants identified routine prenatal screening tests as being most helpful in reducing their anxiety during pregnancy, even though prenatal diagnostic tests cannot guarantee that there are no risks of abnormality but only that the specific abnormalities tested for are present or absent.[7]

CURRENT PROCEDURES FOR PRENATAL SCREENING AND DIAGNOSIS

Advances in diagnostic techniques, coupled with the advent of legal abortion in most countries over the last twenty-five years, have increased pressures to extend the availability of prenatal screening and diagnostic testing. Prenatal screening differs from prenatal diagnosis in that screening is applied to individuals without any known physical risk to identify those at a higher-than-average risk of a genetic defect or carrier status, whereas diagnostic testing is done to detect a genetic disease in individuals with a known risk.[8]

Maternal Serum Screening

Maternal serum screening (MSS) is now frequently offered to pregnant women between fifteen and eighteen weeks gestation to detect pregnancies that are screen-positive, that is, with a risk of one in 220 to 300 or greater of being affected with Down syndrome, trisomy 13, trisomy 18, or a neural tube defect.[9] MSS checks the levels of three or four markers that may be indicative of the presence of these specific defects. A diagnostic test is required to obtain a definitive result after MSS.

Quadruple MSS is now commonly used and can have a detection rate of about 75 percent for Down syndrome pregnancies if gestational age is estimated by a scan, with a 5 percent false-screen-positive rate.[10] The 1997 retest rate for women in Victoria, Australia, prompted by raised MSS was 414 in 7248, that is, 5.7 percent.[11] Of 485 Victorian women who had a diagnostic test, 13 (2.7 percent) major chromosome abnormalities were detected. In other words, more than 94 percent of Victorian women screened did not need to have a diagnostic test and were unnecessarily made anxious for days awaiting definitive results and in fear of a possible miscarriage.

Nuchal Translucency Screening or Scan

All fetuses develop some fluid under the skin of the nuchal or neck area. An abnormal nuchal thickening appears black on a U/S screen from ten to thirteen weeks gestation. This has been found to be associated with

Down syndrome. This early U/S scan is known as a nuchal translucency scan (NTS) and it can be indicative of an increased risk of Down syndrome. Its sensitivity is good with an expert operator: it can identify up to 84 percent of pregnancies at an increased risk with a false-positive rate of 6 percent.[12] In a recent NTS of 3550 pregnancies, only 172 (4.9 percent) were at a higher risk of Down syndrome and needed a diagnostic test for confirmation.[13]

Ultrasound

The most widely used prenatal diagnostic technique is U/S imaging or scanning, which is noninvasive, safe, and usually performed at sixteen to twenty weeks' gestation. Images of the deep structures of the fetus are projected on a screen. This is done by recording the echoes of sound impulses which reflect off the planes and densities of the various tissues. U/S can determine accurate pregnancy dates and detect multiple pregnancy and many anatomical fetal defects.[14] Difficulties may arise in interpreting the results, especially by less expert operators, leading to failure of detection of abnormalities or the diagnosis of an abnormality where there is none. U/S results may indicate prenatal diagnostic tests for definitive results and so increase anxiety for mothers.[15]

Amniocentesis

From the fifteenth to sixteenth week of gestation amniocentesis can be performed, and its results are definitive.[16] It involves the withdrawal of 15 to 20 milliliters of fluid from the amniotic cavity by means of a syringe under U/S guidance. This fluid contains cells shed from the fetus's skin. They are genetically identical to the rest of the fetus's cells and can be grown in culture for chromosomal examination and testing. It may take up to two weeks to obtain definitive results of amniocentesis tests, and this usually increases pregnant women's anxiety.

There is a risk of a miscarriage within two to three weeks of the test which is generally deemed to be due to the procedure, for example through infection. Though it is hard to prove that amniocentesis causes fetal loss, it is generally agreed that there is a procedure-related fetal loss rate of about 0.5 percent for expert operators above the natural or background miscarriage rate of 1 percent for the relevant stage of gestation.[17]

The test may be medically indicated for mothers over 35 or 37 years of age, according to local practice; for those who have had a previous child with a congenital defect; for carriers of an inheritable disorder, when there is a history of chromosomal problems; and when there are unexpected results from U/S, NTS, or MSS. The frequency of detection of a fetus with Down syndrome for a woman aged 37 years at the time when amniocentesis is usually performed is 1 in 190 (0.53 percent), about the same as the risk of miscarriage due to the procedure.[18]

Chorionic Villus Sampling

Chorionic villus sampling (CVS) is usually performed between ten and thirteen weeks of gestation and involves taking a sample of placental tissue, which is usually genetically identical to the cells of the fetus. This is done using U/S guidance with little discomfort. CVS is almost as accurate as amniocentesis if enough tissue is sampled to allow for several cultures to be examined for double-checking. CVS results can be available in six to ten days if cells need to be cultured or within forty-eight hours by short-term analysis.[19]

CVS is performed before fetal movements are felt. Some women may be more reluctant to have abortion after they have felt fetal movements or seen their fetus move on the U/S screen. This may also give rise to what is known as the "tentative pregnancy," where a woman is reluctant to bond with her fetus and acknowledge she is carrying her baby before she receives results showing that her baby will not be born with the defect for which a test was done.[20] It is not surprising that a survey found that 76 percent of eighty-three women who participated in a midtrimester screening program would have preferred to have had the test in the first trimester mainly because early termination of pregnancy is easier and/or because of the earlier reassurance.[21] Bearing in mind the background rate of 1 percent for spontaneous losses at that stage, CVS poses an additional risk to the life of the fetus of 1 to 2 percent.[22]

Sex-Selected Insemination

Some success has been achieved in selecting X and Y sperm based on their relative size because statistical analysis shows that human X sperm are larger and longer than Y sperm. Males are conceived by Y sperm and females by X sperm because the egg always has an X chromosome. This raises the possibility of having sex-selected insemination with X sperm to prevent the conception of male offspring at risk of an X-linked disease. Since females have two X chromosomes, if one is defective the other makes up for the deficiency.[23] Recent reports indicate a separation success rate of about 90 percent for X and 65 to 70 percent for Y sperm has been achieved by staining sperm heads with a fluorescent dye.[24]

Preimplantation and Preconception Genetic Diagnosis

Preimplantation genetic diagnosis (PGD) is an experimental technique whereby cells from cleaving embryos are tested for disease. Unaffected embryos are implanted and the rest discarded.[25] Preconception diagnosis for defective genes might one day be done on the egg's polar bodies, which do not contribute any genetic material to the formation of the embryo; an analysis of their genetic makeup could detect up to two thirds of chromosomally abnormal embryos, leaving further detection to PGD.[26]

Therapeutic Benefits of Prenatal Diagnosis

The results of prenatal tests reduce the anxiety and fears of most pregnant women, and this is a benefit for the mother and fetus. If an untreatable defect is detected, the parents, with appropriate genetic and pastoral counseling, may be able to prepare themselves mentally for the birth of a disabled baby. Prenatal information may also be of benefit to obstetricians for the better management of the pregnancy and the prevention of an unnecessary cesarean delivery. It provides an indication for specialist neonatal staff to be at hand in a well-equipped hospital in case urgent treatment is needed after birth.

Unless certain fetal defects are specifically sought, routine prenatal diagnosis does not usually provide information that could lead to fetal therapy. But it can give information that may help provide medication or therapies for the benefit of some fetuses, with the exception of chromosomal and neural tube defects.[27] It is possible to provide fetal therapies for a few conditions discovered by amniocentesis. This information enables an evaluation of pulmonary maturity, fetal hemolysis, and anemia for therapy *in utero*. Where necessary, intrauterine blood transfusions can be made to treat fetal conditions such as anemia or Rh incompatibility. Various anatomical and developmental fetal defects (e.g., variations in heartbeat) discovered by U/S fetal monitoring can be treated by drugs or repaired by means of fetal surgery.[28] Better detection rates of treatable abnormalities could lead to fewer neonatal deaths by arranging for delivery at a tertiary-care center or by means of appropriate fetal therapy.[29]

Ethical Evaluation of Prenatal Screening and Diagnosis

Ethical Acceptability of Prenatal Screening and Diagnosis

Pregnant women have a right to seek and be given accurate information on the state of the health of their fetuses for reassurance, whether the prospects are good or not. Fetuses, too, may benefit from the allaying of maternal anxiety. There is also scope for intrauterine fetal therapy for some pathologies by the maternal administration of medication, intrauterine blood transfusions or fetal surgery. U/S is safe and useful for verifying dates. Prenatal screening and diagnosis are ethically permissible per se, under certain conditions, if their purpose and the methods used are respectful of the life and dignity of the pregnant woman and her fetus, and provided the requirements of informed consent, due sensitivity for persons with disabilities, and social justice are satisfied.[30] But this does not imply that pregnant women have a duty to undergo prenatal screening or diagnostic tests.

Due Respect for the Life of the Fetus

An ethical concern arises over prenatal diagnostic tests that involve risks to the life or health of the fetus. Fetal losses due to prenatal tests are greater when amniocentesis and CVS operators perform too few tests per year to acquire and maintain their expertise.[31] Prenatal tests should be performed by expert operators.[32] Amniocentesis and CVS come under moral scrutiny on account of their procedure-related fetal loss rates of 0.5 to 1 percent and 1 to 2 percent respectively. This increased risk of miscarriage is not intended but results indirectly as a side effect of the test due to infection, membrane rupture, or fetal stress. Every precaution should be taken to prevent it. Unless a prenatal test could proportionately benefit the fetus, either directly or indirectly by alleviating maternal distress, a risk of 0.5 to 1 percent or more of fetal loss would be ethically unacceptable from the sanctity-of-life perspective.[33] If there is no urgency, tests should wait until the risks of miscarriage are less.[34] But in view of their potential therapeutic benefits, tests for the fetus may be justified.

On account of these procedures' invasive nature for the mother and their fetal risks and costs, there is no need, in addition to giving information, to recommend routine amniocentesis or CVS to all pregnant women. The assessment of fetal risks necessarily involves both objective and subjective elements, especially for the mother, who may be enduring great stress and anxiety. A risk that a mother would deem to be unacceptable for her baby may be perceived by a pregnant woman as ethically acceptable for her fetus.

It is the pregnant woman herself, in the midst of her fears and anxieties and in the context of her family situation, who is properly entitled to evaluate the risks involved. The responsible taking of risks is an inevitable part of parents' lives. Many women will have different reactions to the facts and the risks and may, in good faith, make different evaluations. This implies that there may not be only one right answer and that several evaluations could be reasonable without any being unreasonable. It is not right to engender unwarranted feelings of guilt in pregnant women for their conscientious decisions. What is required for ethical action is that the risks involved in prenatal tests must not be judged by pregnant women and their doctors to be out of proportion to the benefits reasonably expected.[35]

Informed Consent, Counseling, and Pastoral Care

Women are under some social pressure to undergo prenatal screening tests and diagnosis. There is evidence that many of those who agree to participate in screening programs do so only in response to an invitation and may not feel free to decline if the request is made by health professionals.[36] Pregnant women and their partners need practical freedom to consent to or decline prenatal screening tests and diagnosis without undue pressure from health professionals or government health departments. There is no ethical duty to have them, nor should women be made to feel guilty if they decline.

Before choosing to have prenatal screening or diagnostic tests, pregnant women have a right to be adequately informed by doctors or qualified genetic counselors about the purpose of the tests, their risks, the available treatment options with their likely outcomes, and the implications for themselves and their children.[37] The autonomy of pregnant women does not require counselors to refrain from alerting them to family and social consequences of prenatal screening and diagnosis so that they may make free, informed, and responsible decisions. Genetic counseling that is dialogical and interlocutory promotes a deliberative process that is aware of alternatives and free of manipulation and coercion.[38] Pregnant women need to consider the implications of both having and not having these tests, including the possibility that their fetuses may or may not be affected by a congenital anomaly. They need to know who will receive the results in addition to their own doctors. Risks, be they genetic or not, are better presented in different forms: "proportional (1 in 4, etc.), by percent (25 percent), and in a verbal form (e.g., higher than the average for the general population)."[39] It is imperative that pregnant women be given the required information on prenatal screening tests and diagnosis in plain language by doctors or genetic counselors.[40]

There is evidence that for many mothers the requirements of informed consent are not always adhered to for MSS and prenatal diagnosis, even if the mothers concerned thought it was sufficient at the time.[41] Doctors and genetic counselors should respect women and their partners as persons, as well as their dignity, integrity, levels of knowledge, autonomy, and privacy. They should help pregnant women make their own decisions, not make decisions for them. They should give the time needed to listen to their concerns. It is their role to point out to women and their partners the relevant scientific and medical facts, the range and degrees of risks involved, and their correct interpretation.[42] It should be made clear to women that they may withdraw from the prenatal screening and diagnosis program at any stage.

It is not usually the role of genetic counselors to give ethical advice. Pregnant women, however, should be made aware that they are free to seek ethical advice from the genetic counselor, other qualified persons on the hospital's staff, or elsewhere. They should be given the opportunity to discuss their ethical concerns with their partners and respective pastors, chaplains, or personal advisers. These advisers should not give advice contrary to their own conscience, while at the same time respecting pregnant women who may choose to make their own, different, informed, conscientious decisions. With pregnant women's consent, genetic counselors and pastoral care workers may at times establish liaison to help women to cope with and accept an abnormal baby.

In secular public hospitals, pro-life medical staff and genetic counselors need not feel they should exclude themselves from practice provided they are prepared, if asked, to present both sides of the ethical debate on selective abortion and they adhere to the relevant guidelines of their hospitals. By replying to pregnant women's requests for information about prenatal diagnosis or test results, they need not compromise their own ethical principles.[43] In a hospital administered by a religious organization, genetic counseling should be given in accord with the hospital's ethical framework, and in one of their first interviews, pregnant women should be informed of the hospital's ethical policy of not providing selective abortion. A woman attending, say, a religious hospital who opts for selective abortion should be treated with every courtesy, be given a copy of her medical record and test results, and be advised to seek further counseling.

A protocol needs to be in place to ensure that pregnant women are informed of positive diagnostic results for an abnormality with great cultural sensitivity and receive appropriate support to cope with their distress.[44] Care is to be taken to avoid directly or indirectly trying to persuade a pregnant woman to have recourse to selective termination of pregnancy if the test for a congenital defect is positive. At the same time, in a secular or a religious hospital, a genetic counselor who is morally opposed to abortion should respect a woman who conscientiously decides to seek an abortion after she finds out that she is carrying a fetus with a disability.

Prenatal Diagnosis and Selective Abortion: a Pro-Life Evaluation

A serious ethical problem dilemma may arise when pregnant women are informed that their fetus is affected by an abnormality, for example, Down syndrome or a neural tube defect. A decision to have an abortion is likely to derive from the fear of being unable to cope with the long-term implications of raising a child with a disability, especially when society condones abortion. Chris Goodey says abortion could not be for the benefit of Down syndrome children since their physical conditions can be treated and they themselves do not seem to suffer mentally.[45] If prenatal diagnosis was designed to help women to choose to abort abnormal fetuses, from a pro-life perspective it would be ethically flawed by this link to abortion.[46] Likewise, it would be unethical for a health professional to try to persuade a woman to have the test for this purpose or to make consent to abortion a condition for having the test.

There is ample evidence to show that there is some link between information about abnormal fetuses and selective abortion. In Victoria in 1997 major chromosomal abnormalities were detected in 4.3 percent of 5,295 prenatal tests.[47] Such results change parents' anxiety of incertitude to the trauma of knowing their fetuses are affected by a defect. A survey at a Melbourne hospital over a five-year period found that three of 179 women decided to continue pregnancy after a diagnosis of Down syndrome—equivalent to a selective termination rate of 98.3 percent of affected fetuses.[48] At the Medical Centre of the Catholic University in Rome, of 304 cases of amniocentesis, all six fetuses (100 percent) diagnosed with abnormalities were aborted.[49] In South Australia, where abortion is legal and MSS is offered to all pregnant women, about 75 percent of women whose fetuses are confirmed to be affected by Down syndrome have an abortion.[50] But there is another side to the coin. In Victoria in 1997, of the 5295 women whose fetuses were given diagnostic tests, 94.1 percent were normal.[51] These results were reassuring for the vast majority of the women concerned and for whom abortion was not the outcome of prenatal diagnosis.

Another ethical problem is whether prenatal diagnosis per se amounts to material cooperation with the subsequent abortion of abnormal fetuses. A survey was conducted of 1,000 participants aged 16 to 44 in a cystic fibrosis (CF) carrier screening program in which they were asked what they would do if their future pregnancies were affected by CF. Their replies to the question of whether they would consider terminating an affected pregnancy were 26 percent yes and 17 percent no, but 57 percent replied they did not know what they would do.[52] This showed that only a minority of participants would consider terminating a future and presumably wanted pregnancy if the fetus was found to be affected by CF. Few would actually form an intention to terminate a wanted pregnancy until it was certain the fetus was found to be abnormal. Prenatal diagnosis, then, is not always nor necessarily linked to abortion. It provides information women have a right to seek. Unless a woman forms an intention to abort a fetus detected with an abnormality, prenatal diagnosis is ethically distinct from a subsequent decision to have an abortion.

There is, however, a remote material link to abortion in at least 4.3 percent of cases of prenatal diagnosis where the results indicate that the fetus is affected by a major chromosome anomaly.[53] Since most women have an abortion once it is confirmed that their fetuses are affected by Down syndrome, one could argue that prenatal diagnosis involves sufficient material cooperation with abortion to constitute collusion with it.

On the other hand, more weight should be given to the fact that in about 95.7 percent of cases, the results of prenatal tests indicate that fetuses are not affected by the major chromosome abnormalities for which they are tested. This shows that in the vast majority of cases, prenatal diagnosis is not per se materially linked to abortion. It is ethical, then, to perform prenatal diagnostic tests to provide accurate information on the health of their fetuses to women who request it. This is not substantially changed by the fact that most prenatal tests that detect Down syndrome do have a material link to abortion.

The above discussion on cooperation applies to institutions as well as to individuals. Hospitals that support the sanctity-of-life principle should not adopt a policy that formally permits or regulates cooperation in abortion after prenatal diagnosis. Their staff should take the necessary steps to guarantee that prenatal screening and diagnosis are guided by ethical principles and do not in practice encourage or condone selective abortion.

Genetic Screening and Morally Responsible Parenthood

The information gained from genetic screening and testing which confirms carrier status for a genetic disease has serious moral implications for responsible parenthood. One may learn of this after one is already married and a parent or before one marries or becomes a parent. There is a difference between screening to enhance the exercise of morally responsible parenthood and screening designed to prevent the birth of babies affected by severe congenital defects.[54] While it would be wrong to put pressure on a couple not to have children once they find out they are both carriers of a serious recessive genetic disease with a one-in-four risk of having an affected child, there are good reasons why such a couple might consider refraining from running this known risk. But I would hesitate to say there was a strict moral duty for such couples to avoid having children.

People who are known carriers of a major recessive genetic defect need to consider carefully the implications of entering committed relationships that could lead to marrying partners who are carriers of the same defect. Genuine seekers of love are not required to enter marriage while turning a blind eye to what could negatively affect their marriage relationships and the well-being of their children. Parents' obligations toward children with disabilities cannot be set aside.[55] If they marry noncarriers, none of their children would be affected but there would be a 50 percent risk of them being carriers. One of the reasons why Catholic Church law forbids marriage between close blood relatives (e.g., cousins) is to prevent inherited illnesses being transmitted to offspring.[56] Without implying a lack of regard for people with disabilities, for the sake of the future children, it would show a mature sense of responsibility for a carrier to marry a partner who does not share the same genetic defect.

People who are already affected by a serious dominant genetic disease or by one with a late onset need to consider whether to choose to assume the responsibility of marrying and having children with a one-in-two risk of having the same disability. It is not a question of it being better to be or not to be. An existing child with a genetic disability is of inestimable value. It is not a question of telling a person with a disability that they should not have been born. This would be extremely offensive. Nor would it be a case of unjust discrimination against a possible disabled child. An act of injustice can be committed only against an existing child, not against a hypothetical, nonexistent child. Morally responsible family planning implies that a couple ought not to risk conceiving a child with a dominant disease unless they are prepared to love and raise their child with dedication. Prior to making their decision, parents in this situation need to be adequately informed of the one-in-two risk of their child being affected, the usual age of onset of the condition, and the degree of likely suffering their future child would have to endure. To chose to have a child with a one-in-two risk of having to endure much suffering for most of their life would involve assuming an awesome responsibility.

Professional Confidentiality and the Common Good

The common good requires a presumption in favor of professional confidentiality until the contrary is proven. At the same time, the social nature of the human person obliges one not to abide by a presumed contract of confidence which would breach one's prior duty of justice to a third party or to the common good. The results of prenatal tests are personal and should be kept strictly confidential between the woman and her doctor and shared with other health professionals only if treatment so requires.

When information is obtained about oneself but which may also be true of another family member, family solidarity morally requires that an offer be made to share this information with the person concerned, especially if this knowledge would influence a relative's decision about becoming a parent with a high risk of the child inheriting a serious disease. The doctor should try to convince the person of his or her duty in such a case. Consider the rare case of a doctor who discovers that a fetus has a dominant, paternally inherited, single-gene defect such as Huntington's disease, who informs the mother, and who then finds out that the mother's partner is not the father of the child. This information should not be shared with other health professionals without the woman's consent. The genetic father ought to be advised of this fact so that he may make morally responsible decisions about fathering more children. The doctor should try to persuade the mother to inform the genetic father, but ultimately it is the mother's responsibility to act, not the doctor's, unless the mother has given informed consent for the doctor to do so.

Sensitivity for People with Disabilities

Persons with a disability are endowed with intact dignity and inviolable rights. They deserve love and care from their families and dedicated service from health professionals. Those who care for persons with disabilities find their own lives enriched as their personalities develop. Some couples, however, are only too acutely aware of the strain put on their marriages and families by the presence of a child with a disability. They are afraid that they might not be able to cope over the years with few prospects of respite. We cannot underrate the stress caused by the realization of the consequences that the long-term presence of a child with a disability may have on the family's daily life, its income, and life plans. These are a few of the reasons why some pregnant women have recourse to prenatal screening and diagnosis. They want to be reassured their fetus is normal, to be able to seek any available fetal therapies, or to prepare themselves to care for a child with a disability. But there is a perception in the community that there is a right to a child without defects. How right was Italy's Health Minister, Pia Garavaglia, when he said: "Desires are not rights. A child is not a consumer good."[57] The myth of the possibility of the perfect child for all must be dismissed.[58] One could hardly disagree with Dorothy Wertz, who said: "the availability of genetic tests must not be allowed to create an illusion that most disabilities are preventable and therefore unacceptable to society."[59]

The long-term impact on the community's consciousness of the social approval of the deliberate taking of the lives of defective fetuses so that subsequently others may be born to enjoy lives free of congenital disease is a cause of concern to people with a disability and cannot be ignored in policy-making in this regard. Liz Hepburn aptly remarked: "Paradoxically, we seem to be prepared to eliminate the very people before birth whom anti-discrimination legislation seeks to protect after birth."[60]

Prenatal screening and diagnosis may be ethical, but attention still needs to be given to how the prevention of disease may be achieved sensitively and without hurting people who are disabled. Priscilla Alderson reports that people with Down syndrome say that: "discrimination is their worst problem in preventing them from contributing to society as they can and want to."[61]

It is up to prenatal health care professionals to ensure that prenatal screening and diagnosis do not send a message of intolerance to people with a disability.[62] Lyn Gillam argues that prenatal diagnosis is not per se discriminatory against disabled people, but the way it is practiced should eliminate or reduce its potential negative effects on people with disabilities.[63] Christopher Newell questions "the dominance of a perspective which believes that a likelihood of a disability is so significant that it acts as a trump card regarding abortion."[64] Referring to the social nature of disability and genetics, he laments: "such technology will perpetuate the oppression and control of people with disability, especially if the knowledge of people with disability is not utilized in bioethical debates."[65]

Acknowledgments

A previous draft of this chapter was published as "Prenatal Screening and Diagnosis," in Norman Ford, *The Prenatal Person. Ethics from Conception to Birth* (Oxford: Blackwell, 2002). Republished with permission.

Notes

1. Michael Connor and Malcolm Ferguson-Smith, *Essential Medical Genetics* (Oxford; Blackwell Science, 1997), 118.
2. Ibid., 118.
3. Ibid., 118.
4. E. Hook, "Prevalence, Risks and Recurrence," in *Prenatal Diagnosis and Screening*, ed. David J. Brock, Charles Rodeck and Malcolm A. Ferguson-Smith (Edinburgh, U.K.: Churchill Livingstone, 1992), 364.
5. C. Bernadette Modell, "Screening as Public Policy," in *Prenatal Diagnosis*, ed. Brock, et al., 612–13.
6. Judith Searle, "Fearing the Worst—Why Do Pregnant Women Feel 'At Risk'?" *Australian and New Zealand Journal of Obstetrics and Gynaecology* 36 (1996): 279–86.
7. Ibid., 284.
8. See Nuffield Council on Bioethics, *Genetic Screening: Ethical Issues* (London: Nuffield Council on Bioethics, 1993), 3.
9. Lachlan de Crespigny, Meg Espie, and Sophia Holmes, *Prenatal Testing. Making Decisions in Pregnancy* (Ringwood, Victoria, Australia: Penguin Books, 1998), 99–104; Nuffield, *Genetic Screening*, 24–25, 112.
10. James E. Haddow, "Antenatal Screening for Down's Syndrome: Where Are We and Where Next?" *Lancet* 352 (1998):336–37; Nicholas J. Wald, et al., "Prenatal Screening for Down's Syndrome Using Inihibin-A as a Serum Marker," *Prenatal Diagnosis* 16 (1996):143–53, at 151; de Crespigny, *Prenatal Testing*, 102.
11. Carole Webley and Jane Halliday, eds., *Report on Prenatal Diagnostic Testing in Victoria 1997* (Melbourne, Australia: Murdoch Institute, 1998), 15.
12. See Report of the Royal College of Obstetrics and Gynaecology Working Party, *Ultrasound Screening for Fetal Abnormalities* (London: Royal College of Obstetrics and Gynaecology, 1997), 14.
13. See George Makrydas and Dimitrios Lolis, Letter, "Nuchal Translucency," *Lancet* 350 (1997):1630–31; de Crespigny et al., *Prenatal Testing*, 104–6.

14. Royal College of Obstetrics and Gynaecology Working Party, *Ultrasound Screening for Fetal Abnormalities*, 5–6.

15. Judith Lumley, "Uncertainty and Ultrasound Diagnosis," in *Ethical Issues in Prenatal Diagnosis and the Termination of Pregnancy*, ed. John McKie (Melbourne, Australia: Monash University Centre for Human Bioethics, 1994), 26–27; Nuffield, *Genetic Screening*, 23, 115.

16. Leonie C. Stranc, Jane A. Evans, and John L. Hamerton, "Chorionic Villus Sampling and Amniocentesis for Prenatal Diagnosis," *Lancet* 349 (1997):711–14.

17. M. E. Pembrey, "Genetic Factors in Disease," in *Oxford Textbook of Medicine*, 3rd. ed., Vol. I, ed. D. J. Weatherall, J. G. G. Ledingham, and D. A. Warrell (Oxford: University Press, 1996), 136; Jane Halliday, et al., "Importance of Complete Follow-up of Spontaneous Fetal Loss after Amniocentesis and Chorionic Villus Sampling," *Lancet* 240 (1992):886–90.

18. Connor and Ferguson-Smith, *Essential Medical Genetics*, 118 and Appendix I; Jane Wheatley, "What If. . . ." *HQ* (Mar-April 1997):44.

19. For more details, see E. Pergament and B. Fine "The Current Status of Chorionic Villus Sampling," in *Preconception and Preimplantation Diagnosis of Human Genetic Disease*, ed. R. G. Edwards (Cambridge, U.K.: Cambridge University Press, 1993), 145–49.

20. Barbara Katz Rothman, *The Tentative Pregnancy: Prenatal Diagnosis and the Future of Motherhood* (New York: Viking Press, 1986), 101–11.

21. L. H. Kornman et al., "Women's Opinions and the Implications of First- versus Second-Trimester Screening for Fetal Down's Syndrome," *Prenatal Diagnosis* 17:11 (1997):1011–18.

22. Pembrey, "Genetic Factors in Disease," 136; N. Silverman and R. Wapner, "Chorionic Villus Sampling," in *Prenatal Diagnosis*, ed. Brock et al., 30–32; E. Hook, "Prevalence, Risks and Recurrence," in *Prenatal Diagnosis*, ed. Brock et al., 364; Stranc et al., "Chorionic Villus Sampling," 711–14; Halliday et al., "Importance of Complete Follow-up of Spontaneous Fetal Loss," 887.

23. C. Matthews and Ke-hui Cui, Letter, *Nature* 366 (1993):117–18.

24. R. G. Edwards and Helen K. Beard, "Sexing Human Spermatozoa to Control Sex Ratios at Birth Is Now a Reality," Editorial, *Human Reproduction* 10:4 (1995):977–78; Amnon Botchan, et al., "Sperm Separation for Gender Preference: Methods and Efficacy," *Journal of Andrology* 10:2 (1997):107–8.

25. Alan H. Handyside, "Preimplantation Genetic Diagnosis: A 10-Year Perspective," in *Towards Reproductive Certainty; Fertility and Genetics beyond 1999*, ed. Robert Jansen and David Mortimer (New York: Parthenon Publishing Group 1999), 389–96; Stuart A. Lavery and R. M. L. Winston, "Clinical Experience with Preimplantation Genetic Diagnosis at

Hammersmith 1989–1998," in *Towards Reproductive Certainty; Fertility and Genetics beyond 1999*, ed. Jansen and Mortimer, 397–404.

26. See Handyside, "Preimplantation Genetic Diagnosis," 390–93; Y. Verlinsky and C. M. Strom, "Preconception Diagnosis of Polar Bodies," in *Preconception*, ed. Edwards, 233–38.

27. See Y. Verlinsky and N. Ginsberg, "A Brief History of Prenatal Diagnosis," in *Preconception*, ed. Edwards, 138–40.

28. For more details, see Joshua A. Copel and Charles S. Kleinman, "Fetal Arrhythmias," in *Fetal Therapy. Invasive and Transplacental*, ed. Nicholas M. Fisk and Kenneth J. Moise Jr. (Cambridge, U.K.: Cambridge University Press, 1997), 184–98.

29. Robin Heise Steinhorn, "Prenatal Ultrasonography: First Do No Harm?" *Lancet* 352 (1998):1568–69.

30. See also A. S. Moraczewski, ed., *Genetic Medicine and Engineering: Ethical and Social Dimensions* (St. Louis, Mo.: Catholic Health Association of USA and Pope John Paul XXIII Medical-Moral Research Centre, 1983), 131–43, where G. M. Atkinson generally favors prenatal diagnosis.

31. Lachlan de Crespigny, "Prenatal Diagnosis: The Australian Clinical Situation," in *Ethical Issues in Prenatal Diagnosis and the Termination of Pregnancy*, ed. McKie, 13.

32. M. d'A. Crawfurd, "Ethical Guidelines in Fetal Medicine," *Fetal Therapies* 2 (1987):178.

33. John Paul II, Encyclical Letter, *Evangelium Vitae, The Gospel of Life*, no. 63: "When they do not involve disproportionate risks for the child and the mother, and are meant to make possible early therapy or even to favor a serene and informed acceptance of the child not yet born, these techniques are morally licit." For more details of Catholic teaching on prenatal diagnosis, see Congregation for the Doctrine of the Faith, *Instruction on Respect for Human Life in its Origin and on the Dignity of Procreation: Replies to Certain Questions of the Day* (Vatican City: Vatican Polyglot Press, 1987), I, 2.

34. See W. Holzgreve et al., "Benefits of Placental Biopsies for Rapid Karyotyping in the Second and Third Trimesters (Late Chorionic Villus Sampling) in High Risk Pregnancies," *American Journal of Obstetrics and Gynaecology* 162 (1990):1188–92, where there were no miscarriages in 498 CVS cases.

35. Pope John Paul II, *Evangelium Vitae*, n. 63.

36. H. Bekker et al., "Uptake of Cystic Fibrosis Testing in Primary Care: Supply Push or Demand Pull?" *British Medical Journal* 306 (1993):1584–86. Of 5529 people invited to take part, only 957 (17 percent) accepted, 70 percent of whom responded to active opportunistic contact and immediate testing. In this trial, twenty-eight carriers and no carrier couples were detected.

37. See Susan Michie and Theresa Marteau, "Genetic Counselling: Some Issues of Theory and Practice," in *The Troubled Helix: Social and Psychological Implications of the New Human Genetics*, ed. Theresa Marteau and Martin Richards (Cambridge, U.K.: Cambridge University Press, 1996), 104-22.

38. See Mary Terrell White, "Making Responsible Decisions: An Interpretative Ethic for Genetic Decision Making," *Hastings Center Report* 29:1 (1999):14-21.

39. Dorothy C. Wertz, et al., *Guidelines on Ethical Issues in Medical Genetics and the Provision of Genetics Services* (Geneva, Switzerland: World Health Organization Hereditary Diseases Program, 1995), 26.

40. Theresa M. Marteau, "Towards Informed Decisions about Prenatal Testing: A Review," *Prenatal Diagnosis* 15 (1995):1215-18.

41. Helen Statham and Josephine Green, "Serum Screening for Down's Syndrome: Some Women's Experiences," *British Medical Journal*, 307 (1993):174; Helen Statham, Josephine Green, and Clare Snowdon, Letter, *British Medical Journal* 306 (1993):858-59.

42. Wertz et al., *Guidelines on Ethical Issues in Medical Genetic*, 17-18; 22.

43. M. d'A. Crawfurd, "Ethical Guidelines in Fetal Medicine," *Fetal Therapies* 2 (1987):176; John M. Thorp, Jr., and Watson A. Bowes Jr., "Pro-Life Perinatologist—Paradox or Possibility?" *New England Journal of Medicine* 326 (1992):1217-19.

44. Statham and Green, "Serum Screening for Down's Syndrome," 174-76.

45. See Chris Goodey, Priscilla Alderson, and John Appleby, "The Ethical Implications of Antenatal Screening for Down's Syndrome," *Bulletin of Medical Ethics* 147 (1999):13-17, at 14.

46. John Paul II, in *Evangelium Vitae*, no. 63, strongly disapproves of prenatal tests whose purpose is to detect abnormal fetuses for abortion: "it not infrequently happens that these techniques are used with a eugenic intention which accepts selective abortion in order to prevent the birth of children affected by various types of anomalies. Such an attitude is shameful and utterly reprehensible, since it presumes to measure the value of a human life only within the parameters of 'normality' and physical well-being, thus opening the way to legitimizing infanticide and euthanasia as well."

47. Webley and Halliday, *Report on Prenatal Diagnostic Testing in Victoria 1997*, 11.

48. de Crespigny, et al., *Prenatal Testing*, 130.

49. Elio Sgreccia, *Bioetica: manuale per medici e biologi* (Milano: Vita e Pensiero, 1986), 151-52.

50. Jane Wheatley, "What If. . . ." 44.

51. Webley and Halliday, *Report on Prenatal Diagnostic Testing in Victoria 1997*, 1.

52. E. Watson, et al., "Screening for Genetic Carriers of Cystic Fibrosis through Primary Health Care Services," *British Medical Journal* 303 (1991):504–7.

53. Webley and Halliday, *Report on Prenatal Diagnostic Testing in Victoria 1997*, 11; see also Fletcher and Evans, "Ethical Issues in Reproductive Genetics," *Seminars in Perinatology* 22:3 (1998):191, where it is stated that positive results do not usually exceed 4 percent.

54. Editorial, "Screening for Cystic Fibrosis," *Lancet* 340 (1992):209–10.

55. Wertz et al., *Guidelines on Ethical Issues in Medical Genetics*, 37.

56. Canon Law Society of Great Britain and Ireland, *The Canon Law, Letter and Spirit* (Alexandria, New South Wales, Australia: E. J. Dwyer, 1995), can. 1091.

57. Quoted in Jonathan M. Berkowitz and Jack W. Snyder, "Racism and Sexism in Medically Assisted Conception," *Bioethics* 12 (1998):43.

58. For a comprehensive discussion of all the relevant ethical questions pertaining to medical genetics and genetic counseling, see Wertz et al., *Guidelines on Ethical Issues in Medical Genetics*, 52–70.

59. Wertz et al., *Guidelines on Ethical Issues in Medical Genetic*, 61.

60. Liz Hepburn, "Genetic Counseling: Parental Autonomy or Acceptance of Limits?" *Concilium* 275 (1998):40.

61. Goodey, Alderson, and Appleby, "The Ethical Implications of Antenatal Screening for Down's Syndrome," 16.

62. Paul Robinson, "Prenatal Screening, Sex Selection and Cloning," in *A Companion to Bioethics*, ed. Helga Kuhse and Peter Singer (Oxford: Blackwell Publishers, 1998), 173–85.

63. See Lynn Gillam, "Prenatal Diagnosis and Discrimination against the Disabled," *Journal of Medical Ethics* 25 (1999):163–71.

64. See Christopher Newell, "The Social Nature of Disability, Disease and Genetics: A Response to Gillam, Persson, Holtug, Draper and Chadwick," *Journal of Medical Ethics* 25 (1999):174.

65. Ibid., 174.

15

Gene "Therapy": A Test Case for Research with Children

M. Therese Lysaught, Ph.D.

THE CULTURAL FACE of gene therapy is that of a child. The image of a child has served as the secular icon of the mystery and promises of human gene transfer since its inception in 1990.[1] The particular face of the child changes, determined by the message to be communicated. It is the face of Ashanti De Silva, her identity long obscured behind the veil of confidentiality, who at four years old was the first subject enrolled in a human gene transfer protocol with therapeutic intent. Her identity revealed after years in the protocol, she stands as a symbol of hope and promise. It is a countenance like that of Jacob Sontag, his worried face gracing the cover of the *New York Times Magazine*, drawing observers attention to the desperation of parents and faith in the salvific powers of human gene transfer.[2] It is the image of an Amish child, floating surreally in a vivid sea of ultraviolet light, pointing to the numinous power of human gene transfer and its ability to transcend the most intransigent of social boundaries.[3]

Those familiar with the history of human subjects research and the controversies surrounding recombinant DNA in the late 1970s and 1980s ought to find this iconography of particular interest. For it is a striking phenomenon: the emergence of an entirely new modality of clinical intervention, methods untested, risks unknown yet publicly feared, whose earliest human subjects were children.[4] Moreover, as iconic, these images often function to forestall critique, to displace argument, to garner public support—and public monies—for human gene transfer research.

Despite the high visibility of children in the development and justification of the field of human gene transfer, the topic remains an "orphan issue."[5] A full examination of the use of child subjects in human gene transfer research remains beyond the scope of this essay.[6] Such an account would attend to questions such as: How did this come to pass? What social, political, and/or rhetorical factors account for the fact that children were enrolled in human gene transfer so early in the development of such a controversial field, in fact in only the second clinical trial? How many children have been enrolled in clinical trials of human gene transfer? What have these trials entailed? How have these protocols fitted with the federal regulations and federal and professional policies governing the use of child subjects in research?

Two recent events suggest the timeliness of the investigation. First, there is increasing public pressure to enroll children as subjects in clinical trials. In November 1998, the Food and Drug Administration (FDA) announced a major change in the rules governing the process by which pharmaceutical companies will receive approval for new medications. As will be discussed below, this change requiring companies to submit data on the safety and effectiveness of new drugs and vaccines in children occurred largely as an effect of public, political, and professional pressure for a more aggressive and, some argued, a more just approach to developing therapies for fatal childhood diseases.[7] At the same time, however, questions are being raised about compliance with established guidelines and the effectiveness of public oversight of human gene transfer research. Following in the wake of the tragic and troubling death of Jesse Gelsinger at the University of Pennsylvania in September 1999, emerging findings suggest lack of compliance with enrollment criteria, failure to report of adverse events, and a deeply problematic informed-consent process.[8]

As a first step toward a broader examination of the role of child subjects in human gene transfer research, this essay will examine the tensive interface between federal and professional guidelines governing research with child subjects and research practice. Guidelines and oversight mechanisms exist; how are they interpreted, operationalized, and implemented? These questions will be examined by displaying one particular human gene transfer protocol—one that proposed to enroll HIV-positive children as subjects—as it negotiated the process of oversight and approval. Focusing on a particular protocol provides an opportunity for concrete display of the scientific, clinical, and political dimensions of human gene transfer research involving children. At the same time, it highlights both ambiguities in the guidelines themselves as well as the tendency of research practice to challenge straightforward criteria. In order to provide a framework with which to examine the protocol, I begin with a brief history and overview of current regulations and policies governing research with children as human subjects.

CHILDREN AS RESEARCH SUBJECTS: FEDERAL REGULATIONS AND POLICY

Two competing and legitimate tensions govern the question of whether and how to conduct research with children. On the one hand, the historical witness of the abuse of vulnerable human subjects within the U.S. medical system, especially in conjunction with government involvement, has justifiably led to a protectionist stance toward research involving children.[9] Conservative, cautious safeguards are necessary to protect children, who cannot consent and who may experience the burdens of medical intervention as more frightening and excruciating than adults, from being unduly exploited by the interests of researchers, the desperation of their parents, and the relentless momentum of the research imperative. As the recent FDA ruling suggests, however, the legitimate need to safeguard particular children must be balanced with the interests of justice, particularly the need to advance the medical care of children. These tensions inform current federal and professional regulations and policy concerning research involving children.[10]

The history of the federal regulations governing research on children reflects the concern to protect child subjects. Public concern over the cases mentioned above led in part to the congressional institution, in 1974, of the National Commission for the Protection of Human Subjects of Biomedical and Behavioral Research. In the same year, the Department of Health, Education and Welfare (DHEW; now known as the Department of Health and Human Services, DHHS) issued its first regulations for the protection of human subjects in research.[11] The first DHEW/DHHS regulations covering research involving children appeared in 1978.[12] Following the counsel of the 1977 report *Research Involving Children* published by the National Commission for the Protection of Human Subjects in

Biomedical and Behavioral Research,[13] the DHHS published its final regulations, providing "additional protections for children involved as subjects in research" on March 8, 1983. These regulations became effective on June 5, 1983, codified as 45 CFR 46, Subpart D.[14]

The precepts of Subpart D have guided research on children for the past sixteen years. These precepts distinguish five categories of research to clarify when children may be enrolled as research subjects. These categories are:[15]

§46.401 [Exempt research].

§46.404 Research not involving greater than minimal risk.

§46.405 Research involving greater than minimal risk, but presenting the prospect of direct benefit to the individual subjects.

§46.406 Research involving greater than minimal risk and no prospect of direct benefit to individual subjects, but likely to yield generalizable knowledge about the subject's disorder or condition.

§46.407 Research not otherwise approvable, which presents an opportunity to understand, prevent, or alleviate a serious problem affecting the health or welfare of children.

As is evident from even a brief scan of these categories, the anchoring concept of the regulations is the concept of risk, articulated operationally as "minimal risk." Only once the level of risk is established may deliberations regarding the acceptability of enrolling children move to assessing the prospect of benefit; correlatively, should risk be minimal or less, the notion of benefit to individual child subjects becomes irrelevant. As Janofsky and Starfield note: "only if the IRB determines that a proposed project will in fact entail greater than minimal risk must its members additionally address other issues, such as whether there is prospect of direct benefit to individual subjects. . . . Thus, decisions about risk must be made separately from and before any consideration of possible benefit is judged."[16]

"Minimal risk" is defined earlier in the federal regulations as follows: "'Minimal risk' means that the risks anticipated in the proposed research are not greater, considering probability and magnitude, than those ordinarily encountered in daily life or during the performance of routine physical or psychological examinations or tests. (§46.102G)"[17]

The definition for "greater than minimal risk" emerges from §46.406. Research that entails "greater than minimal risk" may be permitted in certain situations, but limits are still set: "The intervention or procedure presents experiences to subjects that are reasonably commensurate with those inherent in their actual or expected medical, dental, psychological, social or educational situations."

As will be argued below, human gene transfer protocols proposed to date fall clearly entail "greater than minimal risk" and so must be considered under §§46.405, 46.406, or 46.407.

But as anyone who has sat on an institutional review board (IRB) knows well, determining how to categorize a specific protocol is more an art than a science. For it is in the concrete negotiations of IRB approval that the concern for protectionism competes with the legitimate need to conduct research on children. Even before Subpart D was issued, pediatricians argued that restricting research with children would compromise pediatric medical care. As early as 1968, H. C. Shirkey coined the term "therapeutic orphan," arguing that a reluctance to test the safety and efficacy of drugs on children could dangerously inhibit the development of needed pharmaceutical interventions for children, handicapping pediatric medicine.[18]

Shirkey's concerns were not unfounded. As the Committee on Drugs of the American Academy of Pediatrics (AAP) recently reported:

> A survey of the 1973 *Physician's Desk Reference* revealed that prescribing information for 78% of medications included a disclaimer or lack of dose information for use by children. A subsequent survey of the 1991 *Physician's Desk Reference* showed that 81% of listed drugs contained language disclaiming use in children or restricting use to certain age groups. A survey of new molecular entities approved by the Food and Drug Administration (FDA) from 1984 through 1989 revealed that 80% were approved without labeling for children. In 1992, 19 new molecular entities with potential use in children were approved by the FDA; 79% of these drugs were not labeled for use in children. In most instances drugs are not labeled for use in children because sufficient studies have not been conducted in children.[19]

Adult studies, however, are not sufficient. As Grodin and Alpert note: "children are not little adults."[20] The AAP concurs:

> Growth, differentiation, and maturation can alter the kinetics, end organ responses, and toxicities of drugs in the newborn, infant, child,

or adolescent compared to the adult. Drug studies in adult humans may not adequately predict the pharmacokinetic, pharmacodynamic, or toxic properties of drugs in children. When drugs have been administered to children without sufficient pharmacology studies to identify the optimal therapeutic approach, children have occasionally suffered severe toxic effects, including death. These toxic effects could have been avoided with some of the drugs if appropriate drug studies had been undertaken before their widespread use in children.[21]

Given these significant physiological issues, coupled with the crisis in the pharmacological armamentarium, the AAP calls for children to be included "in clinical studies of a drug when the drug offers potential benefits to them."[22]

The recommendation of the AAP recently became public policy. Public pressure, channeled through the lobbying power of Congress and the Clinton White House, led to the recent issue of the National Institutes of Health (NIH) Policy and Guidelines on the Inclusion of Children as Participants in Research Involving Human Subjects. Effective October 1, 1998, the purpose of the policy is explicit: "to increase the participation of children in research so that adequate data will be developed to support the treatment modalities for disorders and conditions that affect adults and may also affect children."[23] The policy holds as follows:

It is the policy of the NIH that children (i.e., individuals under the age of 21) must be included in all human subjects research, conducted by the NIH, unless there are scientific and ethical reasons not to include them . . . proposals for research involving human subjects must include a description of plans for including children. If children will be excluded from the research, the application or proposal must present an acceptable justification for the exclusion. . . . The inclusion of children as subjects in research must be in compliance with all applicable subparts of 45 CFR 46 as well as with other pertinent federal laws and regulations.

Following the lead of the NIH, on November 27, 1998, the FDA issued new rules (effective April 1, 1999) requiring pharmaceutical companies seeking FDA approval for new drugs and vaccines to submit data on the safety and effectiveness of these modalities in children "if the product is likely to be used in a substantial number of pediatric patients" or if it provides a "meaningful therapeutic benefit" over existing treatments for children of similar ages.[24] The FDA guidelines, as might be expected, met with strenuous objections from drug companies, who deemed them "impractical and burdensome," costly, and, surprisingly, "unethical, because they might put children at risk."

In the near term, then, it seems that the balance between protectionism and expansionism is shifting toward those who advocate increased involvement of children in biomedical research. Yet even those who recognize the need for research with children recognize that requirements for the inclusion of children might place children at an unacceptable risk. In light of this, a consensus has emerged among federal, professional, and academic commentators alike confirming a long-standing precept in research with child subjects: the necessary priority of research on adults. A survey of statements reveals this consensus.

The 1998 NIH policy outlines a series of seven "justifications for exclusions" by which researchers might legitimately avoid the requirement to conduct studies on children. Of these, the fifth is most pertinent to our discussion.

> 5. Insufficient data are available in adults to judge potential risk in children (in which case one of the research objectives could be to obtain sufficient adult data to make this judgment). While children usually should not be the initial group to be involved in research studies, in some instances, the nature and seriousness of the illness may warrant their participation earlier based on careful risk and benefit analysis.

This caveat first appears in the National Commission's report, where Recommendation 2B states:

> Where appropriate, studies have been conducted first on animals and adult humans, then on older children, prior to involving infants.... Whenever possible, research involving risk should be conducted first on animals and adult humans in order to ascertain the degree of risk and the likelihood of generating useful knowledge. Sometimes this is not relevant or possible, as when the research is designed to study disorders or functions that have no parallel in animals or adults. In such cases, studies involving risk should be initiated on older children to the extent feasible prior to including infants, because older children are less vulnerable and they are better able to understand and to assent to participation. In addition, they are more able to communicate about any physical or psychological effects of such participation.[25]

As Robert Levine notes in his commentary on the report, "Investigators who proposed to do research on children without having first done such research on animals and/or adults will be obligated to persuade the IRB that this is necessary. Suitable justification might be that the disorder or function to be studied has no parallel in animals or adults."[26]

Although this recommendation was not included specifically as a regulation in Subpart D, the DHHS states in the Preamble to the 1983 regulations that it expects this recommendation to be followed.[27] In their 1994 work *Children as Research Subjects*, Michael A. Grodin and Leonard H. Glantz propose a set of "Points to Consider in Proposing or Reviewing Research Involving Children." The first of these points asks the following questions:

1. Is the use of children as research subjects justified in this instance?

    (b) if the research question can be addressed first in adults, has research with adults been conducted?

    (c) has the adult research produced results that would indicate that the proposed research would benefit, or not be harmful to, the children?[28]

Finally, the AAP concurs: "In most cases, studies in children should be preceded by initial clinical trials in adults to provide preliminary pharmacokinetic, safety, and efficacy data. In some instances, drugs intended to treat specific diseases that primarily or exclusively occur in children may be studied initially in children."[29]

This prioritization is reflected in the commentary on the regulations found in the *IRB Guidebook* published by the NIH's Office for Protection from Research Risks (OPRR):

Phase 1 Trials. The issue of Phase 1 drug studies deserves special consideration. The usual approach to designing drug studies involving children as subjects is for appropriate studies to be conducted first in animals, adults, and older children before young children are involved as research subjects. There are some studies, however, in which data may not be entirely generalizable from older populations, and in which the existence of life-threatening conditions for children are important considerations in the IRB's risk/benefit analysis. The requirement for previous testing in adults or older children may thus not be appropriate. Furthermore, some diseases specific to children may require that children be involved without data from older groups (e.g., there is no adult model that mimics the state of HIV-infected newborns; Wilms' tumor and various cancers such as neuroblastoma affect infants who do not survive into older childhood). In some cases "tandem" studies in older populations and children may be justifiable. For example, some Phase 1 studies in children might be based on only pharmacologic safety and toxicity data (completed Phase 1 and ongoing Phase 2) but without complete effectiveness data from trials in adults and older children. If the IRB approves a Phase 1 drug trial, the consent document must specify what is known about the probability that, and the degree to which, an intervention will be of possible benefit based on all of these data.[30]

Note here the meaning of "tandem studies" as used by the OPRR: researchers are to initiate Phase I trials in children when Phase I trials have been concluded in adults and adult research has moved to Phase II trials. As we will see, this is an important caveat vis-à-vis human gene transfer research, given that almost all human gene transfer clinical trials to date have been Phase I trials.

CHILDREN AND HUMAN GENE TRANSFER RESEARCH: THE HIV PROTOCOL

While these guidelines may seem relatively straightforward and seem to provide a well-crafted, prudent, and thoughtful balance between protectionism and the legitimate need to enroll children in clinical trials, in practice their application is much more contested. Key terms and criteria of the guidelines are ambiguous to the point of admitting what seems like almost any interpretation. What counts as "minimal risk" or, if that can be established, a "minor increment" over minimal risk? What constitutes a "prospect" of benefit? Is this different from a "possibility," a "hope," an "intention"? How important must knowledge be to be "vital"? While common sense might provide relatively straightforward answers to these questions in the context of a particular protocol, clinical experience may often lead investigators to assess notions such as "risk," "prospect," and urgency differently. In cases where the guidelines are so clear as to raise questions about a particular research endeavor on ethical grounds, reviewers and committees often find themselves assailed with charges of ignorance or obstructionism, or worse, the ultimate argument stopper, with imperiling the lives of dying children.

Clearly, these issues are not unique to human gene transfer research. The concrete display of the applicability of these guidelines in the context of a human gene transfer protocol, however, simultaneously illumines issues surrounding the guidelines and issues surrounding the conduct of human gene transfer research. For the purposes of this essay, I will focus on a protocol for a Phase I trial entitled: "Transduction of CD34(+) Cells from the Bone Marrow of HIV-1 Infected Children: Comparative Marking by an RRE Decoy Gene and a Neutral Gene."[31] An analysis of the protocol in light of the guidelines as well as a partial narrative of its progress through the review process will highlights the issues identified above.

I was first asked to review this protocol as part of the consolidated review process by the Recombinant DNA Advisory Committee of the National Institutes of Health (RAC) in November 1995.[32] My charge was to determine whether or not the protocol could be approved through consolidated review or whether it warranted review by the entire committee at one of its quarterly meetings.[33]

In this protocol, the investigator proposed to enroll five HIV-1-positive children between the ages of three and eighteen.[34] Through the protocol, the researchers sought to "determine whether the transduced CD34+ cells engrafted and produced peripheral blood leukocytes [T cells and their progenitors] that would have a selective survival advantage." The researchers hypothesized that the transduced gene would make the T cells resistant to HIV infection; they would then produce offspring T cells that would be likewise resistant to infection. Such resistance might, in theory, slow progression of the disease course.

The protocol proposed to subject the child subjects to the following clinical procedures:

- inclusion assessments, including phlebotomy, electroencephalogram, electrocardiogram, and chest X ray;
- bone marrow aspirate to obtain cells for stromal growth;
- preoperative screening assay;
- bone marrow harvest under general anesthesia (10–15 cc/kg);

- cell transduction: CD34+ T lymphocytes were to be isolated and transduced with two different retroviral vectors, one an anti-HIV-1 gene (L-RRE-neo) and the other a neutral marking gene (LN);
- infusion of transduced cells;
- overnight hospitalization;
- post-"treatment" assays, including phlebotomy (approximately ten over a two-year period); and
- a possible additional one or two bone marrow aspirates.

In my (albeit brief) history with the RAC at this point, it seemed clear that this protocol raised certain flags that might ordinarily have triggered the full review process:

It was the first anti-HIV protocol presented to the RAC that proposed to use children as research subjects. Generally, new diseases or new study populations would automatically trigger full RAC review. Four protocols following a similar strategy of intracellular replication inhibition for CD4+ or CD34+ cells had been approved for Phase I clinical trials in adults (9309–057, 9503–103, 9508–117, and 9511–134); three had just been approved in 1995 (March, August, and November respectively). None had yet issued data. None was referenced in the protocol. Hence, while the RAC had reviewed HIV protocols, it had not reviewed an HIV protocol involving children.

None had used this particular retroviral vector construct. Generally, new vectors would trigger full RAC review. A similar RRE decoy vector had been proposed in the adult protocol approved in November (9511–134). Other protocols had used a similar strategy of intracellular replication inhibition of CD34+ or CD4+ lymphocytes of HIV-positive patients. But the vector proposed here was a new construct.

No animal studies were referenced. This would not have been an unusual situation in and of itself. Many human gene transfer protocols at this point cited the lack of availability of good animal models. Human HIV is notorious for lack of a good animal model. However, this issue was not addressed in the protocol.

The first of these two flags would have served as a sufficient trigger for many protocols reviewed by the full committee. However, reading the protocol against the background of the federal guidelines for research on children raised a host of other issues that seemed worthy of fuller discussion. While a complete review of the protocol is beyond the constraints of this essay, three areas are worthy of highlight: the degree of risk presented to child subjects by the protocol; the "prospect" of benefit to the child subjects; and the lack of adult data.

As discussed above, the key concept anchoring the federal guidelines is the nature and degree of risk presented to the child subjects by the protocol. Risk assessment for child subjects must take into account not only risks of physical harm but nonphysical "burdens" the subjects will bear as well. In the case of children, these "burdens" take on greater weight, as children cannot truly consent to bear them. Hence, the determination of risks and burdens for this particular protocol would be required to address at least the following:

- the spectrum of preinclusion, preoperative, and follow-up screening procedures;
- the risks associated with sedation and general anesthesia in children and specifically in HIV-positive children;
- the risks associated with the bone marrow harvest procedure — risks of hemorrhage or infection during or after the procedure; risks of osteomyelitis; the possibility that small children might require postharvest blood transfusions to replace blood and marrow lost during the harvest;
- the burden of additional hospitalization and additional medicalization of the lives of these children, including at least two overnight hospitalizations and at least ten follow-up visits;
- the risks and burdens of one bone marrow aspiration with the possibility of one to two additional aspirations;
- the risks associated with the reinfusion of the genetically modified cells, including (as stated in the consent form, Protocol p. 55) "fever, chills, difficult breathing, and rarely, a severe allergic reaction that can lead to death";
- the standard risks associated with human gene transfer, including unpredicted vector integration leading to cancer and germ-line issues.

Given this list, one begins with the question: Does this protocol entail "greater than minimal risk"? For the risks to be categorized as "minimal," as noted above, "the risks anticipated in the proposed research [must not be] not greater, considering probability and magnitude, than those ordinarily encountered in daily life or during the performance of routine physical or psychological examinations or tests." Would the above present experiences that these particular children would likely encounter in daily life or in the course of routine examinations? Clearly not. Hence a relatively easy assessment is made that this protocol presents "greater than minimal risk." If so, it must be considered under §§45 CFR 405, 406, or 407.

§45 CFR 405, as noted above, raises the question of "prospect" of benefit. If the risks presented by the protocol are deemed "greater than minimal," research is approvable if it can be demonstrated that there is a "prospect of direct benefit to the individual subjects." As the guidelines note: "The risk is justified by the *anticipated* benefit to the subjects; and the relation of the *anticipated* benefit to the risk is at least as favorable to the subjects as that presented by available alternative approaches."

Therefore, to justify the more than minimal risk and burden, the investigators in this case must be able to demonstrate that there is a reasonable "prospect of direct benefit" to the child subjects to be enrolled, that benefits are *anticipated*.

What sort of case did the investigators present? When turning to the text of the protocol itself, one finds some rather clear clues:

> Although *we do not think that this initial study is likely to have a significant medical benefit for the patient*, we believe that it is safe and may provide useful information for progressing to potential beneficial treatments for AIDS. (Emphasis mine; Protocol, nontechnical abstract; similar statement included in the scientific abstract.)

> While this relatively low-risk protocol is *not likely to have significant clinical benefit*, it may provide useful information on the feasibility and potential efficacy of this approach. Positive results in this study may allow consideration of future studies entailing higher risks but higher potential benefits. (Emphasis mine; Protocol, p. 30.)

> Although patients will have routine monitoring of the disease status performed as part of their standard clinical care, it is not a primary endpoint of the study. (Protocol, p. 25.)

The protocol is relatively clear that *no benefit* is anticipated to accrue to the child subjects. Moreover, the protocol was submitted as a Phase I clinical trial, which by definition is not designed to offer benefit.

The audience, however, makes a difference in how the case is presented. When the investigators turn to address the RAC directly, the above disavowals of anticipation and prospect of clinical benefit are offset with a disclaimer about theoretical possibilities:

> While the *possibility* of benefit from the protocol is *unknown*, the *potential* benefits *could be significant*. Slowing or preventing immunologic deterioration could be life-extending or even *life-saving*. (Emphasis mine; Response to NIH "Points to Consider," M-II-A-1, p. 23.)

> In this Phase I study, *it is not known* whether the subjects will receive any benefits. The major goal of the study is to determine if this gene therapy approach is safe and feasible in patients with HIV infection. It is possible that the presence of the RRE decoy gene in some blood cells will allow them to avoid active infection by HIV. This could help maintain immune function and lessen the risk of infection. . . . While we have attempted to not overstate this possibility, it certainly is *hoped* by the investigators that this would be the result. (Emphasis mine; Response to NIH "Points to Consider," M-II-3, p. 26.)

Here, theory takes the place of evidence; hope has replaced warranted anticipation.

A similar shift in rhetoric appears in the Informed Consent form. Here the emphasis on the positive possibilities becomes even more pronounced. The first sentences of the consent form state:

> You/your child [are]/is infected with human immunodeficiency virus (HIV) and are/is eligible to participate in a medical study of gene therapy as a possible *treatment*. A possible form of *treatment* for HIV is gene therapy, where a new gene is put into a patient's cells. . . . If the RRE decoy can bind the HIV protein, *it might prevent the virus from growing in the body*. (Emphasis mine; Protocol, Informed Consent Form, p. 49.)

> It is *not known* whether gene therapy for HIV will be effective. It may be difficult to get the gene into patient's bone marrow stem cells, the gene may not be turned-on in the cells in the body, or the gene may not actually protect the cells. The purpose of this study is to test whether gene therapy with the RRE decoy gene can be performed safely, whether it can get the RRE decoy gene into the cells and whether the RRE decoy gene will inhibit HIV infection of cells. (Emphasis mine; Protocol, Informed Consent Form, p. 50.)

The "Potential Benefits" section of the consent form includes this less optimistic assessment, but alters it significantly with the closing qualification:

> We *do not know* if there will be any direct benefits to you/your child from participating in the study. The major goal of the study is to determine if this gene therapy approach is safe and feasible in patients with HIV infection. . . . If the RRE decoy gene were present in sufficient numbers of cells, *this could help maintain immune function and lessen the risk of infection.* (Emphasis mine; Protocol, Informed Consent Form, p. 51.)

The therapeutic possibilities dangle before the eyes of parents with HIV-stricken children. What parent would not provide their child with this chance?

The institutional IRB noted the ambiguity in this language and that the data did not support the claim to *prospect, probability,* or *likelihood* of benefit. For this reason, upon its initial review, the IRB provided a lengthy critique of the protocol and deferred it.[35]

The investigators made two responses to this critique. First, they pointed to the more positive statements above and argued that "there is *possibility* of direct benefit to the subjects, *if* the hypotheses being tested are found to be correct" (emphasis mine; Protocol, Investigator's Response to IRB, p. 15). The members of the IRB noted that there were a number of "ifs" that would have to occur for possibility to accrue. The investigator notes that he and his colleagues "certainly hope" that this would be the result. However, as Kathryn Whartenby of the FDA noted in her recommendation that this protocol receive full RAC review, there is little proof to substantiate this statement or this hope.[36] In fact, some of the data presented in the protocol seems to argue against clinical benefit.[37]

Secondly, the investigator argued that in the nontechnical abstract, he "thought it best to be pessimistic to not recruit patients with false claims" (Response from Investigator, p. 6).[38] This statement seems difficult to reconcile with the qualified-but-optimistic rhetoric in the informed consent form. In other words, when addressing the scientific community, the investigator seems to be more pessimistic and realistic, acknowledging that the potential benefits of this research will accrue not to the child subjects themselves but to progress in the battle against AIDS. When addressing reviewers and prospective parents, however, realism gives way to hope.

However, even though the "prospect" and "possibility" of direct benefit to enrolled child subjects may have been remote in this case, the research might still have been approvable under §46.406, which allows for "research involving greater than minimal risk and no prospect of direct benefit to individual subjects, but likely to yield generalizable knowledge about the subject's disorder or condition." Again, a number of conditions obtain:

> The risk represents a *minor* increase over minimal risk;
>
> The intervention or procedure presents experiences to subjects that are *reasonably commensurate* with those inherent in their actual or expected medical, dental, psychological, social, or educational situations; *and*
>
> The intervention or procedure is likely to yield generalizable knowledge about the subjects' disorder or condition, which is of *vital importance* for the understanding or amelioration of the subjects' disorder or condition. (Emphasis mine; §46.406.)

Here again, the open-ended nature of the terms used in the guidelines leaves room for debate and disagreement. The easiest of these criteria would be the second. It seems relatively clear that the procedures proposed in the protocol, namely, bone marrow aspirates and bone marrow harvest under general anesthesia, would not normally be encountered by HIV-positive children in the course of treatment for their disease. Thus the procedures would not be "reasonably commensurate."

Would the protocol, however, present only a "minor" increase over minimal risk? Here the guidelines become the most ambiguous and have generated the most debate. The investigators argued, for example, that the procedures did indeed represent only "a minor increase over minimal risk" (Response of Investigators to RAC Review, p. 3) and that "the risks and discomforts are relatively minor" (Response of Investigators, p. 12). In response to critiques, the investigator stated that since the bone marrow aspirations are performed under conscious sedation, they are "not significantly more painful than having a phlebotomy," and that children undergoing bone marrow harvests under general anesthesia "have minimal pain post-operatively" (Response of Investigators, p. 3).

As Janofsky and Starfield have noted, clinical experience certainly shapes one's assessment of the magnitude and possibility of clinical risk.[39] These investigators had performed many bone marrow procedures on children and experienced no adverse events; hence their assessment of the procedures would differ radically from, say, a layperson like myself reading about the procedures.[40]

The third and final criterion of §46.406 is that research entailing greater-than-minimal risk without prospect of benefit to the child subject should provide generalizable knowledge of *vital* importance for the understanding or amelioration of the subject's condition or disorder. But given the improbability of benefit, it was not clear to me how the scientific and molecular findings from this trial would be of "vital" importance to the amelioration of pediatric HIV. How ought this be assessed? According to what criteria? Or is it the case that any findings relative to a terrible disease with certain mortality in children count as "vital," that is, that it is the nature of the disease rather than the nature of the scientific findings that characterizes knowledge as vital?

Clearly, the nature and seriousness of pediatric HIV requires that research be conducted, research that will at some point require enrollment of child subjects, perhaps earlier in the research process than might be otherwise encouraged. However, as the NIH documents on inclusion of children note, "their participation earlier [must be] based on careful risk and benefit analysis."

While physiological differences between children and adults reduce the usefulness of adult models for understanding pediatric HIV, it was not clear that there were no points of contact. Would none of the data from the Phase I adult HIV human gene transfer trials be generalizable to pediatric populations? Or might some of the findings provide at least a modicum of insight? In other words, it seemed that certain research questions might be addressable in adults or other populations.[41] The trials approved by the RAC, however, had not yet had time to be initiated nor had they produced any published findings. This situation seemed to meet well the recent NIH exclusion criterion: "Insufficient data are available in adults to judge potential risk in children (in which case one of the research objectives could be to obtain sufficient adult data to make this judgment)."

These seemed to be the key questions, and questions on which I wished to have more counsel. However, since the conjunction joining the three criteria given for §46.406 is an "and," it seemed questionable to me that this protocol could be justified under this category.[42]

If §46.406 remained an open question, §46.407 remained an option. Here the guidelines provide a mechanism for "Research not otherwise approvable, which presents an opportunity to understand, prevent, or alleviate a serious problem affecting the health or welfare of children." Such research might be approved by "the Secretary [for Health and Human Services], after consultation with a panel of experts in pertinent disciplines (for example: science, medicine, education, ethics, law), and following opportunity for public review and comment." RAC review would be the ideal vehicle for such consultation.

This, then, was the modest conclusion that I reached in my review: since the arguments for justification under §46.405 and §46.406 remained debatable, the protocol seemed a good candidate for the standard process of full RAC review. Extended consideration by a panel of persons with diverse expertise would help to settle the scientific and ethical questions. I consequently recommended, on December 4, that the protocol ought not be exempted from full RAC review. As the RAC was meeting on December 4 and 5, 1995, the protocol would be considered at the RAC's next meeting, in March 1996.

This recommendation, however, was declined, an artifact equally of challenge to the guidelines and of political circumstances.[43] The March 1996 meeting of the RAC was canceled ostensibly because of "lack of protocols to review"; clearly, by early February, the decision had been made to approve the protocol. In early spring, the then-director of the Office of Recombinant DNA Activity (ORDA) announced that he was retiring from the NIH to accept a position with the Institute for Human Gene Therapy at the University of Pennsylvania. The June meeting of the RAC was likewise canceled. In May, Harold Varmus, then-director of the NIH itself, unilaterally disbanded the RAC, an action that met with controversy—approbation from some and outrage from others. In light of public response, the RAC was later reconstituted, but in a significantly reconfigured form. When it met again in December 1996, its authority to recommend approval or disapproval of protocols had been stripped.

Human Gene Transfer Research and Child Subjects: A Critical Juncture

The preceding analysis and narrative is not presented to suggest that the circumstances surrounding this particular protocol have been typical of

RAC approval of human gene transfer trials involving children. Nonetheless, the history of this protocol, when read in light of the guidelines governing human subjects research with children, raises a number of questions for both research with children in general and the field of human gene transfer research specifically.

With regard to the federal guidelines and professional policies, this particular protocol provides additional evidence for the increasingly perceived disjunction between well-crafted research guidelines and their actual implementation in the clinical research setting. Part of this disjunction stems, no doubt, from the fluidity of terms and criteria contained in the guidelines. Certainly, given the open-ended and creative nature of clinical research, guidelines governing research with human subjects ought to admit of some openness to interpretation. But the regression of interpretability ought not be infinite.

As the U.S. research industry moves toward increased enrollment of children in clinical trials, we must seriously consider how useful the federal guidelines will prove in ensuring that the interests of individual subjects are not sacrificed for the good of future children. This is especially urgent for children enrolled in Phase I protocols or children enrolled as controls, where only risks await them. It clearly falls within the purview of the oversight and ethics community to create clarity and consensus as far as possible. The National Bioethics Advisory Commission (NBAC) or OPRR need to provide guidance to IRBs—which are sometimes plagued with inexperience and at other times faced with institutional pressure—with regard to the interpretation and implementation of key terms and criteria. Research ought to be conducted in the near term to elaborate acceptable boundaries for interpreting phrases such as "greater than minimal risk," "minor increment," "prospect of benefit," and "vital knowledge." These agencies ought to provide criteria for distinguishing between a benefit that is "anticipated" or "likely" and one that is simply "possible," "hoped for," or "intended."

Such clarification would also serve to address a problem endemic to the practice of clinical research in general, but especially troubling with regard to research with children—the deeply problematic conflation of "research" and "treatment."[44] Phase I trials, contrary to the language used in the informed consent document of the HIV protocol above, ought not be classified as "treatment." These endeavors are strictly scientific experiments, designed to test scientific end points and to assess the toxicity of the compounds being administered. The situation is clearly different with Phase II and Phase III clinical trials, the kind specifically called for by the recent NIH and FDA rulings. Here, the regulatory agencies require pharmaceutical companies to provide dosing and safety information for moieties already proven efficacious in adults. In these instances, the word "treatment" might more reasonably be used in conjunction with the word "experiment." In both cases, however, the OPRR ought to provide guidance to investigators and IRBs on the appropriate language to use in the informed consent process (both the written and the face-to-face components) as well as how to warrant "prospect" of benefit in the case of radically new interventions. The new push to increase child enrollment in clinical trials, the political pressure exerted by parents on regulatory bodies to allow dying children the "right" to be enrolled, and the increased financial power of the biotechnology industry converge to make these issues urgent; unless they are addressed in the near terms, current guidelines may prove to be little more than formalities sacrificed to expediency or desperation.

These issues remain especially urgent for the field of human gene transfer research. The review of the HIV protocol, however, points to an additional set of questions specific to the field. The HIV protocol and the events surrounding the existence and purview of the RAC from 1996 forward provided preludes to the sobering findings that emerged from the Gelsinger case:

- a cavalier and obstructionist attitude on the part of industry and many researchers toward ethical guidelines and broader oversight of research practice;[45]

- the fact that human gene transfer remains a fledgling field in which much basic research remains to be done;[46]

- the fact that the risks of known methods of human gene transfer have perhaps been underreported or downplayed;[47]

- the fact that risks of effective methods of human gene transfer remain largely unknown;

- the fact that human gene transfer's therapeutic promise remains sadly unfulfilled ten years after initiation of the first protocol.

Much more basic research needs to be done in order to understand the diverse molecular mechanisms governing the array of disorders potentially amenable to gene transfer technologies. Ironically, it was just at the juncture when this conclusion was first publicly stated that the power of public oversight of human gene transfer research was disabled.[48] One of the main reasons given for disbanding the RAC in 1996 was that it attended too much to issues of ethics; as Harold Varmus stated: "the RAC had begun to exhibit a taste for *trivia*: it often got bogged down in debates over the wording of patient consent forms."[49] Perhaps it is the case, however, that issues of patient well-being and the protection of human subjects are not quite so trivial after all.

Moreover, the HIV protocol points to a global question with regard to the field of human gene transfer as a whole. It was a Phase I protocol. Almost all of the human gene transfer protocols initiated to date, whether with adult subjects or child subjects, have likewise been Phase I studies. The first human gene transfer protocol began May 22, 1989. It was a "marking" protocol that accompanied an immunotherapy study in adult patients with metastatic melanoma with a life expectancy estimated at up to 90 days. By definition, long-term risks would be impossible to assess in this population. On February 5, 1990, the investigators presented preliminary data from five of the first six patients treated to that point. The second protocol, the famous SCID-ADA protocol, was approved by the RAC on July 31, 1990, and enrolled its first patient on September 14, 1990.

This situation and the significant participation of children in the early days of human gene transfer return us to one of the few points of agreement noted earlier between those who hold a protectionist stance on the issue of research on children and those who advocate a more expansionist approach: Phase I trials in children ought to be initiated only after Phase I trials in adults have provided important indications on both safety and efficacy. Had sufficient research been conducted by 1990 or 1993 to assess the risks that might be presented to child subjects? Was there sufficient evidence in this phase to substantiate "prospect" of benefit from adult studies?

These are larger questions than can be entertained here. Clearly severe combined immunodeficiency (SCID), neuroblastoma, some of the leukemias studied, as well as the single-gene disorders, are serious childhood illnesses. They have no parallel in adults. But this alone does not warrant their further subjection to experimentation.[50] Human gene transfer as a proposed therapeutic modality is, as the bulk of the protocols and the general rhetoric surrounding genetic intervention witnesses, an intervention with potential application to a broad range of diseases that afflict adults as well as children. A number of the research questions faced by early researchers in human gene transfer could have been answered by adult studies. Hence human gene transfer seems a logical candidate for OPRR's proposal for "tandem" research:

> In some cases "tandem" studies in older populations and children may be justifiable. For example, some Phase 1 studies in children might be based on only pharmacologic safety and toxicity data (completed Phase 1 and ongoing Phase 2) but without complete effectiveness data from trials in adults and older children. If the IRB approves a Phase 1 drug trial, the consent document must specify what is known about the probability that, and the degree to which, an intervention will be of possible benefit based on all of these data.[51]

This, however, was not the route chosen by the field of human gene transfer. Hence we remain faced with an important question: How was it that this new, untried, and controversial research endeavor moved so quickly to the use of children as subjects?

The preliminary answer, I will hazard at this point, is as much a matter of the sociopolitical context of NIH-funded research as it is a matter of clinical science. Is it a coincidence that the Human Genome Project and the field of human gene transfer research, whose icon was a vulnerable child suffering from an incurable disease, were launched almost simultaneously? The elaboration of that answer must await another day.

Acknowledgments

I would like to thank Andrew Karla, Department of Chemistry, Ohio State University, for his invaluable research assistance with this project, Terrence W. Tilley for his helpful remarks on earlier drafts of this text, and the Office of Recombinant DNA Activity for their assistance in collecting human gene transfer protocols involving children. Research for this essay was supported by a 1999 Research Council Seed Grant from the University of Dayton.

Notes

1. As Larry Churchill, et al., argue in "Genetic Research as Therapy: Implications of 'Gene Therapy' for Informed Consent," *Journal of Law, Medicine, & Ethics* 26:1 (1998):38–47, the phrase gene "therapy" often serves rhetorically to mask the fact that human gene transfer interventions ought properly to be classified as research. See also my response, "Commentary: Reconstruing Genetic Research as Research," *Journal of Law, Medicine, & Ethics* 26:1 (1998):48–54.

2. Michael Winerip, "Fighting for Jacob," *New York Times Magazine* (December 6, 1998), recounts the story of Richard and Jordana Sontag, who fought an emotionally excruciating political and PR battle to enroll Jacob, afflicted with Canavan disease, in a disputed clinical trial of human gene transfer.

3. Denise Grady, "At Gene Therapy's Frontier, the Amish Build a Clinic," *New York Times* (June 29, 1999). Grady reports here on a planned clinical trial that would enroll three Amish children in a human gene transfer protocol designed to address Crigler-Najjar syndrome, a deadly autosomal recessive disorder caused by the lack of a liver enzyme required for removing bilirubin from the blood. Affected children, of whom there is a disproportionately high population within Pennsylvania Amish communities due to their habits of intermarriage, currently combat the syndrome with high-tech light therapy. The images are most striking.

4. A retrospective study of human gene transfer protocols with child subjects (see note 7) reveals that children were the intended subjects of 50 percent of the protocols in 1990 and of 50 to 63 percent of approved protocols in 1991 (depending on the age limits used to define "child"). This percentage dropped in 1992 as the overall number of protocols approved began to climb, but overall, in 1992, the total percentage of protocols enrolling children remained at 33 percent. This figure eventually leveled off and stabilized at around 20 percent.

5. Only two articles to date come close to this issue. The first, by John C. Fletcher, "Ethical Issues in and beyond Prospective Clinical Trials of Human Gene Therapy," *Journal of Medicine and Philosophy* 10 (1985):293–309, questions the wisdom of enrolling terminally ill children in the first trials because desperation may be used to justify unreasonable risk. This article was written, however, five years before the first trial commenced. The second piece seems that it would come close to this issue, given where it appears. This is Nancy Ondrusek, "Ethical Issues in Gene Therapy," in *Ethics in Pediatric Research*, ed. Gideon Koren (Malabar, Fl.: Krieger Publishing, 1993), 155–70. However, Ondrusek's essay is simply a standard

overview of ethical issues in human gene transfer research; she gives little attention to the issue of children as subjects.

6. Such a full examination is under way. Spurred by the events surrounding the RAC in 1996, I embarked on a retrospective study of human gene transfer research involving children. This study is currently ongoing. At this point, we have obtained copies of the approximately thirty protocols involving children approved by the RAC from 1989 to 1996 (the point of its hiatus) and are in the process of conducting a qualitative review of these protocols in light of the federal regulations governing research with children.

7. Robert Pear, "F.D.A. Will Require Companies to Test Drugs on Children," *New York Times* (November 28, 1998). The NIH issued a similar policy, effective October 1998; see *NIH Policy and Guidelines on the Inclusion of Children as Participants in Research Involving Human Subjects*, available from http://grants.nih.gov/grants/guide/notice-files/not98-024.html.

8. An eighteen-year-old subject in a clinical trial targeting ornithine transcarbamylase deficiency, Jesse Gelsinger, being under twenty-one years of age, would have qualified as a "child subject" according to certain guidelines. For an account of the findings of the initial inquiry at the December 8–9, 1999, meeting of the RAC, see Sheryl Gay Stolberg, "F.D.A. Officials Fault Penn Team in Gene Therapy Death," *New York Times* (December 9, 1999).

9. The current list of classic cases includes the Tuskegee Syphilis Study, the injection of live cancer cells into chronically ill patients, the hepatitis studies on mentally handicapped children at Willowbrook State Hospital in New York, Henry Beecher's exposé of ethically suspect research published in respected, peer-reviewed, scientific journals, controversy around research involving live and aborted fetuses, and the U.S. radiation experiments. For an overview of these cases in the context of Nazi experimentation and the resulting Nuremberg and Helsinki Codes, see Gregory Pence, *Classic Cases in Medical Ethics*, 3rd ed. (San Francisco: Harper, 2000).

10. As historians of human subjects research know well, the basic outlines of this tension were initially mapped in the debates between Paul Ramsey and Richard McCormick in the early 1970s. For more on this see, among others, Richard A. McCormick, "Proxy Consent in the Experimentation Situation," *Perspectives in Biology & Medicine* 18 (1974):2–20; Paul Ramsey, "Proxy Consent for Children," *Hastings Center Report* 7 (1977):4ff; Richard A. McCormick, "Experimentation in Children: Sharing in Sociality," *Hastings Center Report* 6 (1976):41–46; Paul Ramsey, "The Enforcement of Morals: Nontherapeutic Research on Children—A Reply to Richard McCormick," *Hastings Center Report* 6 (1976):21–30.

11. *Federal Register* 43:18 (1974), 914.

12. *Federal Register* 30:31 (1978), 786.

13. National Commission for the Protection of Human Subjects of Biomedical and Behavioral Research, *Report and Recommendations: Research involving Children.* (Washington, D.C.: DEHW Publication No. (OS) 77–0004, 1977). In 1979, the commission published the landmark Belmont Report, outlining the ethical framework upon which its recommendations regarding human subjects was premised. See National Commission for the Protection of Human Subjects of Biomedical and Behavioral Research, *The Belmont Report: Ethical Principles and Guidelines for the Protection of Human Subjects of Research.* DHEW Publication No. (OS) 78–0012, Appendix 1, DEHW Publication No. (OS) 78–0013, Appendix II. (Washington, D.C.: DHEW Publication No. (OS) 78–0014, 1978).

14. 45 CFR 46, Subpart D (1983).

15. For those not familiar with the federal guidelines, a summary of the components relevant to the discussion in this essay is provided here. The complete text of §45 CFR 46 is available from http://www.med.umich.edu/irbmed/FederalDocuments/hhs/HHS45CFR46.html.

§46.401 Exempt research

Certain research is exempt for all human subjects following specifications in §46.101. Exemptions include certain educational tests; the collection of existing data, documents, records, pathological specimens, or diagnostic specimens; certain research and demonstration projects; and tests of taste, food quality, and consumer acceptance.

§46.404 Research not involving greater than minimal risk

HHS will conduct or fund research, in which the IRB finds that no greater than minimal risk to children is presented, only if the IRB finds that adequate provisions are made for soliciting the assent of the children and the permission of their parents or guardians, as set forth in §46.408.

§46.405 Research involving greater than minimal risk, but presenting the prospect of direct benefit to the individual subjects

HHS will conduct or fund research, in which the IRB finds that more than minimal risk to children is presented by an intervention or procedure that holds out the prospect of direct benefit for the individual subject, or by a monitoring procedure that is likely to contribute to the subject's well-being, only if the IRB finds that:

The risk is justified by the anticipated benefit to the subjects;

The relation of the anticipated benefit to the risk is at least as favorable to the subjects as that presented by available alternative approaches; and

Adequate provisions are made for soliciting the assent of the children and permission of their parents or guardians, as set forth in §46.408.

§46.406 Research involving greater than minimal risk and no prospect of direct benefit to individual subjects, but likely to yield generalizable knowledge about the subject's disorder or condition.

HHS will conduct or fund research, in which the IRB finds that more than minimal risk to children is presented by an intervention or procedure that does not hold out the prospect of direct benefit for the individual subject, or by a monitoring procedure which is not likely to contribute to the well-being of the subject, only if the IRB finds that:

The risk represents a minor increase over minimal risk;

The intervention or procedure presents experiences to subjects that are reasonably commensurate with those inherent in their actual or expected medical, dental, psychological, social, or educational situations;

The intervention or procedure is likely to yield generalizable knowledge about the subjects' disorder or condition, which is of vital importance for the understanding or amelioration of the subjects' disorder or condition; and

Adequate provisions are made for soliciting assent of the children and permission of their parents or guardians, as set forth in §46.408.

§46.407 Research not otherwise approvable, which presents an opportunity to understand, prevent, or alleviate a serious problem affecting the health or welfare of children

HHS will conduct or fund research that the IRB does not believe meets the requirements of §46.404, §46.405, or §46.406, only if:

The IRB finds that the research presents a reasonable opportunity to further the understanding, prevention, or alleviation of a serious problem affecting the health or welfare of children, and

The Secretary, after consultation with a panel of experts in pertinent disciplines (for example: science, medicine, education, ethics, law), and following opportunity for public review and comment, has determined either:

That the research in fact satisfies the conditions of §46.404, §46.405, or §46.406, as applicable, or the following: (i) the research presents a reasonable opportunity to further the understanding, prevention, or alleviation of a serious problem affecting the health or welfare of children; (ii) the research will be conducted in accordance with sound ethical principles; (iii) adequate provisions are made for soliciting the assent of children and the permission of their parents or guardians, as set forth in §46.408.

16. Jeffrey Janofsky and Barbara Starfield, "Assessment of Risk in Research on Children," *Journal of Pediatrics* 98 (1981):843.

17. U.S. Department of Health and Human Services, "Final Regulations Amending Basic HHS Policy for the Protection of Human Research Subjects," *Federal Register* 46 (January 26, 1981): 8387.

18. See H. C. Shirkey, "Therapeutic Orphans," *Journal of Pediatrics* 72 (1968):119–20; and Robert Levine, *Ethics and the Regulation of Clinical Research*, 2nd ed. (Baltimore, Md.: Urban and Schwarzenberg, 1986), 240–41. The term "therapeutic orphan" has shifted from its initial meaning, now referring more general to "orphan diseases," those so rare that therapies, if developed, would not meet a significant enough market to result in profitability. Hence drug companies do not invest their resources in finding the cause or therapies for these diseases; many single-gene disorders qualify as "orphan diseases." See also Levine, *Ethics and the Regulation of Clinical Research:* "the prevailing practice in the United States is to ignore the orphaning clauses on the package labels. Consequently, we have a tendency to distribute unsystematically the unknown risks of drugs in children and pregnant women, thus maximizing the frequency of their occurrence and minimizing the probability of their detection. Parenthetically, it should be noted that most drugs proved safe and effective in adults do not produce unexpected adverse reactions in children; however, when they do, the numbers of harmed children tend to be much higher than they would be if the drugs had been studied systematically before they were introduced into the practice of medicine."

19. Committee on Drugs, American Academy of Pediatrics, "Guidelines for the Ethical Conduct of Studies to Evaluate Drugs in Pediatric Populations," *Pediatrics* 95 (1995):286.

20. Michael A. Grodin and Joel J. Alpert, "Children as Participants in Medical Research," *Pediatric Clinics of North America* 35 (1988):1391.

21. American Academy of Pediatrics, 286.

22. American Academy of Pediatrics, 287.

23. *NIH Policy and Guidelines on the Inclusion of Children as Participants in Research Involving Human Subjects*.

24. Robert Pear, "F.D.A. Requires Companies to Test Drugs on Children," *New York Times*, November 28, 1998. Pear notes, "If, for example, a new drug is urgently needed to treat a life-threatening disease in children and if there is no adequate therapy on the market, the Government may insist that the manufacturer immediately begin tests in children, before full data on adults are available."

25. The National Commission for the Protection of Human Subjects of Biomedical and Behavioral Research, *Research involving Children: Report and Recommendations* (Washington, D.C.: DHEW, 1977), 2–3. In the late 1970s and early 1980s, this position was likewise echoed by the FDA and the American Academy of Pediatrics. See Paolo L. Morselli and Francois Regnier, "Ethics in Pediatric Research for New Antiepileptic Drugs," in *Antiepileptic Drug Therapy in Pediatrics*, ed. P. L. Morselli, C. E. Pippenger, and J. K. Penry (New York: Raven Press, 1983), 310.

26. Robert J. Levine, "Research involving Children: The National Commission's Report," *Clinical Research* 26 (1978):62.

27. Department of Health and Human Services, "Additional Protections for Children Involved as Subjects in Research," *Federal Register* 48 (1983):9814–20; at 9816.

28. Michael A. Grodin and Leonard H. Glantz, eds. *Children as Research Subjects: Science, Ethics, and Law* (New York: Oxford University Press, 1994), 215.

29. American Academy of Pediatrics, 287. This recommendation is also found in the Canadian National Council on Bioethics in Human Research, "Revised Recommendations of *the NCBHR Report on Research Involving Children*," 4 (1993):11; and the American Medical Association, *Current Opinions of the Council of Ethical and Judicial Affairs on Clinical Investigation* (1992):5. Sujit Choudhry, in his "Review of Legal Instruments and Codes on Medical Experimentation with Children," *Cambridge Quarterly of Healthcare Ethics* 3 (1994), draws as his first conclusion: "(1) There seems to be a general consensus that pediatric research is permissible but should only be conducted when research with adults cannot yield the same information," (570). The AAP also notes: "[The investigator] must strive to obtain as much information as possible about the safety and efficacy of a drug before enrolling children as subjects" (287). "The investigator must be aware of possible conflicts between their own academic, professional, and financial interests; the 'need to know'; and the interests of the child subject" (287).

30. Available from http://ohrp.osophs.dhhs.gov/irb/irb_chapter6.htm (April 18, 2003).

31. Donald B. Kohn, "Transduction of CD34(+) Cells from the Bone Marrow of HIV-1 Infected Children: Comparative Marking by an RRE

Decoy Gene and a Neutral Gene," NIH/RAC Protocol 9602–147. The protocol itself, as well as all relevant documents referred to below, were made available as part of the public record at the December 1996 and March 1997 RAC meetings. Copies can be requested through the Office of Biotechnology Activities (formerly the Office of Recombinant DNA Activities; ORDA) of the National Institutes of Health. The protocol itself, my review, the IRB review, Kathryn Whartenby's letter, the investigator's responses to my review and the IRB review, and the then-director of ORDA's correspondence with the investigators were included in the materials for the December 1996 meeting of the RAC. My response to the director of ORDA's correspondence was included in the meeting materials for the March 1997 meeting of the RAC. Citations to these documents will be included parenthetically in the text.

32. My interest in these questions stems from my work with the Recombinant DNA Advisory Committee (RAC) of the National Institutes of Health. From June 1995 to June 1998, I had the considerable privilege of serving as a member of the RAC.

33. A word may be in order here about the responsibilities of the RAC as well as the history and process of consolidated review. At that time, the RAC's charter outlined a broad scope of responsibilities for the committee, all in an advisory capacity to the Director of the NIH and the Secretary of Health and Human Services. As the 1994 charter stated:

> Function: The Recombinant DNA Advisory Committee shall advise the Secretary (Department of Health and Human Services), the Assistant Secretary for Health, and the Director, National Institutes of Health (NIH), concerning the current state of knowledge and technology regarding DNA recombinants, and recommend guidelines to be followed by investigators working with recombinant DNA.

> As noted in the *NIH Guidelines for Recombinant DNA Research* (51 FR 16958), the Director, NIH, must seek the advice of the Recombinant DNA Advisory Committee before taking the following actions: changing containment levels for types of experiments that are not explicitly considered in the *NIH Guidelines*; certifying new host-vector systems; promulgating and amending a list of classes of recombinant DNA molecules to be exempt from the *NIH Guidelines*; permitting experiments specified by Section III-A of the *NIH Guidelines*; adopting other changes in the *NIH Guidelines*; interpreting and determining containment levels upon request by the Office of Recombinant DNA Activities; revision of the Classification of Etiologic Agents for the purposes of the *NIH Guidelines*.

In the late 1970s and early 1980s, the RAC reviewed all recombinant DNA research performed in institutions receiving federal funds for such research. Over time, its responsibilities shifted. As data accrued regarding the relative safety of such experiments, the RAC rather quickly transferred the bulk of the responsibility for review to local institutional biosafety committees. With the advent of human gene transfer in 1989, the RAC's attention shifted almost exclusively to this new rDNA application. While continuing to advise the NIH director about changes in the NIH guidelines concerning containment levels, new host-vector systems, and the classification of etiologic agents, the majority of its work in the 1990s focused on advising the director whether or not to "permit experiments specified by Section III-A of the NIH Guidelines."

The RAC provided such advice to the NIH director through protocol review. Each human gene transfer protocol submitted to the NIH for funding was likewise submitted to the RAC. Following the practice established by many IRBs, at this time each protocol was assigned to three primary reviewers, one of whom was to be a "nonscientist"; the reviewers would identify questions and issues and lead the discussion of the protocol at one of the RAC's quarterly meetings.

As one can imagine, with the first human gene transfer protocols, this process was quite rigorous and careful. Over time, as a base of knowledge and experience had been laid, it was determined that many protocols raised no new issues. Using similar vectors, similar clinical procedures, treating the same or similar diseases, it seemed unnecessary to devote scarce committee time to these protocols. Consequently, in 1995, the RAC instituted a policy of "consolidated review." Each protocol submitted to the NIH would continue to be carefully reviewed by three primary reviewers (one nonscientist), but if the reviewers deemed that the protocol presented no new issues or raised no concerns, their review was sufficient and the protocol was exempted from review by the full committee. If, on the other hand, the reviewers felt the protocol presented a new methodology or raised clinical, scientific, or ethical issues worthy of further discussion, they were to recommend that the protocol be reviewed by the full committee at its next meeting.

34. The institutional IRB approved the protocol only for children aged seven years or older.

35. Upon its initial review, the IRB at the researcher's home institution had refused approval, citing a number of reservations in addition to the above including:

> This procedure has not been previously performed in humans.

> There is no animal data because there is no animal model of human HIV infection.

The principal investigator states that this approach is not likely to have clinical benefit (Protocol, p. 8).

The IRB further states: "Discussion arose about the bone marrow aspirate(s) in children for research purposes only. These children would not receive this procedure if they were not in the research protocol. The risk is mainly pain, however it does appear to be greater than minimal risk as defined in 45 CFR 46.102. . . . The investigator does not believe that this protocol provides significant clinical benefit to the subjects. Thus, one can not justify the risks of the bone marrow aspirate(s) based on potential benefits to the subject" (Protocol, IRB Review, 10).

36. As Whartenby notes: "The major issues associated with this protocol are of an ethical nature: First, the use of this product in children may not be justified, considering the invasiveness of the required procedures and their associated risks. Although the investigators state that some potential for benefit may exist, there is little proof. The investigators also contend that benefit may not be observed in adults, but since the purpose of a phase I trial should be to evaluate toxicity, this point may not be sufficient to justify testing this product in children" (Memorandum, Katherine A. Whartenby, Ph.D. to Philip Noguchi, November 22, 1995).

37. Scientific indications against prospect of benefit are presented by two preclinical/clinical studies the investigators presented in support of the protocol. The first is an *in vitro* study that demonstrates that CD34+ cells can be isolated from cord blood and normal bone marrow, can be transduced with anti-HIV-1 vectors including the vectors proposed for use in this research, that these vectors do not interfere with the normal function of the cell, and that these vectors significantly suppress the replication of HIV-1 *in vitro*. This data appears strong, except for one point: these *in vitro* cultures have been selected by G418 for the transduced cells. The suppression of replication of HIV-1 occurs in cultures where over 70 percent of the cells have been transduced. The investigator himself notes that "challenge of cultures with 30–40 percent of the cells transduced, as achieved directly after gene transfer, fails to show significant inhibition of HIV-1 probably due to the virus production by the non-transduced cells." One might propose that the latter scenario more closely approximates the *in vivo* situation. Elsewhere the authors estimate that the transduced cells will be present *in vivo* at a level of 1 in 10,000 or 1 percent.

The second clinical study cited was the amended SCID-ADA protocol of Dr. Michael Blase, involving the capture of cord blood from newborn prenatally diagnosed infants with SCID. This cord blood was subjected to a similar protocol to that proposed here: the ADA gene was transduced into hematopoetic stem cells, which were reinfused into the patients. This study demonstrated that the transduced cells would engraft and that from a very small (in fact undetectable) number of transduced stem cells, the immune systems of these patients could be reconstituted. It is this study that demonstrated the presence of transduced PBMCs at a level of 1 in 10,000 or CD34(+) cells at a level of 1 percent after one year. Yet the investigators' conclusions regarding the effects of gene expression at these levels were unclear. While this protocol seemed to alleviate concerns surrounding the risks of reinfusion of transduced cells, only three children had been enrolled at that time.

Furthermore, even if this small number of transduced cells in the SCID-ADA protocol was to be demonstrated to have a clinically significant effect, the mechanism of effect for SCID-ADA and HIV-1 inhibition seemed to be markedly different. If the desired outcome was the production of a missing protein, a cell population of 1 percent might be adequate to produce sufficient quantities of the needed enzyme. However, if the desired outcome was to repopulate the immune system through cell replication, a starting point of 1 percent cell population seemed to have a different significance.

The investigators raised further questions regarding the significance of the SCID-ADA study to support the claim to prospect of benefit in their article entitled "Engraftment of Gene-Modified Umbilical Cord Blood Cells in Neonates with Adenosine Deaminase Deficiency," *Nature Medicine* 1:10 (1995), which was included as supporting material for the protocol. Here they stated: "The frequency of vector-containing progenitor cells exceeds by 100-fold the frequency of vector-containing cells in the mature haematopoetic cell compartments. The explanation for this dichotomy is unknown. Potentially, the expression of the ADA gene is beneficial for progenitor cell proliferation and allows expansion of the committed DC34+ progenitor pool in a fashion similar to that expected for T-lymphoid progenitors. However, the relatively high frequency of progenitor cells containing the vector is not reflected in mature leukocytes. . . . This observation suggests that although primitive progenitor cells may engraft without cytoablative therapy, they fail to undergo complete maturation *in vivo*. Alternatively the presence of the vector may interfere with mature haematopoietic cell production. Some reports have suggested that the neomycin phosphotransferase gene (*neo*) may impair hematopoietic cell function" (1021).

Thus information from the SCID-ADA study, coupled with questions from *in vitro* data on the replication of HIV-1 in a culture of 40 percent transduced cells, seems to argue against the anticipation of "prospect" of benefit for the children enrolled.

38. It appears that it was on the basis of these two responses and the claims outlined in note 43 below that the IRB at the investigator's home institution finally approved the protocol.

39. Janofsky and Starfield, "Assessment of Risk in Research on Children." For further description of the wide variability in the assessment of "minimal risk" with regard to protocols involving children, see Benjamin Freedman, Abraham Fuks, and Charles Weijer, "In Loco Parentis: Minimal Risk as an Ethical Threshold for Research upon Children," *Hastings Center Report* 23 (1993):13–19; and Saul Krugman's defense of the Willowbrook Hepatitis Study in Krugman, "The Willowbrook Hepatitis Studies Revisited: Ethical Aspects," *Reviews of Infectious Diseases* 8 (1986):157–62.

40. I was concerned that I was perhaps allowing the names of the procedures to weigh more heavily in my assessment of their risk than they warranted. Therefore, in the process of my initial review, I consulted a handful of individuals and colleagues in the field of pediatrics familiar with the issues of research on children. When asked whether they would consider bone marrow aspirations and harvests in children a procedure entailing only a "minor increase over minimal risk," the almost unanimous opinion was that the bone marrow procedures themselves entailed significantly "greater than minimal risk." This is certainly not a scientific sample, but it did address the concern about clinical versus lay perceptions.

41. In my review, I suggested possible alternatives. The investigators themselves had noted that certain scientific questions might be addressed in SIV or SCID/nu mice (Protocol, p. 88). They did not, however, elaborate or show that these experiments had been conducted. I also proposed that, given the information they provided about the physiology of pediatric HIV, an argument might be made for initially attempting this approach in HIV-positive newborns using retrieved umbilical cord blood and either proceeding within the neonatal period or waiting to see whether or not these infants seroconvert. This approach would test a number of the same concepts and greatly minimize the risk and burden associated with the bone marrow process.

42. If one concludes that these children will most likely not benefit directly from this research, there are still two additional justifications for research on children that must be considered. The first would be that children are the only population in which this particular condition is found. While that is the case for a disease like SCID or PNP-deficiency, it is not the case for HIV.

Alternatively, one could argue that the physiology of children is sufficiently different from that of other human populations that research on, for example, adults will not provide helpful information. This argument is made by the investigator. He states in response to the IRB critique that pediatric HIV infection must be studied in children for four reasons: "(1) Pediatric AIDS has a number of unique clinical aspects and therefore the efficacy of gene therapy for HIV-1 infected children cannot adequately be studied in adults, (2) the procedure may only be effective or be more effective in young children (e.g., <12 years old) because the level of thymic function which would be needed for the transduced CD34+ cells to become functional T lymphocytes is greatest *earliest in life* (emphasis mine); and (3) gene transfer into CD34+ cells is most effective with younger donors. [Therefore], it may be possible that these methods would fail to show efficacy in adults and yet be beneficial if applied to children" (Protocol, Response to IRB, 15).

A number of problems are presented by this claim. First, the investigator makes this claim *solely* in the context of the response to IRB critique; except for the first point above (which is not elaborated upon), the investigator does not raise these issues in the context of the protocol. Moreover, neither in the response nor in the protocol does the investigator provide any citations or data in support of these claims. He states that pediatric HIV infection is unique, but does not say how. He does address the second point in the protocol, but the claim is different; he states that "thymic functions, which may be necessary for stem cells to undergo differentiation to mature T lymphocytes, is likely to be *best early in the disease course*" (emphasis mine, Protocol, 33). This is a different claim. Nor is data included on the relationship between age and CD34+ cell transduction.

Whartenby also raises concerns relevant to these claims. As she notes: "The investigators would like to test this product in children because of their less developed immune system, but the protocol is written for patients aged 3–18. This age span seems to encompass more than one patient population, since the immune system of an 18-year-old should be more similar to that of an adult than that of a 3-year-old. Use of the older group may be more ethical, but may not provide any more information than use in adults. It is not clear how this group should be divided or how the issue of an appropriate age group should be handled" (Whartenby memorandum). This issue becomes more complicated by the IRB amendment that the patient population be restricted to children aged seven or older. Many of these children might well not be "early in the disease course."

Finally, the investigator's third claim above is problematic as well: it certainly may be the case that a given procedure fails to show efficacy in adults but is beneficial in children. But the investigator does not provide data to indicate that this approach has been tried in adults and has failed. It may in fact be the case that this approach could be tried in adults *and might work*. This would provide clinical data which could be useful in informing a pediatric protocol.

43. The end of the narrative of this particular protocol may be of interest to some. As noted above, I had submitted my review on December 4, 1995. On February 1, 1996, I received via fax, the investigator's response to my questions and critiques. Then, on approximately February 5, RAC members

received notice that the March 1996 meeting had been canceled, ostensibly due to "lack of protocols to discuss." It seemed apparent that my recommendation had been overturned and the decision about the protocol had been made. On February 8, however, I received a phone call from Dr. Nelson Wivel, then-director of ORDA. As quickly became clear, Dr. Wivel was calling to urge me to reconsider my recommendation that the protocol go before the entire committee. Playing the card of the urgency of pediatric HIV research, he made it clear to me that he had greater expertise regarding issues related to research involving children and that my concern was, frankly, out of line. I made it clear to Dr. Wivel that I understood the justifications and parameters for nonbeneficial research involving pediatric subjects, especially in the case of lethal diseases, but I also made it quite clear that I was not convinced that the HIV protocol met established guidelines for the ethical conduct of such research. I reiterated that I had reviewed the investigator's materials and responses thoroughly, that I had conducted a significant amount of outside research into this question specifically in light of this protocol, and that because of this research, I remained convinced that the protocol deserved consideration by the full RAC. I also stated that, as I had just been informed that the March meeting had been canceled, it seemed obvious that the decision about this protocol had already been made and that my comments and position were no longer relevant. I told him, nonetheless, that my own opinion remained unchanged. The phone call ended.

Since the RAC did not meet again until December 1996, it was not until this meeting that documents relevant to this protocol were made part of the public record and circulated to members of the RAC. In those materials, appended to the protocol, was a letter from Dr. Wivel to the principal investigator, dated February 8, 1996, the day of our phone conversation. Wivel herein stated:

> I apologize for the somewhat prolonged delay in responding to your letter and the significant amount of information that you presented in response to the comments of Dr. Lysaught. As I think that you are well aware, the two scientific reviewers of your protocol made the judgment that it should be exempt from RAC review.... Your materials were forwarded to Dr. Lysaught for her review. Subsequently I had the opportunity to speak to her, and it is my assumption that she has a clearer concept of the necessity for Phase I trials involving pediatric research subjects who have lethal diseases, notwithstanding the fact that there may be no benefit to a particular subject. After careful evaluation of all the available information, it has been determined that the protocol... is exempt from RAC review.

I had not been copied on this letter. One of my scientific colleagues on the RAC, who found the tenor of the correspondence rather problematic, brought it to my attention. My written response to this letter was made part of the public record in the materials of the March 1997 RAC meeting.

44. Again, for more on this, see Churchill, et al., "Genetic Research as Therapy"; and Lysaught, "Commentary: Reconstruing Genetic Research as Research."

45. Information emerging from the continued inquiry into the practices surrounding Jesse Gelsinger's death reveal what has been characterized as a "95 percent failure rate" in the reporting of adverse events from investigators conducting clinical trials of human gene transfer involving adenoviral vectors; see Sheryl Gay Stolberg, "Agency Failed to Monitor Patients in Gene Research," *New York Times*, (February 2, 2000). The University of Pennsylvania, in its official response to the RAC/FDA findings released in December 1999, characterized the failures in their program as "little more than 'minor deviations' in bookkeeping"; Sheryl Gay Stolberg, "Scientists Defend Suspended Gene Therapy," *New York Times*, (February 15, 2000).

46. See Stuart H. Orkin and Arno G. Motulsky, National Institutes of Health Ad Hoc Committee Report, *Report and Recommendations of the Panel to Assess the N.I.H. Investment in Research on Gene Therapy* (December 7, 1995), available from http://www.nih.gov/news/panelrep.html.

47. Paul Gelsinger, Jesse's father, reports that "he and his son had no idea there were risks" entailed in the procedure and were not informed of adverse events that had occurred with other subjects. Moreover, he reports that he was told by an investigator involved with the protocol that "the treatment was already working in some patients"; Sheryl Gay Stolberg, "Youth's Death Shaking Up Field of Gene Experiments on Humans," *New York Times*, (January 27, 2000). The FDA/NIH investigation revealed that "the informed consent form that the investigators gave patients deviated from the one the agency [the FDA] had approved, in that it omitted information about the death of monkeys that had received treatment similar to that given Mr. Gelsinger, although much more powerful"; Stolberg, "FDA Officials Fault Penn Team in Gene Therapy Death").

48. The Orkin/Motulsky Report (see note 46 above) was issued in December 1995, at about the time the decision was made to abolish the RAC.

49. Eliot Marshall, "Varmus Proposes to Scrap the RAC," *Science* 272 (1996):94 (emphasis added).

50. This point was also relevant to the HIV protocol. The investigator noted that 80 percent of the pediatric AIDS population is African American or Latino and that 75 percent of the younger patients followed at the investigator's institution were likewise African American or Latino. This raises social

and ethical questions regarding whether children who are possibly already doubly disadvantaged—socioeconomically and by their disease—are being asked to assume an additional undue burden of nonbeneficial research. It is a catch-22 (if research is not conducted, this disadvantaged population is further disadvantaged by this terrible disease) unless clinical trials are designed with a serious prospect of benefit.

51. Available from http://ohrp/osophs.dhhs.gov/irb/irb_chapter6.htm (April 18, 2003).

16

Science, Ethics, and Policy: Relating Human Genomics to Embryonic Stem-Cell Research and Therapeutic Cloning

Gerard Magill, Ph.D.

THIS CHAPTER discusses a value-based connection between the emerging technologies of human genomics, embryonic stem-cell research, and therapeutic cloning. The goal is to provide an ethics analysis that seeks to promote and protect society's interests in the current environment of scientific progress and technological breakthroughs.

To set the scene of the emerging capacity of bioengineering today, I present a case study into the treatment of Molly and Adam Nash, the first documented medical therapy to combine human genomics and embryonic stem-cell research. My argument is that human life constitutes the most basic human value that must permeate an ethical analysis of life sciences research today. The emphasis on the value of human life is evident in both human genomics and embryonic stem-cell research, including therapeutic cloning. The breakthroughs in human genomics raise many ethics concerns. But the first death of a patient in a gene therapy trial in 1999 gave prominence to a profound concern about patient safety in human genomics research. Second, the announcement by President Bush in August 2001 permitting federal funding of research on a limited number of embryonic stem cells generated widespread debate about the meaning of embryonic human life. Moreover, dubious claims about a private cloning company called Clonaid having cloned human babies in late 2002 and early 2003 have heightened the rhetoric of policy discussion about embryonic stem-cell research and human cloning.[1] Despite such claims, science continues to encounter such serious difficulties with cloning techniques in nonhuman primates that some scientists suggest that reproductive cloning in nonhuman (and, by extrapolation, human) primates may be unachievable.[2] Nonetheless, scientists continue their research to accomplish the cloning of human embryos.[3] This chapter argues that human genomics can be connected to embryonic stem-cell research based on respecting the value of human life. However, respecting human life in each area appears to yield bipolar results. That is, respect for the value of human life in genomics research is being articulated in terms of strict regulatory measures to protect patient safety; whereas respect for the value of human life in embryonic stem-cell research entails a policy trajectory that seems tolerant of the destruction of some human embryos, even if Congress closes its regulatory doors to human therapeutic cloning.

Briefly, by undertaking a value-based analysis in human genomics and embryonic stem-cells research, the chapter presents a case for protecting and promoting society's interests in the fast-changing landscape of biotechnology today.

Case Study: Molly and Adam Nash

Molly Nash was a six-year-old girl with Fanconi anemia, a rare genetic disorder that prevents bone marrow from being made by the body. This condition can kill at a very young age. A bone marrow transplant from a matching sibling can offer an 85 percent rate of success for treating this disease. Because Molly did not have a sibling, her parents decided to have another baby, hoping to use the placenta and umbilical cord blood after its birth for a stem-cell transplant for Molly. The parents opted for assisted reproduction and genetic screening from the Reproductive Genetics Institute in Chicago. Using preimplantation genetic diagnosis, the parents ensured that the new baby did not have the same disease as Molly and that there would be a good match for the transplant. On August 29, 2000, baby Adam was born. A few weeks later, and after further screening, his six-year-old sister received transfusion of stem cells from his umbilical cord and placenta.[4] Both baby Adam and Molly flourished.

This therapeutic intervention was the first recorded experiment that merged the technologies of genomics (via preimplantation genetic diagnosis) and stem-cell research (via the transplant). The case of Molly and Adam Nash amply demonstrates how human genomics and stem-cell research can work together. But the therapy was not without ethical controversy. For example, to undertake preimplantation genetic diagnosis, fifteen human embryos (at the eight-cell stage, when one cell is typically removed for genetic testing prior to implantation in the mother's womb) were created via *in vitro* fertilization and some were discarded—only baby Adam was born. The main ethical question here concerned the ethical status of the human embryos created to select an appropriate match for treating Molly, especially with regard to treating them as persons with rights or property for medical research.[5]

As bioengineering develops in human genomics and stem-cell research, similarly complex ethical questions will increasingly arise. Hence the ethics discourse that is likely to permeate the new science and technology of life sciences research should inspire policy debate that both promotes and protects the interests of society.

Human Genomics

The science of genetics examines mechanisms that enable biological traits to pass down generations, being expressed in individuals; a genome is the sum total of genetic information contained within the cells of a living individual.[6] Mapping the human genome requires deciphering and arranging in the correct sequence these three billion chemical letters of DNA across our chromosomes. The human genome map was presented at a White House ceremony as a first draft on June 26, 2000.[7] Then, in February 2001, two competing groups (one public, the other private) published their more complete analyses of the human genome. Dr. Francis Collins, who leads the Human Genome Sequencing Consortium, published his team's results in the journal *Nature*,[8] and Dr. Craig Venter, who was the CEO of the private corporation Celera Genomics, published his team's results in the journal *Science*.[9] These publications were associated closely with other efforts to catalog human variation.[10] These papers indicate that the human species possesses about 30,000 genes (not the 100,000 that many projected), just 11,000 more than the laboratory roundworm

(19,000 genes sequenced in 1998), or just over twice as large as the fruit-fly genome (13,600 genes decoded in March 2000)!

Many anticipate effective gene therapies ahead. High expectations abound among the many patients in the United States afflicted with gene-related dysfunctions, such as cardiovascular disease (50 million), diabetes (15 million), cancer (8 million), and Alzheimer's disease (8 million)—not to mention many other costly ailments, such as psychiatric disorders, multiple sclerosis, and obesity. For example, the National Cancer Institute reported in 2001 that more than 100 so-called designer drugs based on molecular genetic medicine are in process toward clinical trials.[11] Of course, there will be many difficulties, not only in shifting from the map of the human genome to practical therapies in the clinic,[12] but also in transforming health care organizations to field these new technologies,[13] including the integration of new gene technologies into health care delivery systems,[14] and developing a clearer understanding of genetic influences on environmentally associated diseases to develop disease-prevention strategies and programs of intervention (such as via the Environmental Genome Project initiated by the National Institute of Environmental Health Sciences).[15] Significant advances have already been made with regard to how the discovery of genetic causes for common diseases is impacting primary-care medicine.[16]

So the map of the human genome constitutes a sort of "book of life" or "holy grail" of molecular medicine that will enable us to understand the molecular pathogenesis of disorders and to develop treatments for many diseases and debilities at their genetic roots.[17] However, for the sake of precision, therapies are likely to be based on the connection between our estimated 30,000 genes and our estimated 300,000 proteins and the pathways between proteins and human diseases. A gene is a section of DNA containing the instructions for assembling a part of a specific protein or enzyme. Proteins underlie all of the body's structure and functions, ranging from the construction and maintenance of skin, bones, and blood, to the production of antibodies to fight infections. The body's collection of proteins is called the "proteome." Just as "genomics" studies how our genes function, "proteomics" studies how our proteins function. Hence gene therapies are likely to focus on the interaction between genomics and proteomics. From this perspective, the map of the human genome provides basic clues about proteins that constitute our biological building blocks as chemical messengers and mechanisms underlying diseased as well as healthy bodies. By exploring the connections between our genes and our proteins (and even between our genes and lipid analysis in so-called "lipomics"), we are more likely to develop new drug therapies.[18]

With the extraordinary expectations in society for breakthroughs in molecular medicine and genetic research, there is an increasing need to integrate science, ethics, and policy.[19] And the extent of the policy and ethics debate on genomics is vast,[20] ranging from grappling with the meaning of normalcy, disability, and deformity in the face of genetic risks and probabilities,[21] to preimplantation genetic diagnosis and genetic screening of newborns,[22] consent for genetic research,[23] privacy and discrimination with regard to genetic information[24] and tissue samples,[25] responsible counseling for genetic testing and screening,[26] justice (including gender justice) for equality of access in the genomics era,[27] race disparities in research,[28] commodification of the human gene, and gene patenting.[29] The need for professional education about the new era of genomics is greater than ever.[30] And religious discourse contributes vigorously to the policy and ethics debate in human genomics.[31] As research and publications on genetics abound,[32] including those that target public involvement,[33] it is not surprising that technological developments will continue to be accompanied by value disputes that will require shrewd ethical analysis to navigate the way ahead.

The first gene therapy protocol was presented to the Human Gene Therapy Subcommittee on September 14, 1990. This was the first officially sanctioned human gene therapy experiment. Since then, research protocols have multiplied. However, perhaps the most significant policy and ethics developments in human genomics have revolved around the issue of patient safety, after the first death ascribed directly to a gene therapy treatment. In September 1999, Jesse Gelsinger, a teenager, died from a severe immune reaction to an experimental deprogrammed virus being used as a vector to deliver an experimental gene therapy for an inherited liver disease. Although his ailment was being managed by medication, he entered the gene therapy trials designed for babies with a fatal form of the disease. The overreaction of his immune system led to respiratory, liver, and kidney failure, causing his death within a few days.[34]

Federal regulatory agencies intervened immediately, discovering multiple problems in the trial protocol, including the prima facie conflict of interest of the project director, who reportedly owned stock in Genovo, the company funding the research. Subsequently, the National Institutes of Health (NIH) discovered that researchers had substantively underreported adverse events. The family filed a lawsuit alleging fraudulence and negligence in recruiting their son as a patient, and eventually settled out of court.[35] As a result of this case, the government closed all gene research protocols at the University of Pennsylvania, where the death had occurred. This case illustrates the urgency and importance of patient safety in biotechnology research today. It is not difficult to recognize the need for increased scrutiny by regulatory bodies to protect human research subjects.

The development of regulatory protection of human subjects in medical research is well documented.[36] But the death of Jesse Gelsinger in a gene therapy research protocol led to more vigorous oversight of human subjects protection by regulatory bodies. Broadly, the famous Belmont Report, written by the National Commission created by Congress in 1974,[37] now elicits much closer scrutiny than previously. As a result of this report, the Department of Health and Human Service (DHHS) and the Food and Drug Administration (FDA) revised their regulations in January 1981 to focus on informed consent.[38] However, these regulations did not adequately consider or explain the significance of risk/benefit assessment. Hence in 1981 a new commission was formed, the President's Commission for the Study of Ethical Problems in Medicine and Biomedical and Behavioral Research. The White House formed a committee of government agencies to consider protections in human subjects research, leading to publication in the *Federal Register*[39] of the "Common Rule," with regulations for multiple federal departments and agencies to provide some standardization as federal policy to protect human subjects.[40] Oversight continued and expanded, such as by turning the Office for Protection from Research Risks (OPRR) into the Office for Human Research Protections (OHRP).[41]

Compliance with the Code of Federal Regulations is more crucial than ever in the genomics era to assure the protection of human subjects in gene therapy trials. Ongoing review of research involving human subjects is attracting increasing attention in research ethics discourse,[42] not least because of the temporary closure in July 2001 of research programs at Johns Hopkins University after the death of Ellen Roche, a twenty-four-year-old laboratory worker in an asthma study.[43]

In December 2000, the National Bioethics Advisory Commission (NBAC) under the leadership of Eric M. Meslin developed two very important reports, one for national issues and one for international issues, for the protection of human subjects in research. The report on international issues was *Ethical and Policy Issues in International Research: Clinical Trials in Developing Countries.*[44] The report on national issues was *Ethical and Policy Issues in Research Involving Human Participants*, published in August 2001 just prior to the expiration on October 3, 2001, of the NBAC charter, whose purpose was to examine the federal system of oversight (NBAC had previously published an influential report on research involving human biological materials in 1999).[45] The report's thirty recommendations (pages 8–18) included the expansion of an oversight system to enhance protection of research participants and promote research consistent with ethical principles. The proposed changes will generate significant costs and require new resources, such as for the National Office of Human Research Oversight.

It is not a surprise, then, that in the environment of increased regulatory oversight in the wake of the Gelsinger case, the NIH issued new rules in November 2001 to improve oversight and public disclosure of problems related to human safety in gene therapy trials.[46] A national committee (the NIH Gene Transfer Safety Assessment Board as a subcommittee of the Recombinant DNA Advisory Committee) will review safety issues in the approximately 360 gene therapy experiments underway in the United States that involve inserting DNA into patients to repair genes or replace damaged genes. The Gelsinger case was a watershed challenge to research in human genomics. The challenge is to recognize the basic value of human life, articulated here in terms of patient safety, as a necessary condition for promoting and protecting society's interests in life sciences research. Incongruously, respect for human life in stem-cell research can yield a result that contrasts starkly with the prominence of patient safety in human genomics. The next section of the chapter discusses this incongruity as crucial for understanding how biotechnology must promote and protect society's interests.

Embryonic Stem-Cell Research and Human Therapeutic Cloning

The Molly and Adam Nash case described at the start of the chapter provides a glimpse into the relation between human genomics and stem-cell research for the development of medical therapies in the future. In the context of the relation between these technologies, this chapter seeks to highlight a value-based connection between these forms of bioengineering today. This ethics analysis suggests that human life constitutes the most basic value that permeates research in the life sciences. Yet that value seems to yield bipolar results in the distinct yet related fields of human genomics and stem-cell research. The first part of the chapter indicates that concern with the value of human life in the field of human genomics, especially in the wake of a patient death in a gene therapy clinical trial, has generated a strenuous concern among regulatory bodies with patient safety. The second part of the chapter suggests, incongruously, that concern with the same value of human life in stem-cell research yields a bipolar result that appears to be tolerant of the destruction of some human embryos. This tolerance occurs in the name of medical science that seeks to develop therapies specifically to enhance the value of human life in so many suffering patients. In other words, in the link between human genomics and embryonic stem-cell research, the shared concern with the basic value of human life yields bipolar results. Addressing this incongruity is necessary for ethics and policy to promote and protect society's interests in bioengineering today.

Research on embryonic stem cells is becoming increasingly controversial because of its connection with human therapeutic cloning, a process that creates human embryos to harvest embryonic stem cells for medical research.[47] President George Bush appointed a new Council of Bioethics to address many of the emerging issues in biotechnology and specifically to examine human cloning (at the same time as the National Academy of Sciences stood against President Bush on human cloning).[48] Of course, ethical discourse on stem-cell experiments raises other issues of patient safety and efficacy, such as in a failed experiment on patients with Parkinson's disease that involved sham surgery.[49] And debate about human cloning also pertains to other issues, such as the genetic traits of our future progeny.[50]

However, the focus in this chapter is to examine embryonic stem-cell research and the related technology of human therapeutic cloning from the value perspective of respecting human life.

The ethics and policy debate on embryonic stem-cell research is very recent. In 1996 a congressional ban seemed to uphold the value of embryonic human life, insofar as the appropriations bill for the Department of Health and Human Services prevented federal funds from being used for research in which embryos are destroyed, discarded, or knowingly subjected to risk or injury. But in late 1998, scientists at the University of Wisconsin-Madison, led by Professor James Thomson (using spare embryos from infertility laboratories) in collaboration with Geron Corporation (Menlo Park, California), and scientists at the Johns Hopkins University, led by John Gearhart (using aborted fetal tissue) announced separate successful experiments in isolating and culturing embryonic human stem cells using somatic-cell nuclear transfer.[51] This process replicated the technology pioneered by Ian Wilmut at the Roslin Institute in Edinburgh, Scotland, to clone the famous ewe Dolly in 1997.[52] Then, in the summer of 2000, another team of U.S. scientists announced their success in culturing bone marrow (adult) human stem cells. While there is no substantive controversy about research on adult human stem cells (conditional upon appropriate consent and risk-benefit analysis), ethics and policy debate has flared on embryonic stem cells because harvesting these stem cells currently entails the destruction of the human embryo.

The intensity of this debate reflects the immense promise of embryonic stem-cell research.[53] Basically, the goal of stem-cell research is to discover and control how stem cells differentiate into human body's 210 cell types (heart, nerve, and blood cells, etc.). Scientific breakthroughs in this research promise to revolutionize both drug development and the testing process used by pharmaceuticals and biotechnology companies, especially as they develop therapies for many diseases and debilities, including neuron cells for Parkinson's, Huntington's, Alzheimer's, ALS, and spinal cord injuries; chondrocytes for arthritis; cardiomyocytes to replace damaged heart tissue; insulin-producing pancreatic cells for diabetes; cancer treatments; the regeneration of vital organs for transplant patients, and so on. The hopes are high to bring relief to vast numbers of patients. In the United States, for example, there are 4000 deaths annually of patients awaiting organ transplants; there are between 400,000 and 800,000 patients with diabetic foot ulcers; there are between 700,000 and 1 million Parkinson's patients; there are 2.5 million bone/reconstructive surgeries annually; there are 3 million cardiovascular procedures and 4.7 million patients with congestive heart failure. With these daunting statistics, it is no wonder that ethics and policy discourse on stem-cell research is now so prominent.

Responding to public expectation for urgent scientific progress, two important reports were presented by the NBAC and by the NIH. In 1999, the NBAC presented a lengthy report, *Ethical Issues in Human Stem Cell Research*; and in August 2000 the NIH published its revised guidelines on embryonic stem-cell research with the support of President Clinton.[54] The NIH guidelines permitted the use of federally funded research to derive human pluripotent stem cells from fetal tissue (derived by agencies not supported by federal funding), provided there was informed consent by the donors and no financial inducements. The NIH guidelines sought to establish a distance between the derivation of embryonic stem cells (involving a process that destroys the embryo) and federal funded research on embryonic stem cells. This particular distinction recurred in the subsequent policy of President George W. Bush.

One year later, on August 9, 2001, President Bush announced his policy that would permit public funding of embryonic stem-cell research, but in a more restrictive fashion than his predecessor.[55] Just before he announced his policy, the NIH provided a comprehensive report to encourage policy and ethics debate.[56] President Bush approved federal funding for more than sixty genetically diverse stem-cell lines. He acknowledged the lines came from stem cells that involved the destruction of human embryos, but emphasized that ongoing research on these immortalized cell lines did not require the subsequent destruction of embryos. In other words, he sought to avoid taxpayer funding that would sanction or encourage further destruction of human embryos. Moreover, he encouraged aggressive federal funding of research on umbilical cord, placenta, adult, and animal stem cells that does not involve the ethical dilemma of destroying human embryos. Subsequently, the NIH established a Web-based human embryonic stem-cell registry to accept applications for federal grants and also released a list of stem-cell colonies approved for federally funded research.[57]

In this policy position President Bush seemed to rely on the same ethical distinction used by President Clinton, yet he sought a very different conclusion. The ethical distinction was to create a distance between federally funded research on embryonic stem cells and the actual harvesting of those stem cells (via a process that involves the destruction of the embryo). The divergent conclusion in each policy is evident: President Clinton sought to permit federal funding for embryonic stem-cell research that tolerated the ongoing destruction of human embryos (albeit by non–federally funded agencies), whereas President Bush sought to permit federal funding for embryonic stem-cell research that prevented the future destruction of human embryos.

To understand how policy might develop on this crucial research topic for the life sciences, it is helpful to grasp the underlying ethical distinction used by both presidents. At first glance, there may not appear to be much difference between their two positions, but in reality there is a very significant ethical difference that impacts future policy development. By creating a distance between federal funding of embryonic stem-cell research and the harvesting of those stem cells (in a process that involves the destruction of human embryos), both presidents seemed to rely on the ethical principle of cooperation, though neither actually acknowledged using the principle.

The ethical principle of cooperation has a long history, beginning in the seventeenth century.[58] The principle seeks to distinguish between material connection and moral complicity with perceived wrongdoing.[59] For example, when an innocent bank teller hands over money to a robber, there is a material connection by the teller that does not entail moral complicity with the robbery; in contrast, the getaway driver for the bank robber may not actually touch the stolen money but is morally complicit with the robbery. With regard to the policy decision by these presidents, it is evident that both established a distance between federal funding of embryonic stem-cell research and the harvesting of the embryonic stem cells—the rationale seems to have been to avoid complicity with the destruction of the human embryo in harvesting the embryonic stem cells. On the one hand, the policy of President Clinton tolerated the ongoing destruction of human embryos (albeit by non–federally funded agencies) to provide the embryonic stem cells for federally funded research. On the other hand, the policy of President Bush permits federally funded research only on immortalized embryonic stem-cell lines to avoid the need for ongoing harvesting of embryonic stem cells with the concomitant destruction of embryos. The significance of this distinction is because complicity with perceived wrongdoing in stem-cell research warrants ethical scrutiny.[60] The debate continues on whether the policy of President Bush effectively avoids moral complicity with the destruction of the donor embryos in the process of harvesting the original embryonic stem cells that were developed into permanent stem-cell colonies.[61]

Underlying the policies of both presidents there is a divergent view of the value of embryonic human life. This divergence reflects the public debate on such a sensitive issue. On the one hand, proponents of embryonic stem-cell research argue that respect for the basic value of human life in so many suffering patients mandates ongoing research to facilitate medical research for effective therapies to address their suffering. This approach tolerates the concomitant destruction of human embryos to harvest the required stem cells. On the other hand, opponents of embryonic stem-cell research that entails the destruction of human embryos argue that respect for all human life from its very beginning mandates the protection of human embryos. This standoff may seem reminiscent of the abortion debate. But it seems unlikely that the abortion paradigm will help to resolve the emerging debate on embryonic stem-cell research. The reason is because *Roe v. Wade* was resolved (in 1973 with Justice Harry A. Blackmun writing the seven-person majority for the Supreme Court) as a matter of the mother's right to privacy founded in the Fourteenth Amendment's view of personal liberty.[62] The right to the mother's privacy does not pertain in the debate on embryonic stem cells insofar as the embryos in question exist independently.

Rather, the ethics and policy debate on embryonic stem-cell research is more likely to be resolved in a manner akin to the debate on fetal tissue research.[63] The U.S. policy debate about research on human embryos goes back at least to 1979, when the U.S. Ethics Advisory Board indicated that it was ethical to create research embryos to investigate safety issues with *in vitro* fertilization technology. This occurred in the wake of *in vitro* fertilization technology pioneered by Robert Steptoe and Patrick Edwards with the birth of Louise Brown in 1978.[64] Policy debate subsequently flared, as in the famous Warnock Report (1985) and approval in Britain and Canada for creating and using research embryos, and in the NIH's *Report of the Human Embryo Research Panel*

(1994).[65] In turn, this debate on fetal-tissue research is connected with the ethical debate on the status of the early embryo. The ethical debate on the status of the early embryo has been vibrant for a considerable time,[66] and it seems likely that this debate on the early embryo will provide the crucial terrain for regulatory policies on embryonic stem-cell research,[67] including the perspectives of religious ethics.[68] The stakes in the ethical debate on the status of the early embryo, with the concomitant policy implications for embryonic stem-cell research, have risen dramatically since recent breakthroughs in human therapeutic cloning, including options to patent human cloning.[69]

In June 1997, the NBAC published its report on human cloning, focusing on the physiological risks to cloned children. The report concluded that because safety concerns posited undue risk to potential offspring, there should be a federal ban against both reproductive and therapeutic cloning, pending ongoing review by Congress.[70] However, Congress did not act on this recommendation. When the debate on stem-cell research surged, the NBAC published another report, in September 1999, recommending that the destruction of human embryos for medical research be permitted, limiting its recommendation to spare embryos only and not applying its analysis to therapeutic cloning.[71] Then in June 2000, Liam Donaldson, Britain's chief medical officer, presented a report recommending therapeutic cloning in Britain to facilitate medical research on human embryonic stem cells. The U.K. Parliament accepted the recommendations by introducing legislation to amend the Human Fertilisation and Embryology Act of 1990.[72] (The 1990 Act permitted some targeted research on embryonic tissues obtained via aborted fetuses or spare embryos from *in vitro* fertilization techniques.)[73] This U.K. law in particular and European perspectives on therapeutic cloning in general are likely to increase the pressure on the U.S. Congress for a policy resolution of the debate.[74]

In the fall of 2001, the debate on human therapeutic cloning and its role in embryonic stem-cell research took a dramatic step forward. On November 25, Advanced Cell Technology in Worcester, Massachusetts, announced that it had successfully cloned human embryos. A frenzy ensued in the media about the ethical and policy implications of therapeutic cloning, even though the experiment had been unable to develop and harvest embryonic stem cells.[75] The ethicist Ronald Green, who is chair of the ethics advisory board for

Advanced Cell Technology (and director of the Ethics Institute at Dartmouth College) shrewdly delineated the challenge that therapeutic cloning presents for ethics and policy discourse.[76] He explained that most members of the ethics board did not consider the human organism produced by therapeutic cloning as equivalent to any ordinary human embryo because it did not result from the normal egg/sperm fertilization process. The board members claimed that the human organism did not merit the same degree of respect and protection as a fertilized human embryo. In other words, the board's perception of the moral status of the early (cloned) embryo appears to have been crucial for this experiment. The development of Green's previous work in the debate on the status of the early embryo obviously influenced his approach to determining the moral status of the cloned embryo.[77]

The proverbial genie is now out of the bottle. Science has passed another milestone by achieving success in human cloning. Undoubtedly, there will be many ethical and policy arguments on both therapeutic cloning for medical research and reproductive cloning for fertility treatment. Already a clear demarcation of opposing viewpoints is emerging, especially on using therapeutic cloning to harvest embryonic stem cells for medical research versus seeking alternatives to human cloning.[78] With the genie out of the bottle, Congress is under more pressure than ever to resolve the policy debate on embryonic stem-cell research and human therapeutic cloning.[79] In particular, the distinction between human reproductive cloning and human therapeutic cloning creates a specific challenge for policy determination.[80] Science policy needs to be shaped in a democratic, pluralistic society, and vigorous ethics discourse must inform such policy.[81]

Conclusion

Congress may have been let out of an awkward corner when the Supreme Court resolved the policy controversy on abortion in 1973. It seems unlikely that the Supreme Court can rescue Congress today on pressing matters in life sciences research. There is urgent need for a policy resolution of the current debate on embryonic stem-cell research and human therapeutic cloning. Both the nation at large and the science community in particular need clarification about what research may be pursued legitimately in our democratic and pluralistic society. The stakes in the modern era of biotechnology are high, and great expectations accompany them. And these stakes are accompanied by grave risks—scientific, ethical, and cultural.

This chapter has considered a value-based connection between the specialties of human genomics, embryonic stem-cell research, and therapeutic cloning. The goal has been to provide an ethics analysis that seeks to promote and protect society's interests in the current environment of scientific progress and technological breakthroughs. While the introductory case of Molly and Adam Nash illustrates the immense potential of life sciences technologies, the chapter has raised crucial questions about how respect for human life seems to yield bipolar results in human genomics and in embryonic stem-cell research, including therapeutic cloning. The chapter presents a challenge to seek a more coherent policy resolution that both promotes and protects society's interests.

Acknowledgments

An earlier version of this chapter was published as "The Ethics Weave in Human Genomics, Embryonic Stem Cell Research, and Therapeutic Cloning: Promoting and Protecting Society's Interests," *Albany Law Review* 65:3 (2002):701–28. I am grateful to the journal for permission to use the material in that article in this chapter.

NOTES

1. Anne Marie Borrego, "Cloned-Baby Claim Renews Calls for Legislation Scientists Fear," *Chronicle of Higher Education* (January 10, 2003):A20.

2. Calvin Simerly, et al., "Molecular Correlates of Primate Nuclear Transfer Failures," *Science* 300 (2003):297.

3. Antonio Regaldo, "Human Embryo Is Cloned, Expert to Say in Paper," *Wall Street Journal* (April 9, 2003):D2.

4. See Frank Clancy, "A Perfect Match," *Medical Bulletin* (Minnesota Medical Foundation) (2001):15–17. For recent developments with preimplantation diagnosis, see Yury Verlinsky, et al., "Preimplantation Diagnosis for Fanconi Anemia Combined with HLA Matching," *Journal of the American Medical Association* 285 (2001):3130–33.

5. See Paul Lauritzen, "Neither Person nor Property. Embryo Research and the Status of the Early Embryo," *America* (March 26, 2001):20–23; Michael J. Meyer and Lawrence J. Nelson, "Respecting What We Destroy: Reflections on Human Embryo Research," *Hastings Center Report* (2001):16–23.

6. Bruce R. Korf, *Human Genetics: A Problem-Based Approach*, 2nd ed. (Malden, Mass: Blackwell Science, 2000), xiii; D. Peter Snustad, Michael J. Simmons, and John B. Jenkins, *Principles of Genetics* (New York: John Wiley and Sons, 1997), 180–379. For a clear introduction for the layperson to the function of DNA and genes, see LeRoy Walters and Julie Gage Palmer, *The Ethics of Human Gene Therapy* (New York: Oxford University Press, 1997), chap. 1.

7. Arielle Emmett, "The Human Genome," *The Scientist* 14:15 (2000):1, 17–19.

8. Francis Collins, et al., "Initial Sequencing and Analysis of the Human Genome," *Nature* 409 (2001):860–921.

9. Craig Venter, et al., "The Sequence of the Human Genome," *Science* 291:5507 (2001):1304–1351. Craig Venter resigned as president of Celera Genomics in January 2002; see Nicholas Wade, "Thrown Aside, Genome Pioneer Plots a Rebound," *New York Times* (April 30, 2002): D1; and Scott Hensley, "Genetics' Venter Uses His Profit For New Causes," *Wall Street Journal* (April 30, 2002):B1.

10. The International SNP Map Working Group, "A Map of Human Genome Sequence Variation Containing 1.42 Million Single Nucleotide Polymorphisms," *Nature* 409 (2001):928–33. Icelandic scientists claim that "more than 100 large-scale corrections are needed" in the human genome sequence from the consortium of the Human Genome Project; see Nicholas Wade, "Human Genome Sequence Has Errors, Scientists Say," *New York Times* (June 11, 2002): D4; and Vanessa Fuhrmans, "Human Genome Detailed by Decode," *Wall Street Journal* (June 10, 2002), B3.

11. Francis S. Collins and Alan E. Guttmacher, "Genetics Moves into the Medical Mainstream," *Journal of the American Medical Association* 286 (2001):2323.

12. See the entire issue on opportunities for medical research and treatment in genetics in *Journal of the American Medical Association* 285:5 (2001), especially F. S. Collins and V. A. McKusick, "Implications of the Human Genome Project for Medical Science," 540–44. Also see Elizabeth A. Eisenhauer, Editorial, "From the Molecule to the Clinic—Inhibiting HER2 to Treat Breast Cancer," *New England Journal of Medicine* 344:11 (2001):841–42; Eric S. Lander, "Genomics: Launching a Revolution in Medicine," *Journal of Law, Medicine & Ethics, Special Supplement* 28:4

(2000):3–14; Philip R. Reilly, "The Human Genome Project: Recent Genetic Advances Will Have Far-Reaching Implications for Catholic Health Care," *Health Progress* (2001):24–27.

13. Chris Myers, Nicole Paulk, and Christine Dudlak, "Genomics: Implications for Health Systems," *Frontiers of Health Services Management* 17:3 (2001):3–16. Also see Maxwell J. Mehlmann, "The Effect of Genomics on Health Services Management: Ethical and Legal Aspects," *Frontiers of Health Services Management* 17:3 (2001):17–26; Associate Justice Stephen G. Breyer, "Genetic Advances and Legal Institutions," *Journal of Law, Medicine & Ethics, Special Supplement* 28:4 (2000):23–29.

14. See Joan Stephenson, "As Discoveries Unfold, a New Urgency to Bring Genetic Literacy to Physicians," *Journal of the American Medical Association* 278 (1997):1225–26; Francis S. Collins, "Preparing Health Professionals for the Genetic Revolution," *Journal of the American Medical Association* 278:15 (1997):1285–86; Joan Stephenson, "Group Drafts Core Curriculum for 'What Docs Need to Know About Genetics,'" *Journal of the American Medical Association* 279:10 (1998):735–36; John Bell, "The New Genetics: The New Genetics in Clinical Practice," *British Medical Journal* 316:7131 (1998):618–20; Ann Louise Kinmonth, et al., "The New Genetics: Implications for Clinical Services in Britain and the United States," *British Medical Journal* 316:7133 (1998):767–70; Thomas H. Murray, Mark A. Rothstein, and Robert J. Murray, Jr., eds., *The Human Genome Project and the Future of Health Care* (Bloomington and Indianapolis, Ind.: Indiana University Press, 1996).

15. See Richard R. Sharp and J. Carl Barrett, "The Environmental Genome Project and Bioethics," *Kennedy Institute of Ethics Journal* 9:2 (1999):175–88.

16. See, for example, Francis S. Collins and Alan E. Guttmacher, "Genetics Moves into the Medical Mainstream," *Journal of the American Medical Association* 286 (2001):2322–23. Also see Francis S. Collins and V. A. McKusick, "Implications of the Human Genome Project for Medical Science," *Journal of the American Medical Association* 285:5 (2001):540–44. Also see Barton Childs and David Valle, "Genetics, Biology and Disease," *Annual Review of Genomics and Human Genetics* 01 (2000):1–19.

17. For example, Barton Childs and David Valle, "Genetics, Biology, and Disease," *Annual Review of Human Genetics* 1 (2000):1–19.

18. See the issue of *The Scientist* 15:7 (2001) on recent developments in the human proteome project. Also see Barry A. Palevitz, "Deciphering Protein Evolution," *The Scientist* (2001):18. For a study on the analysis of the proteome in relation to the human genome, see G. Subramanian, et al., "Implications of the Human Genome for Understanding Human Biology

and Medicine," *Journal of the American Medical Association* 286 (2001):2296–2307 (specifically the section on "analysis of the predicted protein set," at 2229–2301); Jim Kling, "Pushing Proteomics," *The Scientist* (April 15, 2002):43; Laura DeFrancesco, "Probing Protein Interactions," *The Scientist* (April 15, 2002):28–30; Carol Ezzell, "Proteins Rule," *Scientific American* (April 2002):41–47; Deborah A. Fitzgerald, "Lipids + Genomics = Lipomics," *The Scientist* (February 4, 2002):42.

19. See Donald S. Fredrickson, *The Recombinant DNA Controversy, A Memoir: Science, Politics, and Public Interest, 1974–1981* (Washington, D.C.: ASM Press, 2001) (Dr. Fredrickson was director of the NIH); Steve Jones, *Genetics in Medicine: Real Promises, Unreal Expectations* (New York: Millbank Memorial Fund, 2000). Also see Jonathan Michael Kaplan, *The Limits and Lies of Human Genetic Research: Dangers for Social Policy* (New York: Routledge, 2000); Michael H. Shapiro, "Is Bioethics Broke? On the Idea of Ethics and Law 'Catching Up' with Technology," *Indiana Law Review* 3 (1999):17ff; Henry T. Greely, "The Revolution in Human Genetics," *South Carolina Law Review* 52 (2001):377ff.

20. For the policy debate, see British Medical Association, *Human Genetics: Choice and Responsibility* (New York: Oxford University Press, 1998), especially chaps. 1 and 8; for the ethics debate, see the doctoral dissertation of R. H. M. V. Hoedemaekers, *Normative Determinants of Genetic Screening and Testing: An Examination of Values, Concepts and Processes Influencing the Moral Debate* (Nijmegen, Netherlands: University of Nijmegen, 1998); for the impact of the debate upon insurance, see five valuable chapters on "Genetics and Insurance" in A. K. Thompson and R. F. Chadwick, *Genetic Information: Acquisition, Access, and Control* (New York: Kluwer, 1999), 31–84; for an application of the debate to Alzheimer's disease, see Stephen G. Post and Peter J. Whitehouse, eds., *Genetic Testing for Alzheimer's Disease: Ethical and Clinical Issues* (Baltimore, Md.: Johns Hopkins University Press, 1998); and Jennifer S. Geeter, "Coding for Change: The Power of the Human Genome to Transform the American Health Insurance System," *American Journal of Law & Medicine* 28 (2000):1–76.

21. For a scientific study of using microarrays for gene expression profiling, including the definition of *normal*, see Hadley C. King and Animesh A. Sinha, "Gene Expression Profile Analysis by DNA Microarrays: Promise and Pitfalls," *Journal of the American Medical Association* 286 (2001):2280–88. For ethical discourse on genetic normalcy, etc., see Bernard Gert, et al., *Morality and the New Genetics* (Boston: Jones and Bartlett, 1996). Also see D. M. Bartels, B. S. LeRoy, and A. L. Caplan, *Prescribing Our Future: Ethical Challenges in Genetic Counseling* (De Gruyter, 1993); Barbara Katz Rothman, *The Book of Life: A Personal and Ethical Guide to Race, Normalcy,*

and the Ethical Implications of the Human Genome Project (Boston: Beacon Press, 1998); Roger A. Willer, ed., *Genetic Testing and Screening: Critical Engagement at the Intersection of Faith and Science* (Minneapolis, Minn.: Kirk House, 1998); Maureen Junker-Kenny and Lisa Sowle Cahill, *The Ethics of Genetic Engineering*, 2 (1998) *Concilium* (Maryknoll: Orbis Books, 1998); Helga Kuhse, "Preventing Genetic Impairments: Does It Discriminate against People with Disabilities?" in Thompson and Chadwick, *Genetic Information*, 17–30; Magnus Reindal Solveig, "Disability, Gene Therapy and Eugenics," *Journal of Medical Ethics* 26 (2000):85–88.

22. For example, Sherman Elias, "Preimplantation Genetic Diagnosis by Comparative Genomic Hybridization," *New England Journal of Medicine* 345 (2001):1569–71; Harvey L. Levy and Simone Albers, "Genetic Screening of Newborns," *Annual Review of Human Genetics* 1 (2000):139–77; Marian D. Damewood, "Ethical Implications of a New Application of Preimplantation Genetic Diagnosis," *Journal of the American Medical Association* 285 (2001):3143–44; Dena Towner and Roberta Springer Loewy, "Ethics of Preimplantation Diagnosis for a Woman Destined to Develop Early-Onset Alzheimer Disease," *Journal of the American Medical Association* 287:8 (2002):1038–40; Yury Verlinsky, et al., "Preimplantation Diagnosis for Early-Onset Alzheimer Disease Caused by V717L Mutation," *Journal of the American Medical Association* 287:8 (2002):1018–21; Germain Kopaczynski, "Preimplantation Genetic Diagnosis," *Ethics & Medics* 27:5 (2002):1–3.

23. For example, George J. Annas, "Reforming Informed Consent to Genetic Research," *Journal of the American Medical Association* 286 (2001):2326–28. Also see George J. Annas and M. A. Grodin, eds., *The Nazi Doctors and the Nuremberg Code: Human Rights in Human Experimentation* (New York: Oxford University Press, 1992); American Society of Human Genetics, "Statement on Informed Consent for Genetic Research," *American Journal of Human Genetics* 59 (1996):471–74; L. O. Gostin and J. G. Hodge, "Genetic Privacy and the Law: An End to Genetic Exceptionalism," *Jurimetrics Journal* 40 (1999):21–58; L. M. Besklow, et al., "Informed Consent for Population-Based Research involving Genetics," *Journal of the American Medical Association* 286 (2001):2315–21; S. Joffee, et al., "Quality of Informed Consent: A New Measure of Understanding among Research Subjects," *Journal of the National Cancer Institute* 93 (2001):139–47.

24. For example, Susan M. Wolf, "Beyond 'Genetic Determinism': Toward the Broader Harm of Geneticism," *Journal of Law, Medicine, & Ethics* 23 (1995):345–53; John F. Kilner, et al., *Genetic Ethics: Do the Ends Justify the Genes?* (Grand Rapids, Mich.: Eerdmans, 1997); M.A. Rothstein,

Genetic Secrets: Protecting Privacy and Confidentiality in the Genetic Era (New Haven, Conn.: Yale University Press, 1997); Ted Peters, *Genetics: Issues of Social Justice* (Cleveland, Ill.: Pilgrim Press, 1998); American Society of Human Genetics, "ASGH Statement: Professional Disclosure of Familial Genetic Information," *American Journal of Human Genetics* 62:2 (1998):474–83; Mark A. Hall and Stephen S. Rich, "Genetic Privacy Laws and Patients' Fear of Discrimination by Health Insurers: The View from Genetic Counselors," *Journal of Law, Medicine & Ethics* 28 (2000):245ff; Mark A. Hall and Stephen S. Rich, "The Impact of Genetic Discrimination of Laws Restricting Health Insurers' Use of Genetic Information," *American Journal of Human Genetics* 66 (2000):293–307; P. A. Roche and G. J. Annas, "Protecting Genetic Privacy," *National Review of Genetics* 2 (2001):392–96; J. R. Botkin, "Protecting the Privacy of Family Members in Survey and Pedigree Research," *Journal of the American Medical Association* 285 (2001):207–11.

25. For example, Ellen Wright Clayton, et al., "Informed Consent for Genetic Research on Stored Tissue Samples," *Journal of the American Medical Association* 274 (1995):1786–87. Also see American Society of Human Genetics, "Statement on Informed Consent for Genetic Research"; and American College of Medical Genetics, "Statement of Storage and Usage of Genetic Materials," *American Journal of Human Genetics* 57 (1995):1499–1500; Henry T. Greely, "Breaking the Stalemate: A Prospective Regulatory Framework for Unforeseen Research Issues of Human Tissue Samples and Health Information," *Wake Forest Law Review* (Genetic Technology issue) 34 (1999):737ff; J. F. Merz, et al., "IRB Review and Consent in Human Tissue Research," *Science* 283 (1999):1647–48.

26. For example, President's Commission for the Study of Ethical Problems in Medicine and Biomedical and Behavioral Research, *Screening and Counseling for Genetic Conditions* (Washington D.C.: Government Printing Office, 1983); Nuffield Council on Bioethics, *Genetic Screening: Ethical Issues* (London: Nuffield Council, 1993); British Medical Association, *Human Genetics: Choice and Responsibility*; World Health Organization, *Proposed International Guidelines on Ethical Issues in Medical Genetics and Genetic Services* (Geneva, Switzerland: WHO, 1998); Ruth Chadwick, Henk A. M. J. ten Have, et al., "Genetic Screening and Ethics: European Perspectives," *Journal of Medicine and Philosophy* 23:3 (1998):255–73; Ruth Chadwick, Henk ten Have, et al., *The Ethics of Genetic Screening* (New York: Kluwer, 1999).

27. Allen Buchanan, Dan W. Brock, Norman Daniels, Daniel Wikler, *From Chance to Choice: Genetics and Justice* (Cambridge, U.K.: Cambridge University Press, 2000); Mary Briody Mahowald, *Genes, Women, and*

Equality (New York: Oxford University Press, 2000); Mary Briody Mahowald, "Genes, Clones, and Gender Equality," *DePaul Journal of Health Care Law* 3 (2000):495–526; Maxwell J. Mehlman and Jeffrey R. Botkin, *Access to the Genome: The Challenge of Equality* (Washington, D.C.: Georgetown University Press, 1998); also see Arti K. Rai, "The Information Revolution Reaches Pharmaceuticals: Balancing Innovation Incentives, Cost, and Access in the Post-Genomics Era," *University of Illinois Law Review* (2001):173ff.

28. For example, S. S. Lee, et al., "The Meaning of 'Race' in the New Genomics: Implications for Health Disparities Research," *Yale Journal of Health Policy, Law, Ethics* 1 (2001):33–75.

29. See David Resnik, "The Morality of Human Gene Patents," *Kennedy Institute of Ethics Journal* 7 (1997):43–61; M. Cathleen Kaveny, "Genetics and the Future of American Law and Policy," *Concilium* 225 (1998):69–70; Mark Hanson, "Biotechnology and Commodification within Health Care," *Journal of Medicine and Philosophy* 24 (1999):267–87; Baruch Brody, "Public Goods and Fair Prices: Balancing Technological Innovation with Social Well-Being," *Hastings Center Report* (1996):5ff; Rebecca S. Eisenberg, "Public Research and Private Development: Patents and Technology Transfer in Government-Sponsored Research," *Virginia Law Review* 82 (1996):1663–67; Michael A. Heller and Rebecca S. Eisenberg, "Can Patents Deter Innovation?" *Science* 280 (1998):698ff; Leon R. Kass, "Triumph or Tragedy: The Moral Meaning of Genetic Technology," *American Journal of Jurisprudence* 45 (2000):1–16; Lisa Sowle Cahill, "Genetics, Commodification, and Social Justice in the Globalization Era," *Kennedy Institute of Ethics Journal* 11 (2001):221–38; Susan Cartier Poland, "Genes, Patents, and Bioethics—Will History Repeat Itself?" *Kennedy Institute of Ethics Journal* 10 (2000):265–72, with history of relevant U.S. patent law.

30. See National Coalition for Health Professional Education in Genetics, "Recommendations of Core Competencies in Genetics Essential for All Health Professionals," *Genetic Medicine* 3 (2001):155–59.

31. Ronald Cole-Turner, "Religion and the Human Genome," *Journal of Religion and Health* 31 (1992):161–73; Ronald Cole-Turner, *The New Genesis: Theology and the Genetic Revolution* (Louisville, Ky.: Westminster John Knox Press, 1993); Ronald Cole-Turner, "Religion and Gene Patenting," *Science* 270 (1995):52; James M. Gustafson, "Genetic Therapy: Ethical and Religious Reflections," *Journal of Contemporary Health Law and Policy* 8 (1992):183–200; James M. Gustafson, "A Christian Perspective on Genetic Engineering," *Human Gene Therapy* 5 (1994):747–54; Mark Hanson, "Religious Voices in Biotechnology: The Case of Gene Patenting," *Hastings Center Report* 26:7 (1997):1–21; M. Cathleen Kaveny, "Jurisprudence and

Genetics," *Theological Studies*, 60 (1999):135-47; James F. Keenan, "What Is Morally New in Genetic Manipulation?" *Human Gene Therapy* 1 (1990):289-98; James F. Keenan, "Genetic Research and the Elusive Body," in *Embodiment, Morality and Medicine*, L. S. Cahill and Margaret A. Farley (Boston: Kluwer Academic Publishers, 1995), 59-73; James F. Keenan, "Christian Perspectives on the Human Body," *Theological Studies* 55 (1994):330-46; Gerard Magill, "Ethical Challenges in Life Science Research," *MCC Messenger* (April 27, 2001):1-4; Ted Peters, "'Playing God' and Germline Intervention," *Journal of Medicine and Philosophy* 20 (1995):365-86; Ted Peters, *For the Love of Children: Genetic Technology and the Future of the Family* (Louisville, Ky.: Westminster John Knox Press 1996); Thomas A. Shannon, "Ethical Issues in Genetics," *Theological Studies* 60 (1999):111-23; Thomas A. Shannon, *Made in Whose Image? Genetic Engineering and Christian Ethics* (Amherst, N.Y.: Humanity Books, 2000); Lisa Sowle Cahill, "The Genome Project: More than a Medical Milestone," *America* 183:4 (2000):7-13; James J. Walter, "Theological Issues in Genetics," *Theological Studies*, 60 (1999):124-134. Also see entire issue on "Ethics and Genetics" of *National Catholic Bioethics Quarterly* 1:4 (2001); "Genetics and Ethics," entire issue of *Health Progress* (March-April, 2001).

32. For example, a new journal, *Journal of Genetic Medicine*, has a cross-disciplinary focus on gene transfer and its clinical applications; it competes with several other prominent journals, such as *Gene Therapy, Human Gene Therapy*, and *Molecular Therapy*; see Michael Hauser, et al., "Gene Medicine," *Journal of the American Medical Association* 286 (2001):2333.

33. For example, Aubrey Milunsky, *Your Genetic Destiny: Know Your Genes, Secure Your Health, and Save Your Life* (Cambridge, Mass: Perseus, 2001).

34. See LeRoy Walters, "The Oversight of Human Gene Transfer Research," *Kennedy Institute of Ethics Journal* 10 (2000):171-74; Joan Stephenson, "Studies Illuminate Cause of Fatal Reaction in Gene-Therapy Trial," *Journal of the American Medical Association* 285 (2001):2570; Julian Savulescu, "Harm, Ethics Committees and the Gene Therapy Trial," *Journal of Medical Ethics* 27 (2001):148-50 (evaluated with regard to quality-adjusted life years); and Donna Shalala, "Protecting Research Subjects—What Must Be Done" *New England Journal of Medicine* 343 (2001):808-10 (commenting on the case of Jesse Gelsinger).

35. "Family Settles Suit Over Teen's Death During Gene Therapy," *Wall Street Journal* (November, 6, 2000): B6.

36. For example, Department of Health and Human Services, *Scientific Research: Continued Vigilance Critical to Protecting Human Subjects* (Washington, D.C.: Department of Health and Human Services, 1996);

Department of Health and Human Services, *Institutional Review Boards: A Time for Reform* (Washington, D.C.: Department of Health and Human Services, 1998); Department of Health and Human Services, *Protecting Human Research Subjects: Status of Recommendations* (Washington, D.C.: Department of Health and Human Services, 2000); Harold Y. Vanderpool, *The Ethics of Research involving Human Subjects* (Frederick, Md.: University Publishing Group, 1996); Cynthia Dunn McGuire and Gary Chadwick, *Protecting Study Volunteers in Research* (Boston, Mass.: CenterWatch Inc, 1999); Sana Loue, *Textbook of Research Ethics: Theory and Practice* (New York: Kluwer, 1999); Institute of Medicine, *Preserving Public Trust: Accreditation and Human Research Participant Protection Programs* (Washington, D.C.: National Academy Press, 2001).

37. National Commission for the Protection of Human Subjects of Biomedical and Behavioral Research, *The Belmont Report. Ethical Principles and Guidelines for the Protection of Human Subjects of Research*, DHEW Publication No. (OS) 78-0012 (Washington, D.C.: US Government Printing Office, 1979). For a critical review of how the report has functioned, see Eric J. Cassell, "The Principles of the Belmont Report Revisited," *Hastings Center Report* 30 (2000):12–21.

38. 45 CFR 46; 21 CFR 50, 56; see 46 FR 8366; 46 FR 8942. On the history of informed consent, see R. R. Faden and T. L. Beauchamp, *A History and Theory of Informed Consent* (New York: Oxford University Press, 1986).

39. See http://ohrp.osophs.dhhs.gov/humansubjects/guidance/45cfr46.htm. Also note that as revised, Subpart A of the DHHS regulations incorporates the Common Rule (Federal Policy) for the Protection of Human Subjects (56 FR 28003).

40. For a succinct explanation of the *Common Rule* adopted by deferral agencies, see Jeremy Sugarman, Anna C. Mastroianni, and Jeffery P. Kahn, *Ethics of Research with Human Subjects: Selected Policies and Resources* (Frederick, Md.: University Publishing Group, 1998), 35–52.

41. 65 FR 37136-37.

42. For example, Carol Mason Spicer, "Federal Oversight and Regulation of Human Subjects Research—An Update," *Kennedy Institute of Ethics Journal* 10 (2000):261–64; Lisa Eckenwiler, "Moral Reasoning and the Review of Research involving Human Subjects," *Kennedy Institute of Ethics Journal* 11 (2001);37–69, with a helpful bibliography.

43. See George J. Annas, "Reforming Informed Consent to Genetic Research," *Journal of the American Medical Association* 286 (2001):2327; Dan Curry, "US Restricts Research at Johns Hopkins after a Volunteer's Death," *Chronicle of Higher Education* (Aug. 3, 2001):A25; Gina Kolata, "Johns Hopkins Death Brings Halt to US Funded Human Research," *New*

York Times (July 20, 2001):A1, A16; subsequently, the family reached an out-of-court settlement, see "Family of Fatality in Study Settles with Johns Hopkins," *New York Times* (Oct. 12, 2001):A14.

44. National Bioethics Advisory Commission, *Ethical and Policy Issues in International Research: Clinical Trials in Developing Countries*, 2 vols. (Bethesda, Md.: U.S. Government Printing Office, 2001).

45. National Bioethics Advisory Commission, *Ethical and Policy Issues in Research Involving Human Participants* (Bethesda, Md.: National Bioethics Advisory Commission, 2001). Also see National Bioethics Advisory Commission, *Research involving Human Biological Materials: Ethical Issues and Policy Guidance*, 2 vols. (Rockville, Md.: U.S. Government Printing Office, 1999), and National Commission for the Protection of Human Subjects of Biomedical and Behavioral Research, *Report and Recommendations: Research on the Fetus* (1975), see 40 FR 33, 530 (1976).

46. See Jeffrey Brainard, "NIH Issues New Rules on Reporting Safety Problems in Gene-Therapy Trials," *Chronicle of Higher Education* (Nov. 30, 2001):A20. Also see *Federal Register*, Nov. 19, 2001.

47. For example, J. E. S. Hansen, "Embryonic Stem Cell Production through Therapeutic Cloning," *Journal of Medical Ethics* 28:2 (2002):86–88.

48. Ron Southwick, "President Bush Names Members of New Council on Bioethics," *Chronicle of Higher Education* (January 25, 2002):A20; Sheryl Gay Stolberg, "Bush's Advisers on Ethics Discuss Human Cloning," *New York Times* (January 18, 2002):A19 and "Science Academy Supports Cloning to Treat Disease," *New York Times* (January 19, 2002):A1, A12.

49. C. R. Freed, et al., "Transplantation of Embryonic Dopamine Neurons for Severe Parkinson's Disease," *New England Journal of Medicine* 344 (2001):710–19. The experiment involved forty Parkinson's patients, aged thirty-four to seventy-five. Either substantia nigra cells from four aborted fetuses (six to ten weeks) were transplanted into a patient's brain *or* the patient received sham surgery. No overall benefit resulted, but significant side effects such as serious writhing occurred in some younger patients. For a report on genomic screening of Parkinson disease, see William K. Scott, et al., "Complete Genomic Screen in Parkinson Disease: Evidence for Multiple Genes," *Journal of the American Medical Association* 286 (2001):2239–50. For critical ethical reviews, see Ruth Macklin, "Ethical Problems in Sham Surgery in Clinical Research," *New England Journal of Medicine* 341:13 (1999):992–96; and Wim Dekkers and Gerard Boer, "Sham Neurosurgery in Patients with Parkinson's Disease: Is It Morally Acceptable," *Journal of Medical Ethics* 27 (2001):151–56. Also see M. J. Friedrich, "Research Yields Clues to Improving Cell Therapy for Parkinson Disease," *Journal of the American Medical Association* 287:2 (2002):175–76.

50. For example, see Christine Willgoos, "FDA Regulations: An Answer to the Questions of Human Cloning and Germline Gene Therapy," *American Journal of Law and Medicine* 27 (2001):101–24.

51. James A. Thompson, et al., "Embryonic Stem Cell Lines Derived from Human Blastocysts," *Science* 282 (1998):1145–47; Michael J. Shamblot, et al., "Derivation of Human Pluripotent Stem Cells from Cultured Primordial Germ Cells," *Proceedings of the National Academy of Sciences* 95 (1998):13,726–31.

52. Ian Wilmut, et al., "Viable Offspring Derived from Fetal and Adult Mammalian Cells," *Nature* 385 (1997):810–13. Wilmut voices open concern about using the technique of cloning for human medicine, see Ian Wilmut, "Cloning for Medicine," *Scientific American* 279 (1998):58–63.

53. For a scientific introduction to stem-cell biology in general and embryonic stem cells in particular, see Daniel R. Marshak, Richard L. Gardner, and David Gottlieb, eds., *Stem Cell Biology* (Cold Spring Harbor, N.Y.: Cold Spring Harbor Laboratory Press, 2001), especially chaps. 1 and 10. Also see "Symposium: Human Primordial Stem Cells," *Hastings Center Report* 29:2 (1999):30–48; G. Keller and H. R. Snodgrass, "Human Embryonic Stem Cells: The Future Is Now," *Nature Medicine* 5 (1999):151–52; A. R. Chapman, et al., *Stem Cell Research and Applications: Monitoring the Frontiers of Biomedical Research* (Washington D.C.: American Association for the Advancement of Science and Institute for Civil Society, 1999); John R. Meyer, "Human Embryonic Stem Cells and Respect for Life," *Journal of Medical Ethics* 26 (2000):166–70; Janis L. Abkowitz, "Can Human Hematopoietic Stem Cells Become Skin, Gut, or Liver Cells?" *New England Journal of Medicine* 346:10 (2002):770–72; Irving L. Weissman, "Stem Cells—Scientific, Medical, and Political Issues," *New England Journal of Medicine* 346:20 (2002):1576–79.

54. National Bioethics Advisory Commission, *Ethical Issues in Human Stem Cell Research*, 3 vols. (Rockville, Md.: Government Printing Office, 1999); Department of Health and Human Services, "National Institute of Health Guidelines for Research Using Human Pluripotent Stem Cells," *Federal Register* 65:166 (2000):51975–81. Also see John C. Fletcher, "The National Bioethics Advisory Commission's Report on Stem Cell Research: A Review," *ASBH Exchange* 3:1 (2000):8–11; National Bioethics Advisory Commission, *Ethical Issues in Human Stem Cell Research: Vol. I. Report and Recommendations* (Rockville, Md.: National Bioethics Advisory Committee, 1999); National Bioethics Advisory Commission, *Ethical Issues in Human Stem Cell Research: Vol. II. Commissioned Papers* (Rockeville, Md.: National Bioethics Advisory Committee, 2000). A related, earlier document is Department of Health, Education and Welfare, *Report and*

Conclusions: Support of Research involving Human in vitro Fertilization and Embryo Transfer (Washington, D.C.: Government Printing Office, 1979). For some scholarly commentary, see Kevin W. Wildes, "The Stem Cell Report," *America* (October 16, 1999), 12–16; Eugene Rosso, "Panel Supports Use and Derivation of Stem Cells," *The Scientist* (October 11, 1999), 4; Charles Marwick, "Funding Stem Cell Research," *Journal of the American Medical Association* 281:8 (1999):692–93; R. Mikes, "NBAC and Embryo Ethics, *National Catholic Bioethics Quarterly* 1:2 (2001):163–87.

55. For the text of the television address by President George Bush, see "Address on Federal Funding of Embryonic Stem Cell Research," *Origins* 31 (2001):213–15. Also see R. Alto Charo, "Bush's Stem Cell Compromise: A Few Mirrors?" *Hastings Center Report* (2001):6–7.

56. National Institutes of Health, "Stem Cells: Scientific Progress and Future Research Directions" (July 2001), available from http://www.nih.gov/news/stemcell/scireport.htm.

57. See Ted Agres, "News Notes: Human Embryonic Stem Cell Registry Opens,' *The Scientist* (Nov. 26, 2001):10. Also see Ron Southwick, "NIH Releases List of Stem-Cell Colonies Approved for Federal Research," *Chronicle of Higher Education* (Nov. 23, 2001):A22.

58. See Charles E. Curran, "Cooperation: Toward a Revision of the Concept and Its Application," *Linacre Quarterly* 41:3 (1977):152–65, especially the section on "history of the concept" (155–57).

59. For examples of the ethical principle of cooperation in health care debates, see James F. Keenan, "Prophylactics, Toleration, and Cooperation: Contemporary Problems and Traditional Principles," *International Philosophical Quarterly* XXIX, 2: 114 (1989):205–20; James F. Keenan and Thomas R. Kopfensteiner, "The Principle of Cooperation," *Health Progress* (April 1995):23–27; and M. Cathleen Kaveny and James F. Keenan, "Ethical Issues in Health-Care Restructuring," *Theological Studies* 56: 1 (1995):136–50. For a modern reinterpretation of the principle, see M. Cathleen Kaveny, "Appropriation of Evil: Cooperation's Mirror Image," *Theological Studies* 61 (2000):280–313.

60. For an analysis on different types of complicity in this debate, see John A. Robertson, "Ethics and Policy in Embryonic Stem Cell Research," *Kennedy Institute of Ethics Journal* 9 (1999):109–36.

61. See, for example, "Reaction to President Bush's Decision on Embryonic Stem-Cell Research," *Origins* 31:12 (2001):206–12.

62. 410 U.S. 113 (1973). See Laurence H. Tribe, *American Constitutional Law* (Mineola, N.Y.: Foundation Press, 1988), 1341–42. For a legal critique, see the work of John T. Noonan, Jr. (Ninth Circuit Court of Appeals), "Raw Judicial Power," *National Review* (March 1973):259–63. The

Court's decision created a virtually unconditioned right of abortion, especially when read together with the Court's decision on the same day in *Doe v. Bolton*. For a more detailed analysis of this debate, see Gerard Magill and R. Randall Rainey, *Abortion & Public Policy* (Omaha, Nebr.: Creighton University Press, 1996), especially the introduction.

63. For a historical review of fetal tissue research and its connection with stem-cell research, see Jason H. Cassell, "Lengthening the Stem: Allowing Federally Funded Researchers to Derive Human Pluripotent Stem Cells from Embryos," *University of Michigan Journal of Law Reform* 34 (2001):547–72.

64. Department of Health, Education and Welfare, Ethics Advisory Board, "Report and Conclusions: HEW Support of Research Involving Human IVF and Embryo Transfer," *Federal Register* 44:118, June 18, 1979. See Robert Edwards and Patrick Steptoe, "Birth after Reimplantation of a Human Embryo," *Lancet* 2 (1978):336; Robert Edwards and Patrick Steptoe, *A Matter of Life* (New York: William Morrow, 1980).

65. See Mary Warnock, *A Question of Life: The Warnock Report on Fertilisation and Embryology* (Oxford: Basil Blackwell, 1985); Mary Warnock, "The Warnock Report," *British Medical Journal* 291 (1985):187–89; Royal Commission on New Reproductive Technologies, *Proceed with Care. Report of the Royal Commission on New Reproductive Technologies* (Ottowa, Canada: Ministry of Government Services, 1993); Human Fertilisation and Embryology Authority (Great Britain), *Code of Practice*, 4th ed. (1998). Also see National Institutes of Health, *Report of the Human Embryo Research Panel* (Bethesda, Md.: National Institutes of Health, 1994), and National Institutes of Health, *Papers Commissioned for the Human Embryo Research Panel*, 2 vols. (Bethesda, Md.: National Institutes of Health, 1994).

66. For example, Norman Ford, *When Did I Begin?* (New York: Cambridge University Press, 1988); Norman Ford, "The Human Embryo as a Person in Catholic Teaching," *National Catholic Bioethics Quarterly* 1:2 (2001):155–60; Antoine Suarez, "Hydatidiform Moles and Teratomas Confirm the Human Identity of the Preimplantation Embryo," *Journal of Medicine and Philosophy* 15 (1990):630–36; Thomas Shannon and Allan B. Wolter, "Reflections on the Moral Status of the Pre-Embryo," *Theological Studies* 51 (1990):603–26; Richard A. McCormick, "Who or What Is the Preembryo?" *Kennedy Institute of Ethics Journal* 1 (1991):1–15; Bonnie Steinbock, *Life before Birth: The Moral and Legal Status of Embryos and Fetuses* (New York: Oxford University Press, 1992); Lisa Sowle Cahill, "The Embryo and the Fetus: New Moral Contexts," *Theological Studies* 54 (1993):124–42; Mark Johnson, "Reflections on Some Catholic Claims for

Delayed Hominisation," *Theological Studies* 56 (1995):743–63; Jean Porter, "Individuality, Personal Identity, and the Moral Status of the Preembryo: A Response to Mark Johnson," *Theological Studies* 56 (1995):763–70; Carol Tauer, "Embryo Research and Public Policy," *Journal of Medicine and Philosophy* 22 (1997):423–39; Thomas A. Shannon, "Fetal Status: Sources and Implications," *Journal of Medicine and Philosophy* 22 (1997):415–222; Lisa Sowle Cahill, "The Status of the Embryo and Policy Discourse," *Journal of Medicine and Philosophy* 22 (1997):407–14; Glen McGee and Arthur Caplan, "The Ethics and Politics of Small Sacrifices in Stem Cell Research," *Kennedy Institute of Ethics Journal* 9 (1999):151–58; Peter Cataldo, "Human Rights and the Human Embryo," *Ethics & Medics* 26 (2001):1–2.

67. For example, Eric Juengst and Michael Fossel, "The Ethics of Embryonic Stem Cells," *Journal of the American Medical Association* 284 (2000):3180–84; and Paul Lauritzen, ed., *Cloning and the Future of Human Embryo Research* (New York: Oxford University Press, 2001), sec. 1 with five insightful essays: Bonnie Steinbock, "Respect for Human Embryos," (21–33); Courtney S. Campbell, "Source or Resource? Human Embryo Research as an Ethical Issue" (34–49); Maura A. Ryan, "Creating Embryos for Research: On Weighing Symbolic Costs" (50–66); James Keenan, "Casuistry, Virtue, and the Slippery Slope: Major Problems with Producing Human Embryonic Life for Research Purposes" (67–81), and R. Alta Charo, "Every Cell Is Sacred: Logical Consequences of the Argument from Potential in the Age of Cloning" (82–91); Thomas A. Shannon, "Human Cloning," *America* (February 18, 2002):15–18.

68. For example, Richard Doerflinger, "The Ethics of Funding Embryonic Stem Cell Research: A Catholic Viewpoint," *Kennedy Institute of Ethics Journal* 9 (1999):137–50; Richard Doerflinger, "The Policy and Politics of Embryonic Stem Cell Research," *National Catholic Bioethics Quarterly* 1:2 (2001):135–43; Sharon M. Parker, "Bringing the 'Gospel of Life' to American Jurisprudence: A Religious, Ethical, and Philosophical Critique of Federal Funding for Embryonic Stem Cell Research," *Journal of Contemporary Health Law and Policy* 17 (2001):771–808; Alfred Cioffi, "Reproductive and Therapeutic Cloning," *Ethics & Medics* 27:3 (2002):1–3; U.S. Catholic Conference, "Harvesting Embryonic Stem Cells for Research: Response to NIH Draft Guidelines," *Origins* 29:35 (2000):566–71.

69. For example, Goldie Blumenstyk, "U. of Missouri Disavows Any Use of Patent to Clone Humans," *Chronicle of Higher Education* (May 31, 2002):A29; Ronald Cole-Turner, *Human Cloning: Religious Responses* (Louisville, Ky.: Westminster Press, 1997); Jonathan R. Cohen, "Creation and Cloning in Jewish Thought," *Hastings Center Report* (1999):7–12.

70. National Bioethics Advisory Commission, *Cloning Human Beings: Report and Recommendations of the National Bioethics Advisory Commission* (Rockville, Md.: NBAC, 1997). For a critique of the NBAC decision, see Ronald Green, *The Human Embryo Research Debates: Bioethics in the Vortex of Controversy* (Oxford University Press, NY: 2001), 113–19.

71. National Bioethics Advisory Commission, *Ethical Issues in Stem Cell Research, Vol. I, Report and Recommendations* (Rockville, Md.: National Bioethics Advisory Commission, 1999). For a critique of the NBAC decision see Ronald Green, *The Human Embryo Research Debates*, 158–63.

72. U.K. Department of Health, "Stem Cell Research: Medical Progress with Responsibility. A Report from the Chief Medical Officer's Expert Group Reviewing the Potential of Developments in Stem Cell Research and Cell Nuclear Replacement to Benefit Human Health" (June 2000). See Her Majesty's Stationery Office, "The Human Fertilisation and Embryology (Research Purposes) Regulations 2001" and Britain's Human Reproductive Cloning Act (2001), which was passed in response to a High Court judgment on November 15, 2001, that "embryos created by cell nuclear replacement were not governed by the Human Fertilisation and Embryology Act 1990." The current situation in Britain is that legislation permits human therapeutic cloning while banning human reproductive cloning. See Gautam Naik, "Britain Is Poised to Ban Cloning Babies to Term," *Wall Street Journal* (Nov. 23, 2001):B5 (explaining the passage of British legislation to ban reproductive cloning was intended to close a loophole in previous legislation that approved human therapeutic cloning). Also see Alex Sleator, *Stem Cell Research and Regulations under the Human Fertilisation and Embryology Act 1990* (London: House of Commons, 2000); Documentation Column, "The Status of the Embryo: The House of Lords Select Committee Report," *The Tablet* (March 9, 2002):31–32; Arlene Judith Klotzko, "A Cloning Emergency in Britain?" *The Scientist* (January 7, 2002):12.

73. See Sheila Dziobon, "Germ-Line Gene Therapy: Is the Existing UK Norm Ethically Valid?" in Thompson and Chadwick, *Genetic Information* (1999), 255–65; see Gene Therapy Advisory Committee, *Third Annual Report* (Health Department of the UK, 1997); *Report of the Committee on the Ethics of Gene Therapy*, The Clothier Report (London: HMSO, 1992).

74. For example, Kathinka Evers, "European Perspectives on Therapeutic Cloning," *New England Journal of Medicine* 346:20 (2002):1579–82; George J. Annas, "Cloning and the U.S. Congress," *New England Journal of Medicine* 346:20 (2002):1599–602.

75. Jose B. Cibelli, et al., "Somatic Cell Nuclear Transfer in Humans: Pronuclear and Early Embryonic Development," *Journal of Regenerative Medicine* 2 (2001):25–31; Jose B. Cibelli, et al., "The First Human Clone,"

Scientific American (January 2002):42-51. Also see Gina Kolata, "Company Says It Produced Embryo Clones," *New York Times* (Nov. 26, 2001):A14; Gautam Naik, "Brave New World: Stem-Cell Researchers Make Cloned Embryos of a Living Human," *Wall Street Journal* (Nov. 26, 2001):A1. Media coverage has ranged from outright support, in Virginia Postreel, "Should Human Cloning Be Allowed? Yes, Don't Impede Medical Progress," *Wall Street Journal* (Dec. 5, 2001):A20, to blatant opposition, in Eric Cohen, "Should Human Cloning Be Allowed? No, It's a Moral Monstrosity," *Wall Street Journal* (Dec. 5, 2001): A20; and from cautionary tales, in Gina Kolata, "In Cloning, Failure Far Exceeds Success," *New York Times* (Dec. 11, 2001):D1, D40, to advocacy support, in Jill Carroll and Antonio Regaldo, "Congress Urged not to Enact Cloning Ban," *Wall Street Journal* (Dec. 5, 2001): B9, and advocacy opposition, in Renato Martino, "Human Cloning Prohibition Urged," *Origins* 31 (Dec. 6, 2001):439-40.

76. For an earlier perspective on related research by a competing corporation, see Geron Ethics Advisory Board, "Research with Human Embryonic Stem Cells: Ethical Considerations," *Hastings Center Report* 29 (1999):31-36. For an earlier essay on human cloning research by several scholars including Ronald Green, see Robert P. Lanza, et al., "The Ethical Validity of Using Nuclear Transfer in Human Transplantation," *Journal of the American Medical Association* 284 (2001):3175-79.

77. See the editorial news report on the role of the advisory ethics board in this experiment, "Vatican and Bush Deplore Cloning of the Human Embryo," in *The Tablet* (Dec. 1, 2001):17722. For the impact of Green's previous works on the moral status of the early embryo on this topic of human cloning, see Ronald M. Green, *The Human Embryo Research Debates: Bioethics in the Vortex of Controversy* (New York: Oxford University Press, 2001). Also see Ronald M. Green, "Much Ado about Mutton: An Ethical Review of the Cloning Controversy," in Paul Lauritzen, ed., *Cloning and the Future of Human Embryo Research*, (New York: Oxford University Press, 2001), 114-31.

78. For example, John A. Robertson, "Ethics and Policy in Embryonic Stem Cell Research," *Kennedy Institute of Ethics Journal* 9 (1999):109-36; Leon R. Kass and James Q Wilson, *The Ethics of Human Cloning* (Washington, D.C.: AEI Press, 1998); Cynthia B. Cohen, "Banning Human Cloning—Then What?" *Kennedy Institute of Ethics Journal* 11 (2001):205-9; Mark J. Hanson, "Cloning for Therapeutic Purposes: Ethical and Policy Considerations," 32 *University of Toledo Law Review* (2001):355-65; Kevin Fitzgerald, "Cloning: Can It Be Good for Us? An Overview of Cloning Technology and Its Moral Implications," *University of Toledo Law Review* 32 (2001):327-36; Paul Lauritzen, ed., *Cloning and the Future of Human Embryo Research* (New York: Oxford University Press, 2001).

79. For insightful essays on public policy issues on cloning, see sec. 3 in Paul Lauritzen, ed., *Cloning and the Future of Human Embryo Research*, including Carol A. Tauer, "Responsibility and Regulation: Reproductive Technologies, Cloning, and Embryo Research" (145–61), Jonathan D. Moreno and Alex John London, "Consensus, Ethics, and Politics in Cloning and Embryo Research" (162–77), Brian Stiltner, "Morality, Religion, and Public Bioethics: Shifting the Paradigm for the Public Discussion of Embryo Research and Human Cloning" (178–200), and Heidi Forster and Emily Ramsey, "The Law Meets Reproductive Technology: The Prospect of Human Cloning" (201–21).

80. For example, Jeffrey Brainard, "Celebrities, Scientists, and Politicians Try to Shape the Debate over Cloning," *Chronicle of Higher Education* (May 17, 2002):A27.

81. See Lori P. Knowles, "Science Policy and the Law: Reproductive and Therapeutic Cloning," *New York University School of Law Journal of Legislation and Public Policy* 4 (2000/2001):13–22.

SELECT BIBLIOGRAPHY

Books

Agich, G., ed. *Responsibility in Health Care.* Dordrecht, Netherlands: D. Reidel Publishers, 1982.

Agius, A., et al., eds. *Germ-Line Intervention and Our Responsibilities to Future Generations.* Boston: Kluwer Academic Publishers, 1998.

American Medical Association. *Current Opinions of the Council of Ethical and Judicial Affairs on Clinical Investigation.* American Medical Association, 1992.

Andrews, L. B. *The Clone Age: Adventures in the New World of Reproductive Technology.* New York: Henry Holt, 2000.

Annas, G. J., and M. A. Grodin, eds. *The Nazi Doctors and the Nuremberg Code: Human Rights in Human Experimentation.* New York: Oxford University Press, 1992.

Annas, G. J., and E. Sherman. *Gene Mapping: Using Law and Ethics as Guides.* New York: Oxford University Press, 1992.

Appleyard, B. *Brave New Worlds: Staying Human in the Genetic Future.* New York: Viking, 1998.

Arney, W. R., and B. J. Bergen. *Medicine and the Management of Living. Taming the Last Great Beast.* Chicago: University of Chicago Press, 1984.

Baldi, P. *The Shattered Self: The End of Natural Evolution.* Boston, Mass.: MIT Press, 2001.

Bankowski, Z., and A. Capron, eds. *Genetics, Ethics and Human Values: Human Genome Mapping, Genetic Screening and Therapy.* CIOMS Round Table Conference, 1991.

Bartels, D. M., B. S. LeRoy, and A. L. Caplan. *Prescribing Our Future: Ethical Challenges in Genetic Counseling.* New York: De Gruyter, 1993.

Beauchamp, T. L., and J. F. Childress. *Principles of Biomedical Ethics.* New York: Oxford University Press, 1994.

Blank, R. H., and A. L. Bonnicksen, eds. *Medicine Unbound: The Human Body and the Limits of Medical Intervention*. New York: Columbia University Press, 1994.

Brahams, D. *Human Genetic Information: Science, Law, and Ethics*. Ciba Foundation Symposium 149. Chichester, U.K.: John Wiley and Sons, 1990.

Brannigan, M. C. *Ethical Issues in Human Cloning: Cross-Disciplinary Perspectives*. New York: Seven Bridges Press, 2001.

British Medical Association. *Our Genetic Future: The Science and Ethics of Genetic Technology*. Oxford: Oxford University Press, 1992.

———. *Human Genetics: Choice and Responsibility*. New York: Oxford University Press, 1998.

Brock, D. J., C. Rodeck, and M. A. Ferguson-Smith, eds. *Prenatal Diagnosis and Screening*. Edinburgh, U.K.: Churchill Livingstone, 1992.

Brodeur, Dennis, and Kevin O'Rourke. *Medical Ethics: Common Ground for Understanding*, 2 vols. St. Louis, Mo.: Catholic Health Association, 1986, 1989.

Brownsword, R., W. R. Cornish, and M. Llewelyn. *Law and Human Genetics: Regulating a Revolution*. Oxford: Hart Publishing, 1998.

Buchanan, A., D. W. Brock, N. Daniels, and D. Wikler. *From Chance to Choice: Genetics and Justice*. Cambridge, U.K.: Cambridge University Press, 2000.

Burley, J., ed. *The Genetic Revolution and Human Rights*. New York: Oxford University Press, 1999.

Cahill, L. S., and M. A. Farley, eds. *Embodiment, Morality and Medicine*. Boston: Kluwer Academic Publishers, 1995.

Campbell, T., et al., eds. *Human Rights: From Rhetoric to Reality*. Oxford: Basil Blackwell, 1986.

Canon Law Society of Great Britain and Ireland. *The Canon Law, Letter and Spirit*. Alexandria, New South Wales, Australia: E. J. Dwyer, 1995.

Caplan, A. L. *If I Were a Rich Man Could I Buy a Pancreas? And Other Essays on the Ethics of Health Care*. Bloomington and Indianapolis: Indiana University Press, 1992.

Chadwick, R., ed.-in-chief. *Encyclopedia of Applied Ethics*. San Diego, Calif.: Academic Press, 1998.

_____. *The Concise Encyclopedia of the Ethics of New Technologies.* San Diego, Calif.: Academic Press, 2001.
Chadwick, R., H. A. M. J. ten Have, et al. *The Ethics of Genetic Screening.* New York: Kluwer, 1999.
Chadwick, R., M. Levitt, and D. Shickle, eds. *The Right to Know and the Right Not to Know.* Aldershot, U.K.: Avebury, 1997.
Chapman, A. R. *Unprecedented Choices: Religious Ethics at the Frontiers Of Genetic Science.* Minneapolis, Minn.: Fortress Press, 1999.
Chapman, A. R., et al. *Stem Cell Research and Applications: Monitoring the Frontiers of Biomedical Research.* Washington, D.C.: American Association for the Advancement of Science and Institute for Civil Society, 1999.
Clothier Report. *Report of the Committee on the Ethics of Gene Therapy.* London: HMSO (1992).
Cole-Turner, R. *The New Genesis: Theology and the Genetic Revolution.* Louisville, Ky.: Westminster John Knox Press, 1993.
_____. *Human Cloning: Religious Responses.* Louisville, Kentucky: Westminster Press, 1997.
Cole-Turner, R., and B. Waters. *Pastoral Genetics: Theology and Care at the Beginning of Life.* Cleveland, Ohio: Pilgrim Press, 1996.
Congregation for the Doctrine of the Faith. *Instruction on Respect for Human Life in Its Origin and on the Dignity of Procreation: Replies to Certain Questions of the Day.* Vatican City: Vatican Polyglot Press, 1987.
Connor, M., and M. Ferguson-Smith. *Essential Medical Genetics.* Oxford: Blackwell Science, 1997.
Cook-Deegan, R. *The Gene Wars: Science, Politics, and the Human Genome.* New York: W. W. Norton, 1994.
Correa, J. D. V., and E. Sgreccia, eds. *Human Genome, Human Person, and the Society of the Future: Proceedings of the Fourth Assembly of the Pontifical Academy for Life, Vatican City, February 23–25, 1998.* Città del Vaticano: Libreria Editrice Vaticana, 1999.
Cranor, C. F., ed. *Are Genes Us? The Social Consequences of the New Genetics.* New Brunswick, N.J.: Rutgers University Press, 1994.
Danish Council of Ethics. *Ethics and Mapping of the Human Genome.* Copenhagen: Danish Council of Ethics, 1993.
Davis, D. S. *Genetic Dilemmas: Reproductive Technology, Parental Choices, and Children's Futures.* New York: Routledge, 2001.

Davis, D. S., and L. Zoloth, eds. *Notes from a Narrow Ridge: Religion and Bioethics.* Hagerstown, Md.: University Publishing. Group, 1999.

De Crespigny, L., M. Espie, and S. Holmes. *Prenatal Testing. Making Decisions in Pregnancy.* Ringwood, Victoria, Australia: Penguin Books, 1998.

Demy, T. J., and G. P. Stewart, eds. *Genetic Engineering: A Christian Response: Crucial Considerations in Shaping Life.* Michigan: Kregel Publications, 1999.

Department of Energy, Human Genome Program. *Primer on Molecular Genetics.* Washington, D.C.: U.S. Department of Energy, 1992.

Department of Health and Human Services. *Scientific Research: Continued Vigilance Critical to Protecting Human Subjects.* Washington, D.C.: Department of Health and Human Services, 1996.

———. *Institutional Review Boards: A Time for Reform.* Washington, D.C.: Department of Health and Human Services, 1998.

———. *Protecting Human Research Subjects: Status of Recommendations.* Washington, D.C.: Department of Health and Human Services, 2000.

Department of Labor, et al. *Genetic Information and the Workplace.* New York: Cornell University, School of Industrial and Labor Relations, 1998.

Dijck, J. van. *Imagination: Popular Images of Genetics.* London: Macmillan Press, 1998.

Dudley, W. *The Ethics of Human Cloning.* San Diego, Calif.: Greenhaven Press, 2001.

Duster, T. *Backdoor to Eugenics.* New York: Routledge, 1990.

Dyson, A., and J. Harris. *Ethics and Biotechnology.* New York: Routledge, 1994.

Edwards, R. G., ed. *Preconception and Preimplantation Diagnosis of Human Genetic Disease.* Cambridge, UK: Cambridge University Press, 1993.

Edwards, R. G., and P. Steptoe. *A Matter of Life.* New York: William Morrow, 1980.

Engelhardt, H. T., Jr. *The Foundations of Bioethics.* Oxford: Oxford University Press, 1996.

Faden, R. R., and T. L. Beauchamp. *A History and Theory of Informed Consent*. New York: Oxford University Press, 1986.

Fisher, N. L., ed. *Cultural and Ethnic Diversity: A Guide for Genetics Professionals*. Baltimore, Md.: John Hopkins University Press, 1996.

Fisk, N. M., and K. J. Moise, Jr. *Fetal Therapy: Invasive and Transplacental*. Cambridge, U.K.: Cambridge University Press, 1997.

Fletcher, J. *The Teaching of Medical Ethics*. Hastings-on-Hudson, N.Y.: Hastings Center, 1973.

———. *The Ethics of Genetic Control: Ending Reproductive Roulette*. Garden City, N.Y.: Anchor Books, 1974.

Fletcher, J., and D. Wertz. *Ethics and Applied Human Genetics: A Cross-Cultural Perspective*. Heidelberg: Springer Verlag, 1988.

Ford, N. M. *Live Out the Truth in Love*. Melbourne, Australia: Catholic Education Office, 1985, 1991.

———. *When Did I Begin? Conception of the Human Individual in History, Philosophy and Science*. Cambridge, UK: Cambridge University Press, 1988.

———. *The Prenatal Person: Ethics from Conception to Birth*. Oxford: Blackwell, 2002.

Fredrickson, D. S. *The Recombinant DNA Controversy, A Memoir: Science, Politics, and Public Interest 1974–1981*. Washington, D.C.: ASM Press, 2001.

Galton, D. *In Our Own Image: Eugenics and the Genetic Modification of People*. London: Little Brown, 2001.

Gene Therapy Advisory Committee. *Report of the Committee on the Ethics of Gene Therapy*. The Clothier Report. London: HMSO, 1992.

———. *Third Annual Report*. UK Department of Health. London: HMSO, 1997.

Gert, B., et al. *Morality and the New Genetics*. Boston: Jones and Bartlett, 1996.

Glannon, W. *Genes and Future People: Philosophical Issues in Human Genetics*. Boulder, Colo.: Westview Press, 2001.

Glasner, P., and H. Rothman, eds. *Genetic Imaginations. Ethical, Legal and Social Issues in Human Genome Research*. Aldershot, U.K.: Ashgate, 1998.

Grant, George. *Technology and Justice*. Notre Dame, Ind.: University of Notre Dame Press, 1986.
Green, R. M. *The Human Embryo Research Debates: Bioethics in the Vortex of Controversy*. New York: Oxford University Press, 2001.
Grobstein, C. *A Double Image of the Double Helix: The Recombinant-DNA Debate*. San Francisco: W. H. Freeman, 1979.
Grodin, M. A., and L. H. Glantz, eds. *Children as Research Subjects: Science, Ethics, and Law*. New York: Oxford University Press, 1994.
Haering, B. *Ethics of Manipulation*. New York: Seabury Press, 1975.
Haldane, J. B. S. *Daedalus: Or Science and the Future*. New York: E. P. Dutton, 1924.
Hamrin, R. R. *Charting New Territory: Legislative Guide to Genetic Privacy and Discrimination*. Lexington, Ky.: Council of State Governments, 1998.
Harris, J. *Clones, Genes, and Immortality: Ethics and the Genetic Revolution*. Oxford: Oxford University Press, 1998.
Hauerwas, S. *Suffering Presence: Theological Reflections on Medicine, the Mentally Handicapped, and the Church*. South Bend, Ind.: University of Notre Dame Press, 1986.
Have, H. A. M. J. ten. *Death and Medical Power*. Bloomington, Ind.: Indiana University Press, 2004.
Have, H. A. M. J. ten, and David Clark. *The Ethics of Palliative Care: European Perspectives*. Buckingham, UK: Open University Press, 2002.
Have, H. A. M. J. ten, and Bert Gordjin, eds. *Bioethics in a European Perspective*. Dordrecht, the Netherlands: Kluwer Academic Publishers, 2001.
Have, H. A. M. J. ten, and R. Janssens, eds. *Palliative Care in Europe. Concepts and Policies*. Amsterdam, the Netherlands: IOS Press, 2001.
Have, H. A. M. J. ten, G. K. Kimsma, and S. F. Spicker, eds. *The Growth of Medical Knowledge*. Dordrecht, Netherlands: Kluwer Academic Publishers, 1990.
Heller, J. C. *Human Genome Research and the Challenge of Contingent Future Persons*. Omaha, Nebr.: Creighton University Press, 1996.

Heller, J. C., Cynthia B. Cohen, Bruce Jennings, E. F. Michael Morgan, David A. Scott, Timothy F. Sedgwick, and David H. Smith. *Faithful Living, Faithful Dying: Anglican Reflections on End of Life Care.* Harrisburg, Penn.: Morehouse Publishing, 2000.

Heller, J. C., and Nick Fotion, eds. *Contingent Future Persons: On the Ethics of Deciding Who Will Live, or Not, in the Future.* Dordrecht, The Netherlands: Kluwer Academic Publishers, 1997.

Heller, J. C., Joseph E. Murphy, and Mark E. Meaney, eds. *Guide to Professional Development in Compliance.* Gaithersburg, Md.: Aspen Publishers, Inc., 2001.

Heyd, D. *Genethics: Moral Issues in the Creation of People.* Berkeley, Calif.: University of California Press, 1992.

Hoedemaekers, R. H. M. V. *Normative Determinants of Genetic Screening and Testing: An Examination of Values, Concepts, and Processes Influencing the Moral Debate.* Nijmegen, Netherlands: University of Nijmegen, 1998.

Holland, S., K. Lebacqz, and L. Zoloth, eds. *The Human Embryonic Stem Cell Debate: Science, Ethics, and Public Policy.* Boston, Mass.: MIT Press, 2001.

Holtzman, N. A., and M. S. Watson. *Promoting Safe and Effective Genetic Testing in the United States: Final Report of the Task Force on Genetic Testing.* Washington D.C.: National Institutes of Health, 1997.

House of Commons Science and Technology Committee. Third Report, Session 1994–95, *Human Genetics: The Science and Its Consequences.* London: HMSO, 1994–1995.

Howard, T., and J. Rifkin. *Who Should Play God? The Artificial Creation of Life and What It Means for the Future of the Human Race.* New York: Delacorte Press, 1977.

Hubbard, R., and E. Wald. *Exploding the Gene Myth: How Genetic Information Is Produced and Manipulated by Scientists, Physicians, Employers, Insurance Companies, Educators, and Law Enforcers.* Boston: Beacon Press, 1993.

Huxley, A. *Brave New World.* London: Chatto and Windus, 1932.

Institute of Medicine. *Assessing Genetic Risks: Implications for Health and Social Policy.* Washington, D.C.: National Academy Press, 1994.

———. *Preserving Public Trust: Accreditation and Human Research Participant Protection Programs.* Washington, D.C.: National Academy Press, 2001.

Jansen, R., and D. Mortimer, eds. *Towards Reproductive Certainty: Fertility and Genetics beyond 1999.* New York: Parthenon Publishing Group, 1999.

Jonas, H. *The Imperative of Responsibility: In Search of an Ethics for the Technological Age.* Chicago: University of Chicago Press, 1984.

Jones, S. *Genetics in Medicine: Real Promises, Unreal Expectations.* New York: Millbank Memorial Fund, 2000.

Junker-Kenny, M., and L. S. Cahill. *The Ethics of Genetic Engineering 2 (1998) Concilium.* Maryknoll, NY: Orbis Books, 1998.

Kaplan, J. M. *The Limits and Lies of Human Genetic Research: Dangers for Social Policy.* New York: Routledge, 2000.

Kass, L. R. *Towards a More Natural Science.* New York: The Free Press, 1985.

Kass, L. R., and J. Q. Wilson. *The Ethics of Human Cloning.* Washington, D.C.: AEI Press, 1998.

Kegley, J. A. K. *Genetic Knowledge: Human Values and Responsibility.* Lexington, Ky.: International Conference on the Unity of the Sciences, 1998.

Kennedy, I. *Treat Me Right: Essays in Medical Law and Ethics.* Oxford: Clarendon Press, 1988, repr. 1994.

Kevles, D. J. *In the Name of Eugenics: Genetics and the Uses of Human Heredity.* Harmondsworth, UK: Penguin, 1985.

Kevles, D. J., and L. R. Hood. *The Code of Codes: Scientific and Social Issues in the Human Genome Project.* Boston: Harvard University Press, 1992.

Kilner, J. F., et al. *Genetic Ethics: Do the Ends Justify the Genes?* Grand Rapids, Mich.: Eedermans, 1997.

Klotzko, A. J. *The Cloning Sourcebook.* New York: Oxford University Press, 2001.

Knoppers, B. M., ed. *Socio-Ethical Issues in Human Genetics.* Cowansville, Québec, Canada: Y. Blais, 1998.

Koechlin, F., and D. Ammann. *Mythos Gen.* Rieden bei Baden, Austria: Utzinger/Stemmle Verlag, 1997.

Koren, G., ed. *Ethics in Pediatric Research.* Malabar, Fla.: Krieger Publishing, 1993.

Korf, B. R. *Human Genetics: A Problem-Based Approach.* 2nd ed. Malden, Mass: Blackwell Science, 2000.

Kristol, W., and E. Cohen, eds. *The Future Is Now: America Confronts the New Genetics.* Lanham, Md.: Rowman and Littlefield, 2002.

Launis, V., J. Pietarinen, and J. Räikkä, eds. *Genes and Morality. New Essays.* Amsterdam: Rodopi, 1999.

Lauritzen, P., ed. *Cloning and the Future of Human Embryo Research.* New York: Oxford University Press, 2001.

Levine, R. *Ethics and the Regulation of Clinical Research.* Baltimore, Md.: Urban and Schwarzenburg, 1986.

Loue, S. *Textbook of Research Ethics: Theory and Practice.* New York: Kluwer, 1999.

Lupton, D. *Medicine as Culture. Illness, Disease and the Body in Western Societies.* London: Sage Publications, 1994.

MacKinnon, B., ed. *Human Cloning: Science, Ethics, and Public Policy.* Urbana, Ill.: University of Illinois Press, 2000.

Magill, G., ed. *Discourse and Context.* Carbondale, Ill.: Southern Illinois University Press, 1993.

———, ed. *Personality and Belief.* Lanham, N.J.: University of America Press, 1994.

Magill, G., and M. D. Hoff. *Values and Public Life: An Interdisciplinary Study.* Lanham, N.Y.: University Press of America, 1995.

Magill, G., and R. R. Rainey. *Abortion & Public Policy.* Omaha, Nebr.: Creighton University Press, 1996.

Mahoney, K. E., and P. Mahoney, eds. *Human Rights in the Twenty-First Century.* Dordrecht, Netherlands: Kluwer Academic Publishers, 1993.

Mahowald, M. B. *Women and Children in Health Care: An Unequal Majority.* New York: Oxford University Press, 1993.

———. *Philosophy of Woman: Classical to Current Concepts,* 3rd ed. (Indianapolis, Ind.: Hackett Publishing Company, 1994.

———. *Genes, Women, and Equality.* New York: Oxford University Press, 2000.

Mahowald, M. B., Anita Silvers, and David Wasserman. *Disability, Difference, Discrimination.* Lanham, Md.: Rowman & Littlefield, 1998.

Mahowald, M. B., Victor McKusick, Angela Scheuerle, and Timothy Aspinwall, eds. *Genetics in the Clinic*. St. Louis, Mo.: Mosby, 2001.
Mangum, J. M., ed. *The New Faith-Science Debate: Probing Cosmology, Technology, and Theology*. Minneapolis, Minn.: Augsburg Press, 1989.
Marshak, D. R., R. L. Gardner, and D. Gottlieb, eds. *Stem Cell Biology*. Cold Spring Harbor, N.Y.: Cold Spring Harbor Laboratory Press, 2001.
Marteau, T., and M. Richards, eds. *The Troubled Helix: Social and Psychological Implications of the New Human Genetics*. Cambridge, U.K.: Cambridge University Press, 1996.
Mattei, J. F., coordinator. *Ethical Eye: The Human Genome*. Strasbourg, France: Council of Europe, 2001.
McLean, Sheila A. M., ed. *Legal Issues in Medicine*. Aldershot, UK: Ashgate Publishing Co., 1981.
———, ed. *Legal Issues in Human Reproduction*. Aldershot, UK: Ashgate Publishing Co., 1989.
———. *A Patient's Right to Know: Information Disclosure, the Doctor and the Law*. Aldershot, UK: Dartmouth Publishing Co., 1990.
———, ed. *Law Reform and Human Reproduction*. Aldershot, UK: Dartmouth Publishing, 1992.
———, ed. *Compensation for Damage: An International Perspective*. Aldershot, UK: Dartmouth Publishing Co., 1993.
———, ed. *Law Reform and Personal Injury Litigation*. Aldershot, UK: Dartmouth Publishing Co., 1995.
———, ed. *Contemporary Issues in Law, Medicine and Ethics*. Aldershot, UK: Dartmouth Publishing Co., 1996.
———, ed. *Death, Dying and the Law*. Aldershot, UK: Dartmouth Publishing Co., 1996.
———. *Old Law, New Medicine*. London: Pandora/Rivers Oram Press, 1999.
———. *Medical Law and Ethics*. Aldershot, UK: Ashgate Publishing Co., 2002.
———, ed. *The Genome Project and Gene Therapy*. Aldershot, UK: Ashgate Publishing Co., 2003.
McLean, Sheila A. M., and Alison Britton. *The Case for Physician Assisted Suicide*. London: Pandora Press, 1997.

McLean, Sheila A. M., and Noreen Burrows, eds. *The Legal Relevance of Gender: Some Aspects of Sex-Based Discrimination*. London: Macmillan, 1988.
McLean, Sheila A. M., and Tom Campbell, et al., eds. *Human Rights: From Rhetoric to Reality*. Oxford: Basil Blackwell, 1986.
McLean, Sheila A. M., and G. Maher. *Medicine, Morals and the Law*. Aldershot, UK: Gower, 1983, repr. 1985.
McLean, Sheila A. M., and John Kenyon Mason. *Legal and Ethical Aspects of Health Care*. London: Greenwich Medical Media Press, 2003.
McLean, Sheila A. M., and N. Pace, eds. *Ethics and the Law in Intensive Care*. Oxford: Oxford University Press, 1996.
McGee, G. *The Perfect Baby: Parenthood in the New World of Cloning and Genetics*. New York: Rowman and Littlefield, 1997.
_____, ed. *The Human Cloning Debate*. Berkeley, Calif.: Berkeley Hills Books, 2000.
McGuire, C. D., and G. Chadwick. *Protecting Study Volunteers in Research* Boston, MA: CenterWatch Inc, 1999.
McKie, J., ed. *Ethical Issues in Prenatal Diagnosis and the Termination of Pregnancy*. Melbourne, Australia: Monash University Centre for Human Bioethics, 1994.
Mehlman, M. J., and J. R. Botkin. *Access to the Genome: The Challenge of Equality*. Washington, D.C.: Georgetown University Press, 1998.
Meilaender, G. C. *Body, Soul, and Bioethics*. South Bend, Ind.: University of Notre Dame Press, 1995.
Meyers, D. *The Human Body and the Law*. Edinburgh, U.K.: Edinburgh University Press, 1971.
Mieth, D., and J. Pohier, eds. *Ethics in the Natural Sciences*. Edinburgh, U.K.: T. and T. Clark Ltd., 1989.
Miller, H. I. *Policy Controversy in Biotechnology: An Insider's View*. Austin, Tex.: R. G. Landes, 1997.
Milunsky, A. *Your Genetic Destiny: Know Your Genes, Secure Your Health, and Save Your Life*. Cambridge, Mass: Perseus, 2001.
Monagle, J. F., and D. Thomasma., eds. *Health Care Ethics: Critical Issues for the 21st Century*. Gaithersburg, Md.: Aspen Publishers, 1998.

Moraczewski, A. S., ed. *Genetic Medicine and Engineering: Ethical and Social Dimensions.* St. Louis, Mo.: Catholic Health Association of USA and Pope John XXIII Medical-Moral Research Centre, 1983.

Morselli, P. L., C. E. Pippenger, and J. K. Penry. *Antiepileptic Drug Therapy in Pediatrics.* New York: Raven Press, 1983.

Muller, H. J. *Out of the Night: A Biologist's View of the Future.* New York: Vanguard Press, 1935.

Murray, T. H., M. A. Rothstein, and R. J. Murray, eds. *The Human Genome Project and the Future of Health Care.* Bloomington and Indianapolis, Ind.: Indiana University Press, 1996.

National Bioethics Advisory Commission. *Cloning Human Beings.* Rockville, Md.: National Bioethics Advisory Committee, 1997.

———. *Report and Recommendations.* Vol. I of *Ethical Issues in Human Stem Cell Research*: Rockville, Md.: National Bioethics Advisory Committee, 1999.

———. *Research Involving Human Biological Materials: Ethical Issues and Policy Guidance.* 2 vols. Rockville, Md.: U.S. Government Printing Office, 1999.

———. *Commissioned Papers.* Vol. II of *Ethical Issues in Human Stem Cell Research.* Rockeville, Md.: National Bioethics Advisory Committee, 2000.

———. *Ethical and Policy Issues in International Research: Clinical Trials in Developing Countries*, 2 vols. Bethesda, Md.: U.S. Government Printing Office, 2001.

———. *Ethical and Policy Issues in Research Involving Human Participants.* Bethesda, Md.: National Bioethics Advisory Commission, 2001.

National Commission for the Protection of Human Subjects of Biomedical and Behavioral Research. *Report and Recommendations: Research on the Fetus.* Washington, D.C.: DEHW Publications, 1975.

———. *Report and Recommendations: Research Involving Children.* Washington, D.C.: DEHW Publications No. (OS) 77–0004, 1977.

———. *The Belmont Report: Ethical Principles and Guidelines for the Protection of Human Subjects of Research.* Washington, D.C.: DEHW Publications No. (OS) 78–0012–14, 1979.

National Institutes of Health. *Report of the Human Embryo Research Panel.* Bethesda, Md.: National Institutes of Health, 1994.

⁣⁣⁣⁣⁣⁣⁣⁣. *Papers Commissioned for the Human Embryo Research Panel.* 2 vols. Bethesda, Md.: National Institutes of Health, 1994.

⁣⁣⁣⁣⁣⁣⁣⁣. *Stem Cells: Scientific Progress and Future Research Directions.* Bethesda, Md.: National Institutes of Health, 2001.

Nelkin, D., and M. S. Lindee. *The DNA Mystique. The Gene as a Cultural Icon.* New York: W. H. Freeman, 1995.

Nelson, J. R. *On the New Frontiers of Genetics and Religion.* Grand Rapids, Mich.: Eerdmans, 1994.

Nordgren, A. *Responsible Genetics: The Moral Responsibility of Geneticists for the Consequences of Human Genetics Research.* Boston: Kluwer Academic Publishers, 2001.

Nuffield Council on Bioethics. *Genetic Screening: Ethical Issues.* London: Nuffield Council on Bioethics, 1993.

Nussbaum, M. C., and C. R. Sunstein, eds. *Clones and Clones: Facts and Fantasies about Human Cloning.* New York: Norton, 1998.

O'Donovan, O. *Begotten or Made?* Oxford: Clarendon Press, 1984.

Parfit, D. *Reasons and Persons.* Oxford: Clarendon Press, 1984.

Paul, D. B. *Controlling Human Heredity, 1865 to the Present.* Atlantic Highlands, N.J.: Humanities Press, 1995.

⁣⁣⁣⁣⁣⁣⁣⁣. *The Politics of Heredity: Essays on Eugenics, Biomedicine, and the Nature-Nurture Debate.* Albany, N.Y.: State University of New York Press, 1998.

Pence, G. E., ed. *Flesh of My Flesh: The Ethics of Cloning Humans: A Reader.* Lanham, Md.: Rowman and Littlefield, 1998.

⁣⁣⁣⁣⁣⁣⁣⁣. *Who's Afraid of Human Cloning?* Lanham, Md.: Rowman and Littlefield, 1998.

⁣⁣⁣⁣⁣⁣⁣⁣. *Classic Cases in Medical Ethics.* San Francisco: Harper, 2000.

Peters, T. *For the Love of Children: Genetic Technology and the Future of the Family.* Louisville, Ky.: Westminster John Knox Press, 1996.

⁣⁣⁣⁣⁣⁣⁣⁣. *Playing God? Genetic Determinism and Human Freedom.* New York: Routledge, 1997.

⁣⁣⁣⁣⁣⁣⁣⁣, ed. *Genetics: Issues of Social Justice.* Cleveland, Ill.: Pilgrim Press, 1998.

Petersen, A., and R. Bunton. *The New Genetics and the Public's Health.* New York: Routledge, 2002.

Peterson, J. C. *Genetic Turning Points: The Ethics of Human Genetic Intervention.* Grand Rapids, Mich.: W. B. Eerdmans Pub., 2001.

Pollack, R. *Signs of Life: The Language and Meanings of DNA.* Boston: Houghton Mifflin, 1994.

Post, S. G., and P. J. Whitehouse, eds. *Genetic Testing for Alzheimer's Disease: Ethical and Clinical Issues.* Baltimore, Md.: Johns Hopkins University Press, 1998.

President's Commission for the Study of Ethical Problems in Medicine and Biomedical and Behavioral Research. *Splicing Life: The Social and Ethical Issues of Genetic Engineering with Human Beings.* Washington, D.C.: U.S. Government Printing Office, 1982.

――――. *Screening and Counseling for Genetics Conditions.* Washington D.C.: U.S. Government Printing Office, 1983.

Purtilo, R. *Ethical Dimensions in the Health Professions,* 4th ed. Philadelphia, Pa.: W. B. Saunders, 2003.

Purtilo, R., and A. Haddad. "Challenges to Patients." In *Health Professional and Patient Interactions.* Philadelphia, Penn.: W. B. Saunders, 1996.

Ramsey, P. *Fabricated Man. The Ethics of Genetic Control.* New Haven, Conn.: Yale University Press, 1970.

Ratcliff, K. S., M. M. Ferree, G. O. Mellow, et al., eds. *Healing Technology: Feminist Perspectives.* Ann Arbor, Mich.: University of Michigan Press, 1989.

Rawson, Beryl, ed. *The Family in Ancient Rome: New Perspectives.* London: Routledge, 1992.

Reilly, P. *Genetics, Law, and Social Policy.* Cambridge, Mass.: Harvard University Press, 1977.

Reiser, S. J., et al., eds. *Ethics in Medicine.* Cambridge, Mass.: MIT Press, 1977.

Rifkin, J. *The Biotech Century: Harnessing the Gene and Remaking the World.* New York: Jeremy P. Tarcher/Putnam, 1998.

Report of the Royal College of Obstetrics and Gynaecology Working Party. *Ultrasound Screening for Fetal Abnormalities.* London: Royal College of Obstetrics and Gynaecology, 1997.

Robertson. *Children of Choice: Freedom and the New Reproductive Technologies.* Princeton, N.J.: Princeton University Press, 1994.

Rolansky, J. D. *Genetics and the Future of Man: A Discussion of the Nobel Conference Organized by Gustavus Adolphus College, St. Peter, Minnesota 1965.* New York: Appleton-Century-Crofts, 1966.

Rothenberg, K. H., and E. J. Thompson, eds. *Women and Prenatal Testing: Facing the Challenges of Genetic Technology.* Columbus, Ohio: Ohio State University Press, 1994.

Rothman, K. B. *The Tentative Pregnancy: Prenatal Diagnosis and the Future of Motherhood.* New York: Viking Press, 1986.

———. *Recreating Motherhood: Ideology and Technology in a Patriarchal Society.* New York: W. W. Norton, 1989.

———. *The Tentative Pregnancy.* New York: W. W. Norton, 1993.

———. *Genetic Maps and Human Imaginations. The Limits of Science in Understanding Who We Are.* New York: Norton, 1998.

———. *The Book of Life: A Personal and Ethical Guide to Race, Normalcy, and the Ethical Implications of the Human Genome Project.* Boston: Beacon Press, 1998.

Rothstein, M. A., ed. *Genetic Secrets: Protecting Privacy and Confidentiality in the Genetic Era.* New Haven, Conn.: Yale University Press, 1997.

Royal Commission on New Reproductive Technologies. *Proceed with Care: Report of the Royal Commission on New Reproductive Technologies.* Ottowa, Canada: Ministry of Government Services, 1993.

Sen, Amartya. *Inequality Reexamined.* Cambridge, Mass.: Harvard University Press, 1992.

Shannon, T. A. *What Are They Saying about Genetic Engineering?* New York: Paulist Press, 1985.

———. *Made in Whose Image? Genetic Engineering and Christian Ethics.* Amherst, N.Y.: Humanity Books, 2000.

Shinn, R. L. *The New Genetics: Challenges for Science, Faith, and Politics.* London: Moyer Bell, 1996.

Sikora, R. I., and B. Barry, eds. *Obligations to Future Generations.* Philadelphia, Pa.: Temple University Press, 1978.

Sleator, A. *Stem Cell Research and Regulations under the Human Fertilisation and Embryology Act 1990.* House of Commons. London: HMSO, 2000.

Snustad, D. P., M. J. Simmons, and J. B. Jenkins. *Principles of Genetics.* New York: John Wiley and Sons, 1997.

Steinbock, Bonnie. *Life before Birth: The Moral and Legal Status of Embryos and Fetuses.* New York: Oxford University Press, 1992.

Stock, G. *Redesigning Humans: Our Inevitable Genetic Future.* Boston: Houghton Mifflin, 2002.

Stock, G., and J. Campbell, eds. *Engineering the Human Germline: An Exploration of the Science and Ethics of Altering the Genes We Pass to Our Children*. New York: Oxford University Press, 2000.

Stewart, G. P., ed. *Genetic Engineering: A Christian Response: Crucial Considerations in Shaping Life*. Michigan: Kregel Publications, 1999.

Sugarman, J., A. C. Mastroianni, and J. P. Kahn. *Ethics of Research with Human Subjects: Selected Policies and Resources*. Frederick, Md.: University Publishing Group, 1998.

Suzuki, D., and P. Knudtson. *Genethics: The Ethics of Genetic Engineering*. Boston, Mass.: Harvard University Press, 1990.

Thompson, A. K., and R. F. Chadwick, eds. *Genetic Information: Acquisition, Access, and Control*. New York: Kluwer, 1999.

Toulmin, S. *Cosmopolis: The Hidden Agenda of Modernity*. Chicago: University of Chicago Press, 1990.

Tribe, L. H. *American Constitutional Law*. Mineola, N.Y.: Foundation Press, 1988.

Tudge, C. *The Engineer in the Garden: Genes and Genetics: From the Idea of Heredity to the Creation of Life*. New York: Hill and Wang, 1993.

UNESCO. *Declaration of the Responsibilities of the Present Generations towards Future Generations*. Paris: United Nations, 1997.

U.K. Department of Health. *Stem Cell Research: Medical Progress with Responsibility. A Report from the Chief Medical Officer's Expert Group Reviewing the Potential of Developments in Stem Cell Research and Cell Nuclear Replacement to Benefit Human Health*. London: HMSO, 2000.

U.S. Congress. House. Committee on Science. Subcommittee on Technology. *The Prohibition of Federal Government Funding of Human Cloning Research: Hearing before the Committee on Science, Subcommittee on Technology, U.S. House of Representatives, One Hundred Fifth Congress, first session, July 22, 1997*. Washington, D.C.: U.S. Government Printing Office, 1998.

———. House. Committee on Commerce. Task Force on Health Records and Genetic Privacy. *Privacy, Confidentiality and Discrimination in Genetics: Task Force on Health Records and Genetic Privacy, July 22, 1997*. Washington, D.C.: U.S. Government Printing Office, 1998.

———. House. Committee on Commerce. Subcommittee on Health and Environment. *Cloning: Legal, Medical, Ethical, and Social Issues. Hearing, February 12, 1998.* Washington, D.C.: U.S. Government Printing Office, 1998.

———. Senate. Committee on Appropriations. Subcommittee on Departments of Labor, Health and Human Services, and Education, and Related Agencies. *Stem Cell Research. Special Hearings, December 2, 1998, January 12, 1999, January 26, 1999.* Washington, D.C.: U.S. Government Printing Office, 1999.

———. Senate. Committee on Appropriations. Subcommittee on Departments of Labor, Health and Human Services, Education, and Related Agencies. *Stem Cell Research: Hearings before a Subcommittee of the Committee on Appropriations, United States Senate, One Hundred Fifth Congress, second session, special hearing, December 2, 1998, Washington DC, January 12, 1999, Washington DC, January 26, 1999, Washington DC.* Washington, D.C.: U.S. Government Printing Office, 1999–2001.

———. House. Committee on Science. Subcommittee on Technology. *Genetics Testing in the New Millennium, Advances, Standards, and Implications: Hearing before the Subcommittee on Technology, of the Committee on Science, U.S. House of Representatives, One Hundred Sixth Congress, first session, April 21, 1999.* Washington, D.C.: U.S. Government Printing Office, 1999.

———. House. Committee on Energy and Commerce. Subcommittee on Oversight and Investigations. *Issues Raised by Human Cloning Research: Hearing before the Subcommittee on Oversight and Investigations of the Committee on Energy and Commerce, House of Representatives, One Hundred Seventh Congress, first session, March 28, 2001.* Washington, D.C.: U.S. Government Printing Office, 2001.

———. House. Committee on the Judiciary. Subcommittee on Crime. *Human Cloning: Hearing before the Subcommittee on Crime of the Committee on the Judiciary, House of Representatives, One Hundred Seventh Congress, first session on H.R. 1644 and H.R. 2172, June 7 and June 19, 2001.* Washington, D.C.: U.S. Government Printing Office, 2001.

———. House. Committee on Energy and Commerce. Subcommittee on Health. *The Human Cloning Prohibition Act of 2001 and the Cloning Prohibition Act of 2001: Hearing before the Subcommittee on Health of the Committee on Energy and Commerce, One Hundred Seventh Congress, first session on H.R. 1644 and H.R. 2172, June 20, 2001.* Washington, D.C.: U.S. Government Printing Office, 2001.

———. House. Committee on the Judiciary. *Human Cloning Prohibition Act of 2001: Report together with Dissenting Views (to accompany H.R. 2505) (including cost estimate of the Congressional Budget Office).* Washington, D.C.: U.S. Government Printing Office, 2001.

———. Senate. Committee on Appropriations. Subcommittee on Departments of Labor, Health and Human Services, Education, and Related Agencies. *Promise of the Genomic Revolution: Hearing before a Subcommittee of the Committee on Appropriations, United States Senate, One Hundred Seventh Congress, first session, special hearing, July 11, 2001.* Washington, D.C.: U.S. Government Printing Office, 2002.

———. House. Committee on Government Reform. Subcommittee on Criminal Justice, Drug Policy, and Human Resources. *Opportunities and Advancements in Stem Cell Research: Hearing before the Subcommittee on Criminal Justice, Drug Policy, and Human Resources of the Committee on Government Reform, House of Representatives, One Hundred Seventh Congress, first session, July 17, 2001.* Washington, D.C.: U.S. Government Printing Office, 2002.

U.S. Department of Health, Education, and Welfare. *Report and Conclusions: Support of Research Involving Human In Vitro Fertilization and Embryo Transfer.* Washington, D.C.: Government Printing Office, 1979.

U.S. Patent and Trademark Office. *Animals—Patentability.* Washington, D.C.: U.S. Government Printing Office, 1987.

Vanderpool, H. Y. *The Ethics of Research Involving Human Subjects.* Frederick, Md.: University Publishing Group, 1996.

Walters, L., and J. G. Palmer. *The Ethics of Human Gene Therapy.* New York: Oxford University Press, 1997.

Warnock, Mary. *A Question of Life: The Warnock Report on Fertilisation and Embryology.* Oxford: Basil Blackwell, 1985.

Waters, Brent. *Dying and Death: A Resource for Christian Reflection.* Cleveland, Ohio: Pilgrim Press, 1996.

———. *Reproductive Technology: Towards a Theology of Procreative Stewardship.* Cleveland, Ohio: Pilgrim Press, 2001.

Waters, Brent, and Ronald Cole-Turner. *Pastoral Genetics: Theology and Care at the Beginning of Life.* Cleveland, Ohio: Pilgrim Press, 1996.

Weatherall, D. J., J. G. G. Ledingham, and D. A. Warrell, eds. *Oxford Textbook of Medicine.* Oxford: Oxford University Press, 1996.

Webley, C., and J. Halliday, eds. *Report on Prenatal Diagnostic Testing in Victoria 1997.* Melbourne, Australia: Murdoch Institute, 1998.

Weir, R. F., S. C. Lawrence, and E. Faces, eds. *Genes and Human Self-Knowledge: Historical and Philosophical Reflections on Modern Genetics.* Iowa City: University of Iowa Press, 1994.

———. *Controlling Human Heredity, 1865 to the Present.* Atlantic Highlands, N.J.: Humanities Press, 1995.

Wertz, D. C., et al. *Guidelines on Ethical Issues in Medical Genetics and the Provision of Genetics Services.* Geneva, Switzerland: World Health Organization Hereditary Diseases Program, 1995.

Wheale, P., Schomberg, R. V., and P. Glasner, eds. *The Social Management of Genetic Engineering.* Brookfield, Vt.: Ashgate, 1998.

Wilkie, T. *Perilous Knowledge: The Human Genome Project and Its Implications.* London: Faber and Faber, 1993.

Willer, R. A. *Genetic Testing and Screening: Critical Engagement at the Intersection of Faith and Science.* Minneapolis, Minn.: Kirk House, 1998.

Williams, P. N., ed. *Ethical Issues in Biology and Medicine: Proceedings of a Symposium on the Identity and Dignity of Man.* Cambridge, Mass.: Schenkman Publishing, 1973.

Williams, S. J., and M. Calnan, eds. *Modern Medicine. Lay Perspectives and Experiences.* London: UCL Press, 1996.

Wingerson, L. *Unnatural Selection: The Promise and the Power of Human Gene Research.* New York: Bantam Books, 1998.

World Health Organization. *Proposed International Guidelines on Ethical Issues in Medical Genetics and Genetic Services.* Geneva, Switzerland: WHO, 1998.

Wright, S. *Molecular Politics: Developing American and British Regulatory Policy for Genetic Engineering 1972–1982.* Chicago: University of Chicago Press, 1994.

Zilinskas, R. A., and P. J. Balint, eds. *The Human Genome Project and Minority Communities: Ethical, Social, and Political Dilemmas.* Westport, Conn.: Praeger, 2001.

Articles

Abbot, A. "Stem-Cell Research Delayed by German Ethics Council." *Nature* 411:6840 (2001):875.

Abkowitz, J. L. "Can Human Hematopoietic Stem Cells Become Skin, Gut, or Liver Cells?" *New England Journal of Medicine* 346:10 (2002):770–72.

Agres, T. "News Notes: Human Embryonic Stem Cell Registry Opens." *The Scientist* (Nov. 26, 2001):10.

Agius, E. "Germ-Line Cells: Our Responsibilities for Future Generations." In *Ethics in the Natural Sciences*, edited by Dietmar Mieth and Jacques Pohier. Edinburgh, U.K.: T. and T. Clark Ltd., 1989.

Alper, J. S. "Genetic Complexity in Single Gene Diseases: No Simple Link between Genotype and Phenotype." *British Medical Journal* 312 (7025) (1996):196–97.

American College of Medical Genetics. "Statement of Storage and Usage of Genetic Materials." *American Journal of Human Genetics* 57 (1995):1499–1500.

American Society of Human Genetics. "Statement on Informed Consent for Genetic Research." *American Journal of Human Genetics* 59 (1996):471–74.

———. "ASHG Statement: Professional Disclosure of Familial Genetic Information." *American Journal of Human Genetics* 62:2 (1998):474–83.

Amnon, G. "Informed Consent in the Genetic Age." *Cambridge Quarterly of Healthcare Ethics* 8 (1999):393–400.

Andejeski, Yvonne, Erica S. Breslau, Elizabeth Hart, Ngina Lythcott, Linda Alexander, Irene Rich, Isabelle Bisceglio, Helene S. Smith, and Fran M. Visco, "Benefits and Drawbacks of Including Consumer Reviewers in the Scientific Merit Review of Breast Cancer Research." *Journal of Women's Health & Gender-Based Medicine* 11 (2002):119–36.

Anderson, W. F. "Human Gene Therapy: Scientific and Ethical Considerations." *Journal of Medicine and Philosophy* 10 (1985):275–91.
Andrews, L. B., and N. Elster. "Regulating Reproductive Technologies." *Journal of Legal Medicine* 21:1 (2000):35–65.
Annas, G. "Why We Should Ban Human Cloning." *New England Journal of Medicine* 339:2 (1998):122–25.
_____. "Reforming Informed Consent to Genetic Research." *Journal of the American Medical Association* 286 (2001):2327.
_____. "Cloning and the U.S. Congress." *New England Journal of Medicine.* 346:20 (2002):1599–1602.
Annas, G., A. Caplan, and S. Elias. "Stem Cell Politics, Ethics and Medical Progress." *Nature Medicine* 5:12 (1999):1339–41.
Antoniou, M. "Embryonic Stem Cell Research. The Case Against." *Nature Medicine* 7:4 (2001):397–99.
Badura-Lotter, G. "Ethical, Biological and Legal Aspects in the Use of Human Embryonic Stem Cells in Germany." *Human Reproduction & Genetic Ethics: An International Journal* 7:2 (2001):38–44.
Baird, P. A. "Cloning of Animals and Humans: What Should the Policy Response Be?" *Perspectives in Biology & Medicine* 42:2 (1999):179–94.
Balint, J. A. "Ethical Issues in Stem Cell Research." *Albany Law Review* 65:3 (2002):729–42.
Baschetti, R. "Use of Stem-Cells in Creation of Embryos." *Lancet* 358:9298 (2001):2078.
Bates, A. W. "Ethics of Human Reproductive Cloning." *Journal of the Royal Society of Medicine* 92:3 (1999):157–58.
Baylis, F. "Our Cells/Ourselves: Creating Human Embryos for Stem Cell Research." *Women's Health Issues* 10:3 (2000):140–45.
Bekker, H., et al. "Uptake of Cystic Fibrosis Testing in Primary Care: Supply Push or Demand Pull?" *British Medical Journal* 306 (1993):1584–86.
Bell, J. "The New Genetics: The New Genetics in Clinical Practice." *British Medical Journal* 316:7131 (1998):618–20.
Berger, E. M., and B. M. Gert. "Genetic Disorders and the Ethical Status of Germ-Line Gene Therapy." *Journal of Medicine and Philosophy* 19 (1991):667–83.

Berkowitz, J. M., and J. W. Snyder. "Racism and Sexism in Medically Assisted Conception." *Bioethics* 12 (1998):25–45.

Besklow, L. M., et al. "Informed Consent for Population-Based Research involving Genetics." *Journal of the American Medical Association* 286 (2001):2315–21.

Biotechnology Industry Organization. "The Use of Stem Cells in Biomedical Research. Biotechnology Industry Organization (BIO) Position Statement." *Journal of Biolaw & Business* 311 (1999):40–44.

Blumenthal, D., "Relationships between Academic Institutions and Industry in the Life Sciences—An Industry Survey." *New England Journal of Medicine* 334:6 (1996):368–73.

Blumenthal, D., et al. "Withholding Research Results in Academic Life Sciences: Evidence from a National Survey of Faculty." *Journal of the American Medical Association* 277:15 (1997):1224–28.

Bogardus, S. T., Jr., J. Concato, and A. R. Feinstein. "Clinical Epidemiological Quality in Molecular Genetic Research: The Need for Methodological Standards." *Journal of the American Medical Association* 281:20 (1999):1919–26.

Bonnicksen, A. L. "National and International Approaches to Human Germ-Line Gene Therapy." *Politics and the Life Sciences* 13:1 (1994):39–49.

Bosch, X. "Geneticists Discuss Ethics of Human Genome Project." *Lancet* 352:9138 (1998):1448.

———. "European Parliament Backs Human Cloning Ban." *Lancet* 358:9295 (2001):1785.

———. "European Parliament Debate on Human Cloning Collapses." *Lancet* 358:9297 (2001):1974.

———. "Ethics Group Advises Caution before EC Issues Stem-Cell Line Patents." *Lancet* 359:9320 (2002):1839.

Boshammer, S., M. Kayss, C. Runtenberg, and J. S. Ach. "Discussing HUGO: The German Debate on the Ethical Implications of the Human Genome Project." *Journal of Medicine and Philosophy* 23:3 (1998):324–33.

Botchan, A., et al. "Sperm Separation for Gender Preference: Methods and Efficacy." *Journal of Andrology* 10:2 (1997):107–8.

Botkin, J. R. "Protecting the Privacy of Family Members in Survey and Pedigree Research." *Journal of the American Medical Association* 285 (2001):207–11.

Boyd, K. M. "Little Lamb, Who Made Thee? A Letter from Edinburgh." *Cambridge Quarterly of Healthcare Ethics* 7:2 (1998):199–202.

Boyle, R. J., and J. Savulescu. "Ethics of Using Preimplantation Genetic Diagnosis to Select a Stem Cell Donor for an Existing Person." *British Medical Journal* 323:7323 (2001):1240–43.

Branick, V., and M. T. Lysaught. "Stem Cell Research: Licit or Complicit? Is a Medical Breakthrough based on Embryonic and Fetal Tissue Compatible with Catholic Teaching?" *Health Progress* 80:5 (1999):37–42.

Breithaupt, H. "They Are Moving. Germany Has Started a Broad Debate about Legalizing the Use of Surplus Embryos for Biomedical Research." *EMBO Reports* 2:7 (2001):552–23.

Breyer, S. G. "Genetic Advances and Legal Institutions." *Journal of Law, Medicine & Ethics, Special Supplement* 28:4 (2000):23–29.

Brock, D. W. "Human Cloning and Our Sense of Self." *Science* 296:5566 (2002):314–16.

Brody, B. "Public Goods and Fair Prices: Balancing Technological Innovation with Social Well-Being." *Hastings Center Report* (1996):5–11.

Brown, Theodore. "An Historical View of Health Care Teams." In *Responsibility in Health Care*, edited by George Agich, 3–21. Dordrecht, Netherlands: D. Reidel Publishers, 1982.

Bunk, S. "Researchers Feel Threatened by Disease Gene Patents." *The Scientist* 13:20 (1999):7.

Burgio, G. R., and F. Locatelli. "Ethics of Creating Programmed Stem-Cell Donors." *Lancet* 356:9245 (2000):1868–69.

Burley, J. "The Ethics of Therapeutic and Reproductive Human Cloning." *Seminars in Cell & Developmental Biology* 10:3 (1999):287–94.

Bush, President W. George. "Address on Federal Funding of Embryonic Stem Cell Research." *Origins* 31 (2001):213–15.

Butler, W. J., and J. M. Roberts. "Ethical Issues in Cloning and Embryo Research." *Journal of the South Carolina Medical Association* 94:9 (1998):401–5.

Cahill, L. S. "The Embryo and the Fetus: New Moral Contexts." *Theological Studies* 54 (1993):124–42.

———. "The Status of the Embryo and Policy Discourse." *Journal of Medicine and Philosophy* 22 (1997):407–14.

———. "The New Biotech World Order. Symposium: Human Primordial Stem Cells." *Hastings Center Report* 29:2 (1999):30–48.

———. "The Genome Project: More than a Medical Milestone." *America* 183:4 (2000):7–13.

———. "Social Ethics of Embryo and Stem Cell Research." *Women's Health Issues* 10:3 (2000):131–35.

———. "Genetics, Commodification, and Social Justice in the Globalization Era." *Kennedy Institute of Ethics Journal* 11 (2001):221–38.

Callahan, D. "The Moral Career of Genetic Engineering." *Hastings Center Report* 9 (1979):9.

———. "Manipulating Human Life: Is There No End to It?" In *Medicine Unbound: The Human Body and the Limits of Medical Intervention*, edited by Robert H. Blank and Andrea L. Bonnicksen, 18–31. New York: Columbia University Press, 1994.

———. "Can Nature Serve as a Moral Guide?" *Hastings Center Report* (1996):21–22.

———. "Cloning: Then and Now." *Cambridge Quarterly of Healthcare Ethics* 7:2 (1998):141–44.

Campbell, C. S. "Care and Cultivation. Protestant Thinking on Biotechnology Seeks a Balance between Freedom and Responsibility." *Health Progress* 83:1 (2002):34–38.

———. "In Whose Image? Religion and the Controversy of Human Cloning." *Second Opinion* 1 (1999):24–43.

Campbell, C. S., and J. Woolfrey. "Norms and Narratives: Religious Reflections on the Human Cloning Controversy." *Journal of Biolaw & Business* 1:3) (1998):8–20.

Canadian National Council on Bioethics in Human Research. "Revised Recommendations of *The National Council on Bioethics in Human Research Report on Research Involving Children*" 4 (1993):11.

Caplan, A. L. "What's So Special about the Human Genome?" *Cambridge Quarterly of Healthcare Ethics* 7:4 (1998):422–24.

Caplan, A. L., and G. McGee. "Cloning Human Embryos: Decisions Must Not Be Made by Private Corporations behind Closed Doors." *Western Journal of Medicine* 176:2 (2002):78–79.

Cassell, E. J. "The Principles of the Belmont Report Revisited." *Hastings Center Report* 30 (2000):12–21.

Cassell, J. H. "Lengthening the Stem: Allowing Federally Funded Researchers to Derive Human Pluripotent Stem Cells from Embryos." *University of Michigan Journal of Law Reform* 34 (2001):547–72.

Cataldo, P. "Human Rights and the Human Embryo." *Ethics & Medics* 26 (2001):1–2.

Chadwick, R. "Criteria for Genetic Screening: The Impact of Pharmaceutical Research." *Monash Bioethics Review* 18 (1999):22–26.

Chadwick, R., H. A. M. J. ten Have, et al. "Genetic Screening and Ethics: European Perspectives." *Journal of Medicine and Philosophy* 23:3 (1998):255–73.

Chadwick, R., and M.A. Levitt. "When Drug Treatment in the Elderly Is Not Cost-Effective: An Ethical Dilemma in the Environment of Healthcare Rationing." *Drugs and Aging* 7:6 (1995):416–19.

Chambers, J. E. "Equal Access to Cloning?" *Cambridge Quarterly of Healthcare Ethics* 11:2 (2002):169–79.

———. "May a Woman Clone Herself?" *Cambridge Quarterly of Healthcare Ethics* 10:2 (2001):194–204.

Chapman, A. R. "Ethics and Human Genetics." *Annual of the Society of Christian Ethics* 18 (1998):293–303.

Charo, R. Alto. "Bush's Stem Cell Compromise: A Few Mirrors?" *Hastings Center Report* (2001):6–7.

Childs, Barton, and David Valle. "Genetics, Biology, and Disease." *Annual Review of Human Genetics* 1 (2000):1–19.

Choudhry, S. "Review of Legal Instruments and Codes on Medical Experimentation with Children." *Cambridge Quarterly of Healthcare Ethics* 3 (1994):570.

Churchill, L. R., et al. "Genetic Research as Therapy: Implications of 'Gene Therapy' for Informed Consent." *Journal of Law, Medicine, & Ethics* 26 (1998):38–47.

Cibelli, J. B., et al. "Somatic Cell Nuclear Transfer in Humans: Pronuclear and Early Embryonic Development." *Journal of Regenerative Medicine* 2 (2001):25–31.

Cibelli, J. B., et al. "The First Human Cloned Embryo." *Scientific American* (2002):42–51.

Cioffi, A. "Reproductive and Therapeutic Cloning." *Ethics & Medics* 27:3 (2002):1–3.

Clancy, F. "A Perfect Match." *Medical Bulletin* (Minnesota Medical Foundation) (2001):15–17.

Clark, M. "This Little Piggy Went to Market: Xenotransplantation." *Journal of Law, Medicine, & Ethics* 27:2 (1999):143–44, nn. 82–91.

Clayton, E. W. "What the Law Says about Reproductive Genetic Testing and What It Doesn't." In *Women and Prenatal Testing: Facing the Challenges of Genetic Technology*, edited by Karen Rothenberg and Elizabeth Thomson, 131–78. Columbus, Ohio: 1994.

———. "Genetics Research: Toward International Guidelines" In *Biomedical Research Ethics: Updating International Guidelines—A Consultation*, edited by Robert J. Levine and Samuel Gorovitz with James Gallagher, 152–69. Geneva: CIOMS, 2000.

———. "Genetics, Populations, and Public Health: A Complex Relationship." *Journal of Law, Medicine, & Ethics*. 30 (2002):290–97.

Clayton, E. W., et al. "Informed Consent for Genetic Research on Stored Tissue Samples." *Journal of the American Medical Association* 274 (1995):1786–87.

Cohen, C. B. "Use of 'Excess' Human Embryos for Stem Cell Research: Protecting Women's Rights and Health." *Women's Health Issues* 10:3 (2000):121–26.

———. "Banning Human Cloning—Then What?" *Kennedy Institute of Ethics Journal* 11 (2001):205–9.

———. "Ethical Issues in Embryonic Stem Cell Research." *Journal of the American Medical Association* 285:11 (2001):1439; discussion 1440.

Cohen, J. R. "Creation and Cloning in Jewish Thought." *Hastings Center Report* (1999):7–12.

Cole-Turner, R. "Is Genetic Engineering Co-Creation?" *Theology Today* 44 (1987):338–49.

———. "Genetic Engineering: Our Role in Creation." In *The New Faith-Science Debate: Probing Cosmology, Technology and Theology*, edited by John M. Mangum. Minneapolis: Augsburg Press, 1989.

———. "Religion and the Human Genome." *Journal of Religion and Health* 31 (1992):161-73.

———. "Religion and Gene Patenting." *Science* 270 (1995):52.

———. "Cloning Humans from the Perspective of the Christian Churches." *Science & Engineering Ethics* 5:1 (1999):33-46.

Collins, F. "Medical and Ethical Consequences of the Human Genome Project." *Journal of Clinical Ethics* 2:4 (1991):260-67.

———. "Preparing Health Professionals for the Genetic Revolution." *Journal of the American Medical Association* 278:15 (1997):1285-86.

———. "Shattuck Lecture—Medical and Societal Consequences of the Human Genome Project." *New England Journal of Medicine* 341:1 (1999):28-37.

Collins, F., and A. E. Guttmacher. "Genetics Moves into the Medical Mainstream." *Journal of the American Medical Association* 286 (2001):2323.

Collins, F., and V. A. McKusick. "Implications of the Human Genome Project for Medical Science." *Journal of the American Medical Association* 285:5 (2001):540-544.

Collins, F., et al. "New Goals for the US Human Genome Project: 1998-2003." *Science* (1998):282.

Collins, F., et al. "Initial Sequencing and Analysis of the Human Genome." *Nature* 409 (2001):860-921.

Collmann, J., and G. Graber. "Developing a Code of Ethics for Human Cloning." *Critical Reviews in Biomedical Engineering* 28:3-4 (2000):563-66.

Committee on Drugs, American Academy of Pediatrics. "Guidelines for the Ethical Conduct of Studies to Evaluate Drugs in Pediatric Populations." *Pediatrics* 95 (1995):286.

Conrad, P., and J. Schneider. "Looking at Levels of Medicalization: A Comment on Strong's Critique of the Thesis of Medical Imperialism." *Social Science and Medicine* 14A (1980):75-79.

Cook-Deegan, R. M. "Human Gene Therapy and Congress." *Human Gene Therapy* 1 (1990):163-70.

_____. "Germ-Line Therapy: Keep the Window Open a Crack." *Politics and the Life Sciences* 13:2 (1994):217–20.

Copel, Joshua A., and Charles S. Kleinman. "Fetal Arrhythmias." In *Fetal Therapy. Invasive and Transplacental*, edited by Nicholas M. Fisk and Kenneth J. Moise Jr., 184–98. Cambridge, U.K.: Cambridge University Press, 1997.

Council for Responsible Genetics, Human Genetics Committee. "Position Paper on Human Germ Line Manipulation." *Human Gene Therapy* 4 (1993):35–37.

Council on Ethical and Judicial Affairs, American Medical Association, "Multiplex Genetic Testing." *Hastings Center Report* 28:4 (1998):15–21.

Craft, I. "Source of Research Embryos for Cloning." *Lancet* 357:9265 (2001):1368.

Crawford, R. "Healthism and the Medicalization of Everyday Life." *International Journal of Health Services* 10:3 (1980):365–88.

Crawfurd, M. d'A. "Ethical Guidelines in Fetal Medicine." *Fetal Therapies* 2 (1987):176–78.

Crysdale, C. S. W. "Christian Responses to the Human Genome Project." *Religious Studies Review* 26:3 (2000):236–42.

Curran, C. E. "Cooperation: Toward a Revision of the Concept and Its Application." *Linacre Quarterly* 41:3 (1977):152–165.

Curry, Dan. "US Restricts Research at Johns Hopkins after a Volunteer's Death." *Chronicle of Higher Education* (Aug. 3, 2001):A25.

Daar, J. "Sliding the Slope toward Human Cloning." *American Journal of Bioethics* 1:1 (2001):23–24.

Damewood, M. D. "Ethical Implications of a New Application of Preimplantation Genetic Diagnosis." *Journal of the American Medical Association* 285 (2001):3143–44.

Davis, B. D. "Germ-Line Therapy: Evolutionary and Moral Considerations." *Human Gene Therapy* 3 (1992):361–63.

Davis, D. S. "Informed Consent for Stem Cell Research in the Public Sector." *Journal of the American Medical Women's Association* 55:5 (2000):270–74.

Dayton, L., and G. Vogel. "Stem Cell Research. Reports Give Green Light in Australia, Israel." *Science* 293:5539 (2001):2367–68.

De Berghmans, W. G. R. L., et al. "Ethical Guidance on Human Embryonic and Fetal Tissue Transplantation: A European Overview." *Medicine, Health Care & Philosophy* 5:1 (2002):79–90.

De Crespigny, Lachlan. "Prenatal Diagnosis: The Australian Clinical Situation." In *Ethical Issues in Prenatal Diagnosis and the Termination of Pregnancy*, edited by J. McKie, 13. Melbourne, Australia: Monash University Centre for Human Bioethics, 1994.

DeFrancesco, L. "Probing Protein Interactions." *The Scientist* (April 15, 2002):28–30.

Dekkers, W., and G. Boer. "Sham Neurosurgery in Patients with Parkinson's Disease: Is It Morally Acceptable." *Journal of Medical Ethics* 27 (2001):151–56.

Denker, H. "Embryonic Stem Cells: An Exciting Field for Basic Research and Tissue Engineering, but also an Ethical Dilemma?" *Cells Tissues Organs* 165:3–4 (1999):246–49.

Department of Health and Human Services. "Additional Protections for Children Involved as Subjects in Research." *Federal Register* 48 (1983):9814–20.

———. "National Institute of Health Guidelines for Research Using Human Pluripotent Stem Cells." *Federal Register* 65:166 (2000):51975–81.

Department of Health, Education and Welfare, Ethics Advisory Board, "Report and Conclusions: HEW Support of Research Involving Human IVF and Embryo Transfer." *Federal Register* 44:118 (1979).

Dequeker, E., et al. "Quality Control in Molecular Genetic Testing." *Nature Reviews Genetics* 2:9 (2001):717–23.

De Wachter, M. "Ethical Aspects of Human Germ-Line Gene Therapy." *Bioethics* 7 (1993):166–77.

———. "The European Convention on Bioethics." *Hastings Center Report* 27:1 (1997):13–23.

Dickson, D. "European Panel Rejects Creation of Human Embryos for Research." *Nature* 408:6810 (2000):277.

———. "UK Ethicists Back Use of Stem Cells." *Nature* 404:6779 (2000):697.

Doerflinger, R. "The Ethics of Funding Embryonic Stem Cell Research: A Catholic Viewpoint." *Kennedy Institute of Ethics Journal* 9 (1999):137–50.

——. "The Policy and Politics of Embryonic Stem Cell Research." *National Catholic Bioethics Quarterly* 1:2 (2001):135–143.

Dork, T., B. Dworniczak, C. Aulehla-Scholz, et al., "Distinct Spectrum of CFTR Gene Mutations in Congenital Absence of Vas Deferens." *Human Genetics* 100:3–4 (1997):365–77.

Dresser, R. "Ethical Issues in Embryonic Stem Cell Research." *Journal of the American Medical Association* 285:11 (2001):1439–40.

Dziobon, S. "Germ-Line Gene Therapy: Is the Existing UK Norm Ethically Valid?" In *Genetic Information: Acquisition, Access, and Control*, edited by A. K. Thompson and R. F. Chadwick, 255–65. New York: Kluwer, 1999.

Editorial. "Screening for Cystic Fibrosis." *Lancet* 340 (1992):209–10.

——, Documentation Column. "The Status of the Embryo: The House of Lords Select Committee Report." *The Tablet* (2002):31–32.

——. "Reaction to President Bush's Decision on Embryonic Stem-Cell Research." *Origins* 31:12 (2001):206–212.

Eckenwiler, L. "Moral Reasoning and the Review of Research involving Human Subjects." *Kennedy Institute of Ethics Journal* 11 (2001):37–69.

Edwards, Robert, and Patrick Steptoe, "Birth after Reimplantation of a Human Embryo." *Lancet* 2 (1978):336.

Edwards, Robert, and H. K. Beard. "Sexing Human Spermatozoa to Control Sex Ratios at Birth Is Now a Reality." *Human Reproduction* 10:4 (1995):977–78.

——. "How Identical Would Cloned Children Be? An Understanding Essential to the Ethical Debate." *Human Reproduction Update* 4:6 (1998):791–811.

Eisenberg, L. "Would Cloned Humans Really Be Like Sheep?" *New England Journal of Medicine* 340:6 (1999):471–75.

Eisenberg, R. S. "Public Research and Private Development: Patents and Technology Transfer in Government-Sponsored Research." *Virginia Law Review* 82 (1996):1663–67.

Eisenhauer, E. A. Editorial. "From the Molecule to the Clinic—Inhibiting HER2 to Treat Breast Cancer." *New England Journal of Medicine* 344:11 (2001):841–842.

Elias, S. "Preimplantation Genetic Diagnosis by Comparative Genomic Hybridization." *New England Journal of Medicine* 345 (2001):1569-71.

Elias, S., and G. C. Annas. "Generic Consent for Genetic Screening." *New England Journal of Medicine* 330:22 (1994):1611-13.

Emanuel, E. "Assistance from the Family Members, Friends, Paid Care Givers, and Volunteers in the Care of Terminally Ill Patients." *New England Journal of Medicine* 341:13 (1999):956-63.

Emmett, A. "The Human Genome." *The Scientist* 14:15 (2000):1, 17-19.

Engelhardt, H. T., Jr. "Persons and Humans: Refashioning Ourselves in a Better Image and Likeness." *Zygon* 19:3 (1984):281-96.

———. "Human Nature Technologically Revisited." *Social Philosophy and Policy* 8:1 (1990):180-91.

———. "Germ-Line Genetic Engineering and Moral Diversity: Moral Controversies in a Post-Christian World." *Social Philosophy and Policy* 14 (1996):47-62.

Enquete Commission. "A Report from Germany—An Extract from *Prospects and Risks of Gene Technology: The Report of the Enquete Commission to the Bundestag of the Federal Republic of Germany.*" *Bioethics* 2 (1988):256-63.

Enriquez, J. "Genomics and the World Economy." *Science* 281 (5379) (1998):925-26.

European Group on Ethics in Science and New Technologies to the European Commission. "Ethical Aspects of Human Stem Cell Research and Use." *Bulletin of Medical Ethics* 165 (2001):20-22.

Evers, K. "The Identity of Clones." *Journal of Medicine and Philosophy* 24:1 (1999):67-76.

———. "European Perspectives on Therapeutic Cloning." *New England Journal of Medicine* 346:20 (2002):1579-82.

Ezzell, Carol. "Proteins Rule." *Scientific American* (April 2002):41-47.

Fasouliotis, S. J., and J. G. Schenker. "Human Umbilical Cord Blood Banking and Transplantation: A State of the Art." *European Journal of Obstetrics, Gynecology, & Reproductive Biology* 90:1 (2000):13-25.

Fitzgerald, Deborah A. "Lipids + Genomics = Lipomics." *The Scientist* (February 4, 2002):42.

Fitzgerald, J. "Geneticizing Disability: The Human Genome Project and the Commodification of Self." *Issues in Law & Medicine* 14:2 (1998):147–63.

Fitzgerald, K. "Human Cloning: Analysis and Evaluation." *Cambridge Quarterly of Healthcare Ethics* 7:2 (1998):218–22.

———. "Cloning: Can It Be Good for Us? An Overview of Cloning Technology and Its Moral Implications." *University of Toledo Law Review* 32 (2001):327–36.

Fletcher, J. C. "Ethical Aspects of Genetic Controls: Designed Genetic Changes in Man." *New England Journal of Medicine* 285:14 (1971):776–83.

———. "Indicators of Humanhood: A Tentative Profile of Man." *Hastings Center Report* 2:5 (1972):1–4.

———. "Ethical Issues in and beyond Prospective Clinical Trials of Human Gene Therapy." *Journal of Medicine and Philosophy* 10 (1985):293–309.

———. "Evolution of Ethical Debate about Human Gene Therapy." *Human Gene Therapy* 1 (1990):55–68.

———. "The National Bioethics Advisory Commission's Report on Stem Cell Research: A Review." *ASBH Exchange* 3:1 (2000):8–11.

Fletcher, J. C., and W.F. Anderson. "Germ-Line Gene Therapy: A New Stage of the Debate." *Law, Medicine, and Healthcare* 20:1–2 (1992):27–39.

Fletcher and Evans. "Ethical Issues in Reproductive Genetics." *Seminars in Perinatology* 22:3 (1998):191.

Flower, J. "A Brave New Medicine: A Conversation with William Haseltine." *Health Forum Journal* 42:4 (1999):28–30, 61.

———. "Genomics vs Genetics." *Health Forum Journal* 42:4 (1999):61.

———. "The Promise of Genomics." *Health Forum Journal* 42:4 (1999):62–63.

———. "The Way It Is, Is Not the Way It Will Be." *Health Forum Journal* 42:4 (1999):16–27.

———. "The Uncomfortable and the Unanswerable. New Technologies Raise an Unending List of Ethical Questions." *Health Forum Journal* 43:3 (2000):24–27.
Flowers, E. B. "The Ethics and Economics of Patenting the Human Genome." *Journal of Business Ethics* 17:15 (1998):1737–45.
Fogle, T. "Information Metaphors and the Human Genome Project." *Perspectives in Biology & Medicine* 38:4 (1995):535–47.
Ford, N. M. "Fetus." In *Encyclopedia of Applied Ethics*. San Diego, Calif.: Academic Press, 1998.
———. "The Human Embryo as a Person in Catholic Teaching." *National Catholic Bioethics Quarterly* 1:2 (2001):155–60.
Foster, M. W., et al. "Communal Discourse as a Supplement to Informed Consent for Genetic Research." *Nature Genetics* 17 (1997):277–79.
———. "The Role of Community Review in Evaluating the Collective Risks of Human Genetic Variation Research." *American Journal of Human Genetics* 64 (1999):1719–27.
Freed, C. R., et al. "Transplantation of Embryonic Dopamine Neurons for Severe Parkinson's Disease." *New England Journal of Medicine* 344 (2001):710–19.
Freedman, B., A. Fuks, and C. Weijer. "In Loco Parentis: Minimal Risk as an Ethical Threshold for Research Upon Children." *Hastings Center Report* 3 (1993):13–19.
Friedmann, Theodore. "Milestones and Events in the Early Development of Human Gene Therapy." *Molecular Genetic Medicine* 3 (1993):1–32.
Friedrich, M. J. "Debating Pros and Cons of Stem Cell Research." *Journal of the American Medical Association* 284:6 (2000):681–82.
———. "Research Yields Clues to Improving Cell Therapy for Parkinson Disease." *Journal of the American Medical Association* 287:2 (2002):175–76.
Furton, E. J., and M. M. Mathews-Roth. "Stem Cell Research and the Human Embryo. Part Two." *Ethics & Medics* 24:9 (1999):3–4.
Galton, D. J., and L. Doyal. "'Goodbye Dolly?' The Ethics of Human Cloning." *Journal of Medical Ethics* 24:4 (1998):279.
Gannon, P., and C. Villers. "Genetic Testing and Employee Protection." *Medical Law International* 4 (1999):39–57.

Garcia, J. B. P. "The Law and the New Challenges Posed by Genetic Science. Presentation of a Meeting." *Law and Human Genome Review.* 4 (1996):202.

Gardner, W. "Can Human Genetic Enhancement Be Prohibited?" *Journal of Medicine and Philosophy* 20:1 (1995):65–84.

Gaylin, W. "Fooling with Mother Nature." *Hastings Center Report* (1990):17–21.

Geeter, J. S. "Coding for Change: The Power of the Human Genome to Transform the American Health Insurance System." *American Journal of Law & Medicine* 28 (2000):1–76.

Geron Ethics Advisory Board. "Research with Human Embryonic Stem Cells: Ethical Considerations." *Hastings Center Report* 29 (1999):31–36.

Giardiello, F. M., J. D. Brensinger, G. M. Petersen, M. C. Luce, L. M. Hylind, J. A. Bacon, S. V. Booker, R. D. Parker, and S. R. Hamilton. "The Use and Interpretation of Commercial APC Gene Testing for Familial Adenomatous Polyposis." *New England Journal of Medicine* 336:12 (1997):823–27.

Gillam, L. "Prenatal Diagnosis and Discrimination against the Disabled." *Journal of Medical Ethics* 25 (1999):163–71.

Gillon, R. "Human Reproductive Cloning—A Look at the Arguments against It and a Rejection of Most of Them." *Journal of the Royal Society of Medicine* 92:1 (1999):3–12.

Gin, B. R. "Genetic Discrimination, Huntington's Disease and the Americans with Disabilities Act." *Columbia Law Review* 97 (1997):1406–34.

Glannon, W. "The Ethics of Human Cloning." *Public Affairs Quarterly* 12:3 (1998):287–305.

Goodey, C., P. Alderson, and J. Appleby. "The Ethical Implications of Antenatal Screening for Down's Syndrome." *Bulletin of Medical Ethics* 147 (1999):13–17.

Gostin, L. O. "Genetic Privacy." *Journal of Law, Medicine, & Ethics* 23:4 (1995):320–30.

Gostin, L. O., and J. G. Hodge. "Genetic Privacy and the Law: An End to Genetic Exceptionalism." *Jurimetrics Journal* 40 (1999):21–58.

Greely, H. T. "Breaking the Stalemate: A Prospective Regulatory Framework for Unforeseen Research Issues of Human Tissue Samples and Health Information." *Wake Forest Law Review* (Genetic Technology issue) 34 (1999):737–66.

———. "The Revolution in Human Genetics." *South Carolina Law Review* 52 (2001):377–90.

Greenspan, P. S. "Free Will and the Genome Project." *Philosophy and Public Affairs* 22:1 (1993):31–43.

Grodin, M. A., and J. J. Alpert. "Children as Participants in Medical Research." *Pediatric Clinics of North America* 35 (1988):1391–1401.

Groves, J. "Taking Care of the Hateful Patient." *New England Journal of Medicine* 298:4 (1978):883–87.

Gustafson, J. M. "Genetic Engineering and the Normative View of the Human." In *Ethical Issues in Biology and Medicine: Proceedings of a Symposium on the Identity and Dignity of Man*, edited by Preston N. Williams, 46–58. Cambridge, Mass.: Schenkman Pub. Co., 1973.

———. "Genetic Therapy: Ethical and Religious Reflections." *Journal of Contemporary Health Law and Policy* 8 (1992):183–200.

———. "A Christian Perspective on Genetic Engineering." *Human Gene Therapy* 5 (1994):747–54.

Guttmacher, A. E. "Genetic Medicine in the Next Five Years." *Health Progress* 80:5 (1999):43–44.

Haddad, F. F., et al. "The Human Genome Project: A Dream Becoming a Reality." *Surgery* 125:6 (1999):575–80.

Haddow, J. E. "Antenatal Screening for Down's Syndrome: Where Are We and Where Next?" *Lancet* 352 (1998):336–37.

Hall, M. A., and S. S. Rich. "Genetic Privacy Laws and Patients' Fear of Discrimination by Health Insurers: The View from Genetic Counselors." *Journal of Law, Medicine & Ethics* 28 (2000):245–57.

———. "The Impact of Genetic Discrimination Laws Restricting Health Insurers' Use of Genetic Information." *American Journal of Human Genetics* 66 (2000):293–307.

Halliday, J., et al. "Importance of Complete Follow-up of Spontaneous Fetal Loss after Amniocentesis and Chorionic Villus Sampling." *Lancet* 240 (1992):886–90.

Halperin, M. "Human Genome Mapping: A Jewish Perspective." *Jewish Medical Ethics* 3:2 (1998):30–33.

Handyside, Alan H. "Preimplantation Genetic Diagnosis: A 10-Year Perspective." In *Towards Reproductive Certainty; Fertility and Genetics beyond 1999*, edited by Robert Jansen and David Mortimer, 389–96. New York: Parthenon Publishing Group 1999.

Hansen, J. E. "Embryonic Stem Cell Production through Therapeutic Cloning." *Journal of Medical Ethics* 28:2 (2002):86–88.

Hanson, M. J. "Religious Voices in Biotechnology: The Case of Gene Patenting." *Hastings Center Report* 27 (1997):1–21.

———. "Biotechnology and Commodification within Health Care." *Journal of Medicine and Philosophy* 24 (1999):267–87.

———. "Cloning for Therapeutic Purposes: Ethical and Policy Considerations." 32 *University of Toledo Law Review* (2001):355–65.

Hare, R. M. "Possible People." *Bioethics* 2 (1988):279–93.

Harris, J. "Is Genetic Engineering a Form of Eugenics?" *Bioethics* 7:2/3 (1993):178–87.

———. "Cloning and Human Dignity." *Cambridge Quarterly of Healthcare Ethics* 7:2 (1998):163–67.

———. "The Concept of the Person and the Value of Life." *Kennedy Institute of Ethics Journal* 9:4 (1999):293–308.

———. "Clones, Genes, and Reproductive Autonomy. The Ethics of Human Cloning." *Annals of the New York Academy of Sciences* 913 (2000):209–17.

Harris, L. H. "Ethics and Politics of Embryo and Stem Cell Research: Reinscribing the Abortion Debate." *Women's Health Issues* 10:3 (2000):146–51.

Hastings Center. "Genetic Grammar: 'Health,' 'Illness,' and the Human Genome Project." *Hastings Center Report* 22 (1992).

Hauser, Michael, et al., "Gene Medicine." *Journal of the American Medical Association* 286 (2001):2333.

Have, H. A. M. J. ten. "Knowledge and Practice in European Medicine: The Case of Infectious Diseases." In *The Growth of Medical Knowledge,* edited by H. A. M. J. ten Have, G. K. Kimsma, and S. F. Spicker, 15-40. Dordrecht/Boston/London: Kluwer Academic Publishers, 1990.

———. "Living with the Future: Genetic Information and Human Existence." In *The Right to Know and the Right Not to Know,* edited by R. Chadwick, M. Levitt, and D. Shickle, 87-95. Aldershot, U.K.: Avebury, 1997.

Have, H. A. M. J. ten, and M. Loughlin. "Responsibilities and Rationalities: Should the Patient Be Blamed?" *Health Care Analysis* 2:2 (1994):119-27.

Hayry, Heta. "How to Assess the Consequences of Genetic Engineering." In *Ethics and Biotechnology,* edited by Anthony Dyson and John Harris, 144-56. New York: Routledge, 1994.

Hedgecoe, A. "Geneticization, Medicalization and Polemics." *Medicine, Health Care & Philosophy* 1:3 (1998):235-24.

Heller, J. C. "Religious Perspectives on Human Cloning: Revisiting Safety as a Moral Constraint." *Valparaiso University Law Review* 32:2 (1998):661-78.

Heller, Michael A., and Rebecca S. Eisenberg. "Can Patents Deter Innovation?" *Science* 280 (1998):698ff.

Hepburn, L. "Genetic Counseling: Parental Autonomy or Acceptance of Limits?" *Concilium* 275 (1998):40.

Hershenov, D. B. "An Argument for Limited Human Cloning." *Public Affairs Quarterly* 14:3 (2000):245-58.

Hodgson, J., and A. Marshall. "Pharmacogenomics: Will the Regulators Approve?" *Nature Biotechnology* 16 (1998):243-46.

Hoedemaekers, R., and H. A. M. J. ten Have. "Geneticization: The Cyprus Paradigm." *Journal of Medicine and Philosophy* 23:4 (1998):274-87.

———. "Genetic Health and Genetic Disease." In *Genes and Morality. New Essays,* edited by V. Launis, J. Pietarinen, and J. Räikkä, 121-43. Amsterdam/Atlanta: Rodopi, 1999.

Holmes, E. M. "Solving the Insurance/Genetic/Fair/Unfair Discrimination Dilemma in Light of the Human Genome Project." *Kentucky Law Journal* 85 (1996-1997):503-64.

Holzgreve, W., et al., "Benefits of Placental Biopsies for Rapid Karyotyping in the Second and Third Trimesters (Late Chorionic Villus Sampling) in High Risk Pregnancies." *American Journal of Obstetrics and Gynaecology* 162 (1990):1188–92.

Hook, E. "Prevalence, Risks and Recurrence." In *Prenatal Diagnosis and Screening*, edited by David J. Brock, Charles Rodeck and Malcolm A. Ferguson-Smith, 364. Edinburgh, U.K.: Churchill Livingstone, 1992.

Housman, D., and F. D. Ledley. "Why Pharmacogenomics? Why Now?" *Nature Biotechnology* 16 (1998):492–93.

Illich, I. "The Medicalization of Life." *Journal of Medical Ethics* 1:1 (1975):73–77.

International SNP Map Working Group. "A Map of Human Genome Sequence Variation Containing 1.42 Million Single Nucleotide Polymorphisms." *Nature* 409 (2001):928–33.

Janofsky, J., and B. Starfield. "Assessment of Risk in Research on Children." *Journal of Pediatrics* 98 (May 1981):843–46.

Jendusa, J. M. "Pandora's Box Exposed: Untangling the Web of the Double Helix in Light of Insurance and Managed Care." *DePaul Law Review* 49 (1999):161–215.

Joffee, S., et al. "Quality of Informed Consent: A New Measure of Understanding among Research Subjects." *Journal of the National Cancer Institute* 93 (2001):139–47.

John Paul II, Pope. "Biological Research and Human Dignity." *Origins* 12 (1982):342–43.

———. "Ethics of Genetic Manipulation." (Address by Pope John Paul II to Members of the World Medical Association) *Origins* 13 (1983):385, 387–89.

Johnson, M. "Reflections on Some Catholic Claims for Delayed Hominisation." *Theological Studies* 56 (1995):743–63.

Juengst, E. "The NIH 'Points to Consider' and the Limits of Human Gene Therapy." *Human Gene Therapy* 1 (1990):425–33.

———. "Germ-Line Gene Therapy: Back to Basics." *Journal of Medicine and Philosophy* 19 (1991):587–92.

———. "Can Enhancement Be Distinguished from Prevention in Genetic Medicine?" *Journal of Medicine and Philosophy* 22:2 (1997):125–42.

———. "The Ethics of Prediction: Genetic Risk and the Physician-Patient Relationship." In *Health Care Ethics: Critical Issues for the 21st Century*, edited by John F. Monagle and David Thomasma, 212–23. Gaithersburg, Md.: Aspen Publishers, 1998.

———. "Groups as Gatekeepers to Genomic Research: Conceptually Confusing, Morally Hazardous, and Practically Useless." *Kennedy Institute of Ethics Journal* 8:2 (1998):183–200.

Juengst, E., and M. Fossel. "The Ethics of Embryonic Stem Cells." *Journal of the American Medical Association* 284 (2000):3180–84.

Kaji, E. H., and J. M. Leiden. "Gene and Stem Cell Therapies." *Journal of the American Medical Association* 285:5 (2001):545–50.

Kass, Leon. "New Beginnings in Human Life." In *The New Genetics and the Future of Man*, edited by M. Hamilton, 61. Grand Rapids, Mich.: Eerdmans, 1972.

———. "Triumph or Tragedy: The Moral Meaning of Genetic Technology." *American Journal of Jurisprudence* 45 (2000):1–16.

Kaveny, M. C. "Genetics and the Future of American Law and Policy." *Concilium* 225 (1998):69–70.

———. "Jurisprudence and Genetics." *Theological Studies* 60 (1999):135–47.

———. "Appropriation of Evil: Cooperation's Mirror Image." *Theological Studies* 61 (2000):280–313.

Kavka, Gregory S. "The Futurity Problem." In *Obligations to Future Generations*, edited by R. I. Sikora and Brian Barry, 180–203. Philadelphia: Temple University Press, 1978.

———. "The Paradox of Future Individuals." *Philosophy and Public Affairs* 11 (1982):93–112.

Keenan, James F. "Prophylactics, Toleration, and Cooperation: Contemporary Problems and Traditional Principles." *International Philosophical Quarterly* XXIX, 2: 114 (1989):205–220.

———. "What Is Morally New in Genetic Manipulation?" *Human Gene Therapy* 1 (1990):289–98.

———. "Christian Perspectives on the Human Body." *Theological Studies* 55 (1994):330–46.

———. "Genetic Research and the Elusive Body." In *Embodiment, Morality and Medicine*, edited by L. S. Cahill and M. A. Farley, 59–73. Boston: Kluwer Academic Publishers, 1995.

Keenan, James F., and T. R. Kopfensteiner. "The Principle of Cooperation." *Health Progress* (April 1995):23–27.
Keller, G., and H. R. Snodgrass. "Human Embryonic Stem Cells: The Future is Now." *Nature Medicine* 5 (1999):151–52.
King, H. C., and A. A. Sinha. "Gene Expression Profile Analysis by DNA Microarrays: Promise and Pitfalls." *Journal of the American Medical Association* 286 (2001):2280–88.
Kinmonth, A. L., et al. "The New Genetics: Implications for Clinical Services in Britain and the United States." *British Medical Journal* 317:7133 (1998):767–70.
Kirk, M. "Genetics, Ethics and Education: Considering the Issues for Nurses and Midwives." *Nursing Ethics* 7:3 (2000):215–26.
Kletzel, M., E. Morgan, and J. Frader. "Ethical Issues in Stem Cell Transplantation." *Journal of Hematotherapy* 7:3 (1998):197–203.
Kling, Jim. "Pushing Proteomics." *The Scientist* (April 15, 2002):43
Knowles, L. P. "Property, Progeny, and Patents." *Hastings Center Report* 29:2 (1999):38–40.
―――. "Science Policy and the Law: Reproductive and Therapeutic Cloning." *New York University School of Law Journal of Legislation and Public Policy* 4 (2000–2001):13–22.
Kohn, Donald B. "Transduction of CD34(+) Cells from the Bone Marrow of HIV-1 Infected Children: Comparative Marking by an RRE Decoy Gene and a Neutral Gene." NIH/RAC Protocol 9602-147. Recombinant DNA Advisory Committee of the National Institutes of Health.
Kohn, Donald B., Weinberg, K. I., Nolta, J. A., Heiss, L. N., Lenarsky, C., Crooks, G. M., Hanley, M. E., Annett, G., Brooks, J. S., El-Khoureiy, A., Lawrence, K., Wells, S., Moen, R. C., Bastian, J., Williams-Herman, D. E., Elder, M., Wara, D., Bowen, T., Hershfield, M. S., Mullen, C. A., Blease, R. M., and Parkman, R. "Engraftment of Gene-Modified Umbilical Cord Blood Cells in Neonates with Adenosine Deaminase Deficiency." *Nature Medicine* 1:10 (1995):1017–1023.
Kopaczynski, G. "Preimplantation Genetic Diagnosis." *Ethics & Medics* 27:5 (2002):1–3.
Kornman, L. H., et al. "Women's Opinions and the Implications of First- versus Second-Trimester Screening for Fetal Down's Syndrome." *Prenatal Diagnosis* 17:11 (1997):1011–18.

Krugman, S. "The Willowbrook Hepatitis Studies Revisited: Ethical Aspects." *Reviews of Infectious Diseases* 8 (1986):157–62.

Kuhse, Helga. "Preventing Genetic Impairments: Does It Discriminate against People with Disabilities?" In *Genetic Information: Acquisition, Access, and Control*, edited by Thompson, A. K., and R. F. Chadwick, 17–30. New York: Kluwer, 1999.

Kukla, H. J. "Embryonic Stem Cell Research: An Ethical Justification." *Georgetown Law Journal* 90:2 (2002):503–43.

Lammers, A., and T. Peters, "Genethics: Implications of the Human Genome Project." *Christian Century* 107 (1990):868–71.

Lander, E. S. "Genomics: Launching a Revolution in Medicine." *Journal of Law, Medicine & Ethics, Special Supplement* 28:4 (2000):3–14.

Lane, P. A. "Sickle Cell Anemia." *Pediatric Clinics of North America* 43:3 (1996):639–64.

Lanza, R. P., J. B. Cibelli, and M. D. West. "Human Therapeutic Cloning." *Nature Medicine* 5:9 (1999):975–77.

Lanza, R. P., et al. "The Ethical Reasons for Stem Cell Research." *Science* 292:5520 (2001):1299.

Lanza, R. P., et al. "The Ethical Validity of Using Nuclear Transfer in Human Transplantation." *Journal of the American Medical Association* 284 (2001):3175–79.

Larkin, M. "J. Craig Venter: Sequencing Genomes His Way." *Lancet* 353:9171 (1999):2218.

Lassen, Eva Maria. "The Roman Family: Ideal and Metaphor." In *Constructing Early Christian Families: Family as Social Reality and Metaphor*, edited by Halvor Moxnes, 103–20. London and New York: Routledge, 1997.

Lauritzen, P. "Neither Person nor Property. Embryo Research and the Status of the Early Embryo." *America* (March 26, 2001):20–23.

Lavery, Stuart A., and R. M. L. Winston. "Clinical Experience with Preimplantation Genetic Diagnosis at Hammersmith 1989–1998." In *Towards Reproductive Certainty; Fertility and Genetics beyond 1999*, edited by Robert Jansen and David Mortimer, 397–404. New York: Parthenon Publishing Group 1999.

Lebacqz, K. "Research with Human Embryonic Stem Cells: Ethical Considerations." *Hastings Center Report* 29:2 (1999):31–36.

Lebacqz, K., C. Tauer, G. McGee, and A. Caplan. "Stem Cells." *Hastings Center Report* 29:4 (1999):4–5.

Lee, S. "Human Stem Cell Research: NIH Releases Draft Guidelines for Comment." *Journal of Law, Medicine & Ethics* 28:1 (2000):81–83.

Lee, S., et al. "The Meaning of 'Race' in the New Genomics: Implications for Health Disparities Research." *Yale Journal of Health Policy, Law, Ethics* 1 (2001):33–75.

Levine, R. J. "Research Involving Children: The National Commission's Report." *Clinical Research* 26 (1978):61–66.

Levy, H. L., and S. Albers. "Genetic Screening of Newborns." *Annual Review of Human Genetics* 1 (2000):139–77.

Lippman, A. "Prenatal Genetic Testing and Screening: Constructing Needs and Reinforcing Inequities." *American Journal of Law & Medicine* 17 (1991):15–50.

———. "Led (Astray) by Genetic Maps: The Cartography of the Human Genome and Health Care." *Social Science and Medicine* 35:12 (1992):1469–76.

———. "Prenatal Genetic Testing and Geneticization: Mother Matters for All." *Fetal Diagnosis and Therapy*. 8:supplement 1 (1993):175–88.

Lipschutz, J. H. "To Clone or Not to Clone—A Jewish Perspective." *Journal of Medical Ethics* 25:2 (1999):105–7.

Lumley, Judith. "Uncertainty and Ultrasound Diagnosis." In *Ethical Issues in Prenatal Diagnosis and the Termination of Pregnancy*, edited by John McKie, 26–27. Melbourne, Australia: Monash University Centre for Human Bioethics, 1994.

Lysaught, M. T. "Commentary: Reconstructing Genetic Research as Research." *Journal of Law, Medicine, & Ethics* 26 (1998):48–54.

Macer, D. "Whose Genome Project?" *Bioethics* 5:3 (1991):183–211.

Macklin, R. "Ethical Problems in Sham Surgery in Clinical Research." *New England Journal of Medicine* 341:13 (1999):992–96.

———. "Ethics, Politics, and Human Embryo Stem Cell Research." *Women's Health Issues* 10:3 (2000):111–15.

Maddox, J. "New Genetics Means No New Ethics." *Nature* 364 (1993):97.

Magill, G. "The Ethics Weave in Human Genomics, Embryonic Stem Cell Research, and Therapeutic Cloning: Promoting and Protecting Society's Interests." *Albany Law Review* 65:3 (2002):701–28.

Mahowald, M. B., D. Levinson, Christine Cassel, et al. "The New Genetics and Women." *Milbank Quarterly* 74:2 (1996):239–83.

———. "Genes, Clones, and Gender Equality." *DePaul Journal of Health Care Law* 3 (2000):495–526.

Makrydas, G., and D. Lolis. Letter, "Nuchal Translucency." *Lancet* 350 (1997):1630–31.

Malinowski, M. J., and M. A. O'Rourke. "A False Start? The Impact of Federal Policy on the Genotechnology Industry." *Yale Journal on Regulation* 13 (1996):163–254.

Maloney, D. M. "IRBs and Human Stem Cell Research Protocols." *Human Research Report* 15:1 (2000):3.

———. "Institutional Review Boards Have New Role in Stem Cell Research." *Human Research Report* 15:10 (2000):1–2.

Mann, Charles C. "Biotech Goes Wild. Genetic Engineering Will Be Essential to Feed the World's Billions." *Technology Review* (July/August, 1999):36–43.

Marshall, E. "Varmus Proposes to Scrap the RAC." *Science* 272 (1996):945.

———. "NIH to Produce a 'Working Draft' of the Genome by 2001." *Science* 281 (1998):1774–75.

———. "Ethicists Back Stem Cell Research, White House Treads Cautiously." *Science* 285 (1999):502.

Marteau, T. M. "Towards Informed Decisions about Prenatal Testing: A Review." *Prenatal Diagnosis* 15 (1995):1215–18.

Martino, R. "Human Cloning Prohibition Urged." *Origins* 31 (2001):439–40.

Martone, M. "The Ethics of the Economics of Patenting the Human Genome." *Journal of Business Ethics* 17:15 (1998):1679–84.

Marwick, C. "Funding Stem Cell Research." *Journal of the American Medical Association* 281:8 (1999):692–93.

Matthews, C., and Ke-hui Cui. Letter. *Nature* 366 (1993):117–18.

McCormick, R. A. "Proxy Consent in the Experimentation Situation." *Perspectives in Biology & Medicine* 18 (1974):2–20.

———. "Experimentation in Children: Sharing in Sociality." *Hastings Center Report* 6 (1976):41–46.

———. "Who or What Is the Preembryo?" *Kennedy Institute of Ethics Journal* 1 (1991):1–15.

McGee, G. "Gene Patents Can Be Ethical." *Cambridge Quarterly of Healthcare Ethics* 7:4 (1998):417–21.

McGee, G., and M. Arruda. "A Crossroads in Genetic Counseling and Ethics." *Cambridge Quarterly of Healthcare Ethics* 7:1 (1998):97–100.

McGee, G., and A. Caplan. "The Ethics and Politics of Small Sacrifices in Stem Cell Research," *Kennedy Institute of Ethics Journal* 9 (1999):151–58.

McGee, G., and A. L. Caplan. "What's in the Dish?" *Hastings Center Report* 29:2 (1999):36–38.

McLaren, A. "Ethical and Social Considerations of Stem Cell Research." *Nature* 414:6859 (2001):129–31.

McLean, S. A. M. "The Right to Reproduce." In *Human Rights: From Rhetoric to Reality*, edited by Campbell, et al. Oxford: Basil Blackwell, 1986.

———. "The New Genetics: A Challenge to Clinical Values?" *Proceedings of the Royal College of Physicians, Edinburgh* 26 (1996):41.

Mehlmann, M. J. "The Effect of Genomics on Health Services Management: Ethical and Legal Aspects." *Frontiers of Health Services Management* 17:3 (2001):17–26.

Meilaender, G. "Affirming Ourselves to Death." *First Things* 84 (1998):13–14.

Mendiola, M. M., T. Peters, E. W. Young, and L. Zoloth-Dorfman. "Research with Human Embryonic Stem Cells: Ethical Considerations." *Hastings Center Report* 29:2 (1999):31–6.

Mendoza, O. "Patient-Centered Healthcare." *Nature Biotechnology* 17:supplement (1999):BV15.

Merz, J. F. "Disease Gene Patents: Overcoming Unethical Constraints on Clinical Laboratory Medicine." *Clinical Chemistry* 45 (1999):324–30.

Merz, J. F., et al. "IRB Review and Consent in Human Tissue Research." *Science* 283 (1999):1647–48.

Meslin, E. M. "Of Clones, Stem Cells, and Children: Issues and Challenges in Human Research Ethics." *Journal of Women's Health & Gender-Based Medicine* 9:8 (2000):831–41.

Meslin, E. M., E. J. Thompson, and J. T. Boyer. "The Ethical, Legal, and Social Implications Research Program at the National Genome Research Institute." *Kennedy Institute of Ethics Journal* 7:3 (1997):291–98.

Messer, N. "Human Cloning and Genetic Manipulation: Some Theological and Ethical Issues." *Studies in Christian Ethics* 12:2 (1999):1–16.

Meyer, J. R. "Human Embryonic Stem Cells and Respect for Life." *Journal of Medical Ethics* 26 (2000):166–70.

Meyer, M. J., and L. J. Nelson. "Respecting What We Destroy: Reflections on Human Embryo Research." *Hastings Center Report* (2001):16–23.

Michie, Susan, and Theresa Marteau. "Genetic Counselling: Some Issues of Theory and Practice." In *The Troubled Helix: Social and Psychological Implications of the New Human Genetics*, edited by Theresa Marteau and Martin Richards, 104–22. Cambridge, U.K.: Cambridge University Press, 1996.

Micielli, R. "Stem Cell Research: Biology, Ethics, and Policy." *Princeton Journal of Bioethics* 3:1 (2000):99–111.

Mikes, R. "NBAC and Embryo Ethics." *National Catholic Bioethics Quarterly* 1:2 (2001):163–87.

Miller, F. G., A. L. Caplan, and J. C. Fletcher. "Dealing with Dolly: Inside the National Bioethics Advisory Commission." *Health Affairs* 17:3 (1998):264–67.

Modell, C. Bernadette. "Screening as Public Policy." In *Prenatal Diagnosis*, edited by Brock, et al., 612–13. Edinburgh, U.K.: Churchill Livingstone, 1992.

Murphy, T. F. "Entitlement to Cloning." *Cambridge Quarterly of Healthcare Ethics* 8:3 (1999):364–68.

———. "Response to 'May a Woman Clone Herself' by Jean Chambers (CQ Vol 10, No 2). Are There Limits to the Use of Reproductive Cloning?" *Cambridge Quarterly of Healthcare Ethics* 11:1 (2002):83–86.

Myers, Chris, Nicole Paulk, and Christine Dudlak. "Genomics: Implications for Health Systems." *Frontiers of Health Services Management* 17:3 (2001):3–16.

Narveson, J. "Utilitarianism and New Generations." *Mind* 76 (1967):62–72.

National Coalition for Health Professional Education in Genetics. "Recommendations of Core Competencies in Genetics Essential for All Health Professionals." *Genetic Medicine* 3 (2001):155–59.

National Institutes of Health. "Genetic Testing for Cystic Fibrosis," NIH Consensus Statement, *Online* (April 14–16, 1997), 15:4 (1997).

Nevin, N. C. "Experience of Gene Therapy in the United Kingdom." *Annals of the New York Academy of Sciences* 862 (1998):184–87.

Newell, C. "The Social Nature of Disability, Disease and Genetics: A Response to Gillam, Persson, Holtug, Draper and Chadwick." *Journal of Medical Ethics* 25 (1999):172–75.

Nirenberg, M. "Will Society Be Prepared?" *Science* 157 (1967):633.

Noonan, John T., Jr. "Raw Judicial Power." *National Review* (March 1973):259–63.

Nsiah-Jefferson, Laurie, and Elaine J. Hall. "Reproductive Technology: Perspectives and Implications for Low-Income Women and Women of Color." In *Healing Technology: Feminist Perspectives*, edited by K. S. Ratcliff, M. M. Ferree, G. O. Mellow, et al., 93–117. Ann Arbor, Mich.: University of Michigan Press, 1989.

Okarma, T. B. "Human Primordial Stem Cells." *Hastings Center Report* 29:2 (1999):30.

Olynyk, J. K., et al. "A Population-Based Study of the Clinical Expression of the Hemochromatosis Gene." *New England Journal of Medicine* 341:10 (1999):718–32.

Ondrusek, Nancy. "Ethical Issues in Gene Therapy." In *Ethics in Pediatric Research*, edited by Gideon Koren, 155–70. Malabar, Fl.: Krieger Publishing, 1993.

O'Neill, Onora. "Informed Consent and Genetic Information." *Studies in History and Philosophy of Biology and Biomedical Sciences* 32:4 (2001):689–704.

Owens, K., and M. C. King. "Genomic Views of Human History." *Science* 286 (1999):451–53.

Palevitz, Barry A. "Deciphering Protein Evolution." *The Scientist* (2001):18.
Parker, S. M. "Bringing the 'Gospel of Life' to American Jurisprudence: A Religious, Ethical, and Philosophical Critique of Federal Funding for Embryonic Stem Cell Research." *Journal of Contemporary Health Law and Policy* 17 (2001):771–808.
Paul, Diane B. "Eugenic Anxieties, Social Realities and Political Choices." In *Are Genes Us? The Social Consequences of the New Genetics*, edited by Carl F. Cranor, 142–54. New Brunswick, N.J.: Rutgers University Press, 1994.

———. "Is Human Genetics Disguised Eugenics?" In *Genes and Human Self-Knowledge: Historical and Philosophical Reflections on Modern Genetics*, edited by Robert F. Weir, Susan C. Lawrence and Evan Faces eds., 67–83. Iowa City: University of Iowa Press, 1994.
Pauling, L. "Reflections on the New Biology." *UCLA Law Review* 15 (1968):268–72.
Pellegrino, E. D. "Intersections of Western Biomedical Ethics and World Culture: Problematic and Possibility." *Cambridge Quarterly of Healthcare Ethics* 3 (1992):191–96.
Pembrey, M. E. "Genetic Factors in Disease." In *Oxford Textbook of Medicine*, 3rd. ed., Vol. I, edited by D. J. Weatherall, J. G. G. Ledingham, and D. A. Warrell, 136. Oxford: University Press, 1996.
Pencarinha, D. F., N. K. Bell, J. G. Edwards, and R. G. Best. "Ethical Issues in Genetic Counseling: A Comparison of M.S. Counselor and Medical Geneticist Perspectives." *Journal of Genetic Counseling* 1:1 (1992):23.
Pera, M. F. "Scientific Considerations relating to the Ethics of the Use of Human Embryonic Stem Cells in Research and Medicine." *Reproduction, Fertility, & Development* 13:1 (2001):23–29.
Pergament, E., and B. Fine. "The Current Status of Chorionic Villus Sampling." In *Preconception and Preimplantation Diagnosis of Human Genetic Disease*, edited by R. G. Edwards, 145–49. Cambridge, U.K.: Cambridge University Press, 1993.
Persidis, A. "The Business of Pharmacogenomics." *Nature Biotechnology* 16 (1998):209–10.

Peters, T. "'Playing God' and Germline Intervention." *Journal of Medicine and Philosophy* 20 (1995):365–86.

Poland, S. C. "Genes, Patents, and Bioethics—Will History Repeat Itself?" *Kennedy Institute of Ethics Journal* 10 (2000):265–72.

Porter, J. "Individuality, Personal Identity, and the Moral Status of the Preembryo: a Response to Mark Johnson." *Theological Studies* 56 (1995):763–70.

Potter, Ralph. "Labeling the Mentally Retarded. The Just Allocation of Therapy." In *Ethics in Medicine*, edited by S. J. Reiser, et al., 623–30. Cambridge, Mass.: MIT Press, 1977.

Purtilo, R. "Interdisciplinary Health Care Teams and Health Care Reform." *Journal of Law, Medicine & Ethics* 22:2 (1994):121–26.

Rahner, Karl. "The Experiment with Man." In *Theological Investigations IX*, Graham Harrison trans., 205–24. New York: Herder and Herder, 1972.

———. "The Problem of Genetic Manipulation" In *Theological Investigations IX*, Graham Harrison trans., 225–51. New York: Herder and Herder, 1972.

Rai, Arti K. "The Information Revolution Reaches Pharmaceuticals: Balancing Innovation Incentives, Cost, and Access in the Post-Genomics Era." *University of Illinois Law Review* (2001):173ff.

Ramsey, P. "The Enforcement of Morals: Nontherapeutic Research on Children—A Reply to Richard McCormick." *Hastings Center Report* 6 (1976):21–30.

———. "Proxy Consent for Children." *Hastings Center Report* 7 (1977):4–48.

———. "Human Sexuality in the History of Redemption." *Journal of Religious Ethics* 16:1 (1988):56–86.

Ravin, A. J., M. B. Mahowald, and C. B. Stocking. "Genes or Gestation? Attitudes of Women and Men about Biological Ties to Children." *Journal of Women's Health* 6:6 (1997):639–47.

Rayl, A. J. S. "Microgravity and Gene Expression: Early Results Point to Relationship." *The Scientist* 13:18 (1999):1, 8–9.

Regaldo, Antonio. "Mining the Genome." *Technology Review* (September/October 1999):55–63.

Reilly, P. R. "Fear of Discrimination Drives Legislative Interest." *Human Genome News* 8 (1997):3–4.

———. "Laws to Regulate the Use of Genetic Information." In *Genetic Secrets: Protecting Privacy and Confidentiality in the Genetic Era*, edited by Mark A. Rothstein, 369–91. New Haven, Conn.: Yale University Press, 1997.

———. "The Human Genome Project: Recent Genetic Advances Will Have Far-Reaching Implications for Catholic Health Care." *Health Progress* (2001):24–27.

Resnick, D. "The Morality of Human Gene Patents." *Kennedy Institute of Ethics Journal* 7 (1997):43–61.

Revel, M. "Research on Animal Cloning Technologies and Their Implications in Medical Ethics: An Update." *Medicine & Law* 19:3 (2000):527–43.

Richter, G., and M. D. Bacchetta. "Interventions in the Human Genome: Some Moral and Ethical Considerations." *Journal of Medicine and Philosophy* 23:3 (1998):303–17.

Roberts, M. A. "Cloning and Harming: Children, Future Persons, and the "Best Interest" Test." *Notre Dame Journal of Law, Ethics & Public Policy* 13:1 (1999):37–61.

Robertson, J. A. "Liberty, Identity, and Human Cloning." *Texas Law Review* 76:6 (1998):1371–1456.

———. "Ethics and Policy in Embryonic Stem Cell Research." *Kennedy Institute of Ethics Journal* 9 (1999):109–36.

———. "Reproductive Liberty and the Right to Clone Human Beings." *Annals of the New York Academy of Sciences* 913 (2000):198–208.

———. "Human Embryonic Stem Cell Research: Ethical and Legal Issues." *Nature Reviews Genetics* 2:1 (2001):74–78.

Robertson, M. J. "Ethical, Legal and Scientific Issues concerning Human Cloning." *Journal of the Oklahoma State Medical Association* 95:4 (2002):261–64.

Robertson, S. "The Ethics of Human Cloning." *Journal of Medical Ethics* 24:4 (1998):282.

Robinson, Paul. "Prenatal Screening, Sex Selection and Cloning." In *A Companion to Bioethics*, edited by Helga Kuhse and Peter Singer, 173–85. Oxford: Blackwell Publishers, 1998.

Roche, P. A., and G. J. Annas. "Protecting Genetic Privacy." *National Review of Genetics* 2 (2001):392–96.

Roche, P. A., and M. A. Grodin. "The Ethical Challenge of Stem Cell Research." *Women's Health Issues* 10:3 (2000):136–39.

Rogers, A., and H. Ashraf. "UK's Position on Human Cloning Provokes Hostile Reaction in Europe Union." *Lancet* 355:9212 (2000):1344.

Rokosz, G. J. "Human Cloning: Is the Reach of FDA Authority too far a Stretch?" *Seton Hall Law Review* 30:2 (2000):464–515.

Rosenberg, R. "Family History and Genetic Risk Factors, Forward to the Future." *Journal of the American Medical Association* 278:15 (1997):1284–85.

Roses, A. D. "Pharmacogenetics and Future Drug Development and Delivery." *Lancet* 355 (2000):1358–61.

———. "Pharmacogenetics and the Practice of Medicine." *Nature* 405 (2000):857–65.

Ross, Lainie Friedman. "Disclosing Misattributed Paternity." *Bioethics* 10 (1996):114–30.

Rosso, E. "Panel Supports Use and Derivation of Stem Cells." *The Scientist* (11 October 1999):4.

Rothstein, M. "Protecting Genetic Privacy by Permitting Employer Access Only to Job-Related Employee Medical Information: Analysis of a Unique Minnesota Law." *American Journal of Law & Medicine* XXIV:4 (1998):400–1.

Royal Commission on New Reproductive Technologies. *Proceed with Care: Report of the Royal Commission on New Reproductive Technologies.* Ottawa, Canada: Ministry of Government Services, 1993.

Ryan, K. J. "The Politics and Ethics of Human Embryo and Stem Cell Research." *Women's Health Issues* 10:3 (2000):105–10.

Ryan, M. P., et al. "Genetic Testing for Familial Hypertrophic Cardiomyopathy in Newborn Infants." *British Medical Journal* 310 (1995):856–59.

Sanchez-Sweatman, L. R. "Reproductive Cloning and Human Health: An Ethical, International, and Nursing Perspective." *International Nursing Review* 47:1 (2000):28–37.

Savulescu, J. "Should We Clone Human Beings? Cloning as a Source of Tissue for Transplantation." *Journal of Medical Ethics* 25:2 (1999):87–95.

---. "Should Doctors Intentionally Do Less Than the Best?" *Journal of Medical Ethics* 25:2 (1999):121–26.

---. "The Ethics of Cloning and Creating Embryonic Stem Cells as a Source of Tissue for Transplantation: Time to Change the Law in Australia." *Australian & New Zealand Journal of Medicine* 30:4 (2000):492–98.

---. "Harm, Ethics Committees and the Gene Therapy Trial." *Journal of Medical Ethics* 27 (2001):148–50.

Schibecci, R., et al. "Genetic Medicine: An Experiment in Community-Expert Interaction." *Journal of Medical Ethics* 25 (1999):335–39.

Schmidt, K. "Just for You." *New Scientist* 160:2160 (1998):32–36.

Schmidtke, J. "Who Owns the Human Genome? Ethical and Legal Aspects." *Journal of Pharmacy and Pharmacology* 44:supplement 1(1992):205–10.

Schwartz, R. S. "Editorial Comment." *New England Journal of Medicine* 340:3 (1999):244.

Schwartz, Thomas. "Obligations to Posterity." In *Obligations to Future Generations*, edited by R. I. Sikora and Brian Barry, 3–13. Philadelphia: Temple University Press, 1978.

Scott, W. K., et al. "Complete Genomic Screen in Parkinson Disease: Evidence for Multiple Genes." *Journal of the American Medical Association* 286 (2001):2239–50.

Searle, J. "Fearing the Worst—Why Do Pregnant Women Feel 'At Risk'?" *Australian and New Zealand Journal of Obstetrics and Gynaecology* 36 (1996):279–86.

Shalala, D. "Protecting Research Subjects—What Must Be Done" *New England Journal of Medicine* 343 (2001):808–10.

Shamblot, Michael J., et al. "Derivation of Human Pluripotent Stem Cells from Cultured Primordial Germ Cells." *Proceedings of the National Academy of Sciences* 95 (1998):13, 726–31.

Shannon, Thomas A. "Fetal Status: Sources and Implications." *Journal of Medicine and Philosophy* 22 (1997):415–222.

---. "Human Cloning." *America* (2002):15–18.

Shannon, Thomas A., James J. Walter, and M. Cathleen Kaveny. "Ethical, Theological, and Legal Issues in Genetics." *Theological Studies* 60 (1999):109–47.

Shannon, Thomas A., and Allan B. Wolter. "Reflections on the Moral Status of the Pre-Embryo." *Theological Studies* 51 (1990):603–26.

Shapiro, M. H. "Is Bioethics Broke? On the Idea of Ethics and Law 'Catching Up' with Technology." *Indiana Law Review* 3 (1999):17–162.

Sharp, R. R., and J. C. Barrett. "The Environmental Genome Project and Bioethics." *Kennedy Institute of Ethics Journal* 9:2 (1999):175–222.

Shirkey, H. C. "Therapeutic Orphans." *Journal of Pediatrics* 72 (1968):119–20.

Shinn, R. "Christian Perspectives on the Human Body." *Theological Studies* 55 (1994):330–46.

——. "Genetic Research and the Elusive Body." In *Embodiment, Morality and Medicine*, edited by Lisa S. Cahill and Margaret A. Farley, 59–73. Boston: Kluwer Academic Publishers, 1995.

Silver, L. M. "Cloning, Ethics, and Religion." *Cambridge Quarterly of Healthcare Ethics* 7:2 (1998):168–72.

Silverman, N., and R. Wapner. "Chorionic Villus Sampling." In *Prenatal Diagnosis*, edited by Brock et al., 30–32. Edinburgh, U.K.: Churchill Livingstone, 1992.

Solomon, M., et al. "Decisions Near the End of Life." *American Journal of Public Health* 83:1 (1993).

Solveig M. R. "Disability, Gene Therapy and Eugenics." *Journal of Medical Ethics* 26 (2000):85–88.

Spicer, Carol Mason. "Federal Oversight and Regulation of Human Subjects Research—An Update." *Kennedy Institute of Ethics Journal* 10 (2000):261–64.

Statham, H., and J. Green. "Serum Screening for Down's Syndrome: Some Women's Experiences." *British Medical Journal* 307 (1993):174.

Statham, H., J. Green, and C. Snowdon. Letter. *British Medical Journal* 306 (1993):858–59.

Steinbock, B. "What Does 'Respect for Embryos' Mean in the Context of Stem Cell Research?" *Women's Health Issues* 10:3 (2000):127–30.

Steinhorn, R. H. "Prenatal Ultrasonography: First Do No Harm?" *Lancet* 352 (1998):1568–69.

Stephenson, J. "As Discoveries Unfold, a New Urgency to Bring Genetic Literacy to Physicians." *Journal of the American Medical Association* 278:15 (1997):1225–26.

———. "Group Drafts Core Curriculum for 'What Docs Need to Know about Genetics.'" *Journal of the American Medical Association* 279:10 (1998):735–36.

———. "Human Genome Studies Expected to Revolutionize Cancer Classification." *Journal of the American Medical Association* 282:10 (1999):927–28.

———. "Studies Illuminate Cause of Fatal Reaction in Gene-Therapy Trial." *Journal of the American Medical Association* 285 (2001):2570.

Stix, G. "Personal Pills." *Scientific American* 279:4 (1998):10–11.

Stone, Richard, "Religious Leaders Oppose Patenting Genes and Animals." *Science* 268 (1995):1126.

Stranc, L. C., J. A. Evans, and J. L. Hamerton. "Chorionic Villus Sampling and Amniocentesis for Prenatal Diagnosis." *Lancet* 349 (1997):711–14.

Strohman, R. C. "Ancient Genomes, Wise Bodies, Unhealthy People: Limits to a Genetic Paradigm in Biology and Medicine." *Perspectives in Biology & Medicine* 37:1 (1993):112–42.

Strong, C. "Cloning and Infertility." *Cambridge Quarterly of Healthcare Ethics* 7:3 (1998):279–93.

———. "Response to 'May a Woman Clone Herself?' by Jean E. Chambers (CQ Vol 10, No 2) and 'Entitlement to Cloning' by Timothy F. Murphy (CQ Vol 8, No 3). Clone Alone." *Cambridge Quarterly of Healthcare Ethics* 11:1(2002):76–82.

Suarez, A. "Hydatidiform Moles and Teratomas Confirm the Human Identity of the Preimplantation Embryo." *Journal of Medicine and Philosophy* 15 (1990):630–36.

Subramanian, G., et al. "Implications of the Human Genome for Understanding Human Biology and Medicine." *Journal of the American Medical Association* 286 (2001):2296–2307.

Suter, M. S. "Whose Genes Are These Anyway? Familial Conflicts over Access to Genetic Information." *Michigan Law Review* 9 (1993):1855.

Tauer, Carol A. "Embryo Research and Public Policy." *Journal of Medicine and Philosophy* 22 (1997):423–39.

———. "Private Ethics Boards and Public Debate." *Hastings Center Report* 29:2 (1999):43–45.
Tavill, A. S. Editorial. "Clinical Implications of the Hemochromatosis Gene." *New England Journal of Medicine* 341:10 (1999):755–57.
Terry, V., and C. D. Shorey. "Artificial Reproductive Technologies in Australia—Legislation for the New Millennium?" *Medicine & Law* 19:2 (2000):209–35.
Thompson, James A., et al. "Embryonic Stem Cell Lines Derived from Human Blastocysts." *Science* 282 (1998):1145–47.
Thorp, J. M., Jr., and W. A. Bowles, Jr. "Pro-Life Perinatologist—Paradox or Possibility?" *New England Journal of Medicine* 326 (1992):1217–19.
Tobin, B. "Reply to Savulescu: Why We Should Maintain a Prohibition on Destructive Research on Human Embryos." *Australian & New Zealand Journal of Medicine* 30:4 (2000):498–502.
Towner, D., and R. S. Loewy. "Ethics of Preimplantation Diagnosis for a Woman Destined to Develop Early-Onset Alzheimer Disease." *Journal of the American Medical Association* 287:8 (2002):1038–40.
Trepagnier, D. M. "Human Embryonic Stem Cell Research: Implications from an Ethical and Legal Standpoint." *Journal of the Louisiana State Medical Society* 152:12 (2000):616–24.
Trosko, J. E. "Cloning of Human Stem Cells: Some Broad Scientific and Philosophical Issues." *Journal of Laboratory & Clinical Medicine* 135:6 (2000):432–36.
U.S. Catholic Conference. "Harvesting Embryonic Stem Cells for Research: Response to NIH Draft Guidelines." *Origins* 29:35 (2000):566–71.
U.S. Department of Health and Human Services. "Final Regulations Amending Basic HHS Policy for the Protection of Human Research Subjects." *Federal Register* 46 (26 January 1981):8387.
Venter, C., et al. "The Sequence of the Human Genome." *Science* 291:5507 (2001):1304–51.
Verlinsky, Y., and N. Ginsberg. "A Brief History of Prenatal Diagnosis." In *Preconception and Preimplantation Diagnosis of Human Genetic Disease*, edited by R. G. Edwards, 138–40. Cambridge, UK: Cambridge University Press, 1993.

Verlinsky, Y., and C. M. Strom. "Preconception Diagnosis of Polar Bodies." In *Preconception and Preimplantation Diagnosis of Human Genetic Disease,* edited by R. G. Edwards, 233-38. Cambridge, UK: Cambridge University Press, 1993.

Verlinsky, Y., et al. "Preimplantation Diagnosis for Fanconi Anemia Combined with HLA Matching." *Journal of the American Medical Association* 285 (2001):3130-33.

Verlinsky, Y., et al. "Preimplantation Diagnosis for Early-Onset Alzheimer Disease Caused by V717L Mutation." *Journal of the American Medical Association* 287:8 (2002):1018-21.

Wagner, R. P. "Understanding Inheritance: An Introduction to Classical and Molecular Genetics." *Los Alamos Science* 20 (1992).

Wald, N. J., et al. "Prenatal Screening for Down's Syndrome Using Inhibin-A as a Serum Marker." *Prenatal Diagnosis* 16 (1996):143-53.

Wallace, R. W. "The Human Genome Diversity Project: Medical Benefits versus Ethical Concerns." *Molecular Medicine Today* 4:2 (1998):59-62.

Walter, J. J. "Theological Issues in Genetics." *Theological Studies* 60 (1999):124-134.

Walters, L. "Human Gene Therapy: Ethics and Public Policy." *Human Gene Therapy* 2 (1991):115-22.

———. "The Oversight of Human Gene Transfer Research." *Kennedy Institute of Ethics Journal* 10 (2000):171-74.

Warnock, M. "The Warnock Report." *British Medical Journal* 291 (1985):187-89.

———. "The Regulation of Technology." *Cambridge Quarterly of Healthcare Ethics* 7:2 (1998):173-75.

Watson, E., et al. "Screening for Genetic Carriers of Cystic Fibrosis through Primary Health Care Services." *British Medical Journal* 303 (1991):504-7.

Watson, J. D. "The Human Genome Project: Past, Present, and Future." *Science* 248 (1990):44-48.

Watson, J. D., and F. H. C. Crick. "Molecular Structure of Nucleic Acids: A Structure for Deoxyribose Nucleic Acid." *Nature* 171 (1953):737-38.

Weissman, I. L. "Stem Cells—Scientific, Medical, and Political Issues." *New England Journal of Medicine* 346:20 (2002):1576–79.

Werber, S. J. "Cloning: a Jewish Law Perspective with a Comparative Study of Other Abrahamic Traditions." *Seton Hall Law Review* 30:4 (2000):1114–81.

Wertz, D. C. "Society and the Not-So-New Genetics: What Are We Afraid Of? Some Predictions from a Social Scientist." *Journal of Contemporary Health Law and Policy* 13 (1997):299–346.

———. "The Difficulty of Recruiting Minorities to Studies of Ethics and Values in Genetics." *Community Genetics* 1 (1998):175–79.

———. "Guidelines Point the Way on Genetics Ethics." *Nature* 399:6734 (1999):297.

Wertz, D. C., and J. C. Fletcher. "Ethical and Social Issues in Prenatal Sex Selection: A Survey of Geneticists in 37 Nations." *Social Science and Medicine* 46:2 (1998):255–73.

Wertz, D. C., J. C. Fletcher, and J. J. Mulvihill. "Medical Geneticists Confront Ethical Dilemmas: Cross-Cultural Comparisons among 18 Nations." *American Journal of Human Genetics* 46 (1990):1201–13.

Wertz, D. C., and R. Gregg. "Genetics Services in a Social, Ethical, and Policy Context: A Collaboration between Consumers and Providers." *Journal of Medical Ethics* 26 (2000):261–65.

Wheatley, Jane. "What If. . . ." *HQ* (Mar-April 1997):44.

White, M. T. "Making Responsible Decisions: An Interpretive Ethic for Genetic Decision Making." *Hastings Center Report* 29:1 (1999):14–21.

Wikler, Daniel. "Can We Learn from Eugenics?" In *Genetic Information: Acquisition, Access, and Control*, edited by Alison K. Thompson and Ruth F. Chadwick, 1–16. New York: Kluwer, 1999.

Wilcox, A., et al. "Genetic Determinism and the Over-Protection of Human Subjects." *Nature Genetics* 21:4 (1999):362.

Wildes, K. W. "The Stem Cell Report." *America* (16 October 1999):12–16.

Williamson, R. "Human Reproductive Cloning Is Unethical Because It Undermines Autonomy: Commentary on Savulescu." *Journal of Medical Ethics* 25:2 (1999):96–97.

Willgoos, C. "FDA Regulations: An Answer to the Questions of Human Cloning and Germline Gene Therapy." *American Journal of Law & Medicine* 27 (2001):101–24.

Wilmut, I. "Cloning for Medicine." *Scientific American* 279 (1998):58–63.

———. "How Safe Is Cloning?" *Cloning* 3:2 (2001):39–40.

Wilmut, I., et al. "Viable Offspring Derived from Fetal and Adult Mammalian Cells." *Nature* 385 (1997):810–13.

Wils, J. P. "Umbilical Cord Blood Stem Cell Transplantation — Ethical Problems." *Biomedical Ethics* 4:3 (1999):92–101.

Wolf, S. M. "Beyond 'Genetic Determinism': Toward the Broader Harm of Geneticism." *Journal of Law, Medicine, & Ethics* 23 (1995):345–53.

Wolf, C. R., G. Smith, and R. L. Smith, "Science, Medicine, and the Future: Pharmacogenetics." *British Medical Journal* 320 (2000):987–90.

Wright, S. M. "The Social Warp of Science: Writing the History of Genetic Engineering Policy." *Science, Technology, and Human Values* 18:1 (1993):79–101.

Zneimer, S. M. "The Human Genome Project: Exploring Its Progress and Successes and the Ethical, Legal, and Social Implications." *Clinical Leadership & Management Review* 16:3 (2002):151–57.

Zola, I. K. "In the Name of Health and Illness: On Some Socio-Political Consequences of Medical Influence." *Social Science and Medicine* 9 (1975):83–87.

NOTES ON CONTRIBUTORS

CAROL BAYLEY, PH.D., is Vice President for Ethics and Justice Education at Catholic Healthcare West, a forty-two-hospital, not-for-profit health care system based in San Francisco. She earned her doctorate in Philosophy at Georgetown University. She has published in the areas of the philosophy of science, alternative medicine, pharmacy ethics, business ethics, and medical error. Her most recent work is "Medical Mistakes and Institutional Culture," in an edited collection on *Promoting Patient Safety* (forthcoming, 2004).

JOHN BREHANY, PH.D., is Director of Mission Services and Ethics at Mercy Medical Center in Sioux City. He graduated with a Ph.D. degree in health care ethics from Saint Louis University in 2003. He also received an S.T.L. degree from the Pope John Paul II Institute for Studies on Marriage and Family in Washington, D.C., and an M.A. degree from the University of St. Thomas, Houston, Texas. He has served as an Adjunct Professor in Saint Louis University's Schools of Medicine and Public Health and at the School of Pharmacy at the University of Nebraska Medical Center. Prior to this, he served as Associate Professor of Ethics and Theology at Mount Angel Seminary, St. Benedict, Oregon. His current research specialties include issues in human genetics, end-of-life care issues, and ethics of stem-cell research.

DENNIS BRODEUR, PH.D., is the Senior Vice President for Stewardship at SSM Health Care System in St. Louis, Missouri. He graduated with a Ph.D. degree in moral theology from the University of Louvain, Belgium, in 1976. He consults nationally on ethical issues in health care and serves on several boards of major organizations and associations. He has published many scholarly essays on a wide range of issues in health care ethics as book chapters and journal articles, and he coauthored with Kevin O'Rourke *Medical Ethics: Common*

Ground for Understanding, 2 vols. (St. Louis, Mo.: Catholic Health Association, 1986, 1989).

RUTH CHADWICK, D.PHIL., is Professor of Bioethics and Director of the ESRC Centre for Economic and Social Aspects of Genomics (CESAGen) at Lancaster University. She held positions in Liverpool, Cardiff, and Preston before taking up her present post in 2000. She has coordinated a number of projects funded by the European Commission, including the Euroscreen projects (1994–1996; 1996–1999) and coedits the journal *Bioethics*. She is a member of the Human Genome Organisation (HUGO) Ethics Committee (Vice-Chair since 1999), the Food Ethics Council, the Medical Research Council Advisory Committee on Scientific Advances in Genetics, and the Advisory Committee on Novel Foods and Processes. She was editor-in-chief of the award-winning *Encyclopedia of Applied Ethics* (San Diego, Calif.: Academic Press, 1998). Current research grants include (with Oxford and Sheffield) the Wellcome Trust Electronic Bioethics Resource, and she is a partner in the North-West Genetics Knowledge Park (NoWGeN).

ELLEN WRIGHT CLAYTON, M.D., J.D., is the Rosalind E. Franklin Professor and Director of the Center for Genetics and Health Policy, Professor of Pediatrics, and Professor of Law at Vanderbilt University in Nashville, Tennessee. She received her M.D. degree from Harvard Medical School in 1985 and her J.D. from Yale Law School in 1979. Her current research interests include research ethics and the implications of new developments in genomics both in and outside the health care setting. She is currently the Chair of the Ethical, Legal, and Social Implications Working Group of the International Haplotype Map Project. She has published two books and many articles, including "Genetics, Populations, and Public Health: A Complex Relationship," *Journal of Law, Medicine, & Ethics* 30 (2002):290–97; and "Genetics Research: Toward International Guidelines," in Robert J. Levine and Samuel Gorovitz with James Gallagher, eds., *Biomedical Research Ethics: Updating International Guidelines—A Consultation* (Geneva: CIOMS, 2000), 152–69.

NORMAN M. FORD, SDB, S.T.L, PH.D., has been the Foundation Director of the Caroline Chisholm Centre for Health Ethics, East Melbourne, since March 1995. He lectures in Philosophy of the Human Person, Practical Ethics, and Bioethics at the Catholic Theological College for the B.Theol. of the ecumenical Melbourne College of Divinity. He is also an Adjunct Professor of the Australian Catholic University and a Senior Honorary Research Fellow at the Centre for Human Bioethics, Monash University. He completed his Ph.D. in philosophy at the Salesian Pontifical University, Rome (1967). At present he specializes in health ethics, especially on issues at the beginning and end of life. He is a member of several hospital ethics committees. He was the President of the Catholic Moral Theology Association of Australia and New Zealand from 1992 to 1997 and President of the Melbourne College of Divinity from 1992 to 1993. He has published many articles on ethics and bioethics, such as "Fetus" in *Encyclopedia of Applied Ethics* (San Diego, Calif.: Academic Press, 1998), as well as several books, including *Live Out the Truth in Love* (Melbourne: Catholic Education Office, 1985, 1991); *When Did I Begin? Conception of the Human Individual in History, Philosophy and Science* (Cambridge, UK: Cambridge University Press, 1988, paperback 1991), translated into Polish in 1995 and Italian in 1997; and *The Prenatal Person. Ethics from Conception to Birth* (Oxford: Blackwell Publishing, 2002).

HENK A. M. J. TEN HAVE, M.D., PH.D., is Professor of Medical Ethics at the University of Nijmegen, the Netherlands. He studied medicine and philosophy at Leiden University and received his medical degree in 1976 from Leiden University and his philosophy degree in 1983. He worked as researcher in the Pathology Laboratory at the University of Leiden (1976–1977), as practicing physician in the Municipal Health Services of the City of Rotterdam (1978–1979), and as Professor of Philosophy in the Faculty of Medicine and Faculty of Health Sciences at the University of Limburg, Maastricht (1982–1991). Since 1991 he has been Professor of Medical Ethics and Director of the Department of Ethics, Philosophy and History of Medicine in Nijmegen. He is involved in many public debates concerning euthanasia, drug addiction, choices in health care, and resource allocation. His current research focuses on ethical issues in palliative care. He has been coordinator of the European Commission–funded project, "Palliative Care Ethics." He also serves on numerous editorial boards. He is editor-in-chief of the recently established journal *Medicine, Health Care & Philosophy*. He is cofounder and Secretary of the European Society for Philosophy of Medicine and Health Care. He has published *Medische Ethiek* (Houten, the Netherlands: Bohn Stafleu Van Loghum, 1998), a textbook for medical curricula. His other recent books include *Palliative Care in Europe. Concepts and Policies*, editor with R. Janssens (Amsterdam, the Netherlands: IOS Press, 2001); *Bioethics in a European Perspective*, editor with Bert Gordjin (Dordrecht, the Netherlands: Kluwer Academic Publishers, 2001); and *The Ethics of Palliative Care: European Perspectives*, editor with David Clark (Buckingham, UK: Open University Press, 2002). He has just finished a manuscript for a book on euthanasia, *Death and Medical Power*, forthcoming (Bloomington, Ind.: Indiana University Press, 2004).

JAN C. HELLER, PH.D., is System Director of the Office of Ethics and Theology for Providence Health System, based in Seattle. He provides education and consultation in health care ethics throughout the system and participates in regional, national, and international forums that address clinical, organizational, and social issues in health care. He received his Ph.D. in 1995 in Ethics and Society from Emory University. Dr. Heller has published numerous articles in scholarly and popular journals, and has published four books: *Human Genome Research and the Challenge of Contingent Future Persons* (Omaha, Nebr.: Creighton University Press, 1996); *Contingent Future Persons: On the Ethics of Deciding Who Will Live, or Not, in the Future*, editor with Nick Fotion (Dordrecht, The Netherlands: Kluwer Academic Publishers, 1997); *Faithful Living, Faithful Dying: Anglican Reflections on End of Life Care*, coauthor with Cynthia B. Cohen, Bruce Jennings, E. F. Michael Morgan, David A. Scott, Timothy F. Sedgwick, and David H. Smith (Harrisburg, Penn.: Morehouse Publishing, 2000); and *Guide to Professional Development in Compliance*, editor with Joseph E. Murphy and Mark E. Meaney (Gaithersburg, Md.: Aspen Publishers, Inc., 2001). He is a priest in the Episcopal Church and was previously the founding Director of the Center for Ethics in Health Care at Saint Joseph's Health System in Atlanta.

M. THERESE LYSAUGHT, PH.D., is an Associate Professor in the Department of Religious Studies at the University of Dayton. She holds an M.A. in Theology from the University of Notre Dame and earned her doctorate in Theological Ethics at Duke University. She has served as a member of the Recombinant DNA Advisory Committee (RAC) of the National Institutes of Health and as an advisor to the Program on Dialogue between Science, Ethics, and Religion at the American Academy of Arts and Sciences as well as at the Catholic Health Association. After completing her doctoral work, she spent a year with the Program in Molecular and Clinical Genetics at the University of Iowa as a postdoctoral research fellow and fellow of the NIH program on the Ethical, Legal, and Social Implications of the Human Genome Project. Her areas of interest include human gene transfer research, human cloning, and stem-cell research. A frequent speaker and writer, Professor Lysaught's publications have appeared in the *Journal of Law, Medicine & Ethics* and *Health Progress*. Her work has been anthologized in *On Moral Medicine*, edited by S. Lammers and A. Verhey (Grand Rapids, Mich.: Eerdmans Publishing, 1998), and she contributed the entry on human gene transfer research to the forthcoming third edition of the *Encyclopedia of Bioethics* (New York: Macmillan, forthcoming). She is also a frequent contributor to the journal *Commonweal* and has contributed to chapters in over ten books.

GERARD MAGILL, PH.D., is a Professor with tenure at the Center for Health Care Ethics at Saint Louis University (health sciences) in St. Louis, Missouri. He holds the positions of Executive Director of the center and Department Chair of the interdisciplinary Ph.D. program at the center, and he has a secondary appointment as Professor in the Department of Internal Medicine at the University Hospital and Medical School. He graduated with a Ph.D. degree from Edinburgh University in 1987. His current research specialties include the ethics of human genomics, the ethics of stem-cell research, and organizational ethics in health care. He has published many scholarly essays and several books, including *Discourse and Context*, editor (Carbondale, Ill.: Southern Illinois University Press, 1993); *Personality and Belief*, editor (Lanham, N.J.: University of America Press, 1994); *Values and Public Life*, coeditor with Marie D. Hoff

(Lanham, N.J.: University of America Press, 1995); *Abortion and Public Policy*, coeditor with R. Randall Rainey (Omaha, Nebr.: University of Creighton Press, 1996).

MARY BRIODY MAHOWALD, PH.D., is Professor Emerita at the University of Chicago. Her Ph.D. in Philosophy was awarded in 1969 from Marquette University. She subsequently taught in the Philosophy Departments at Villanova University, Indiana University at Indianapolis, Oberlin College, and Case Western Reserve University. Since 1982, her primary appointments have been in medical schools, where she is directly involved in cases that raise ethical questions. She has been principal investigator on grants from the National Institutes of Health and the Department of Energy, has had fellowships from the National Endowment for the Humanities, the Rockefeller Foundation, and the American Council of Learned Societies, and has published over a hundred articles in clinical and philosophy journals. Her books include *Philosophy of Woman: Classical to Current Concepts* (Indianapolis, Ind.: Hackett Publishing Company, 3rd ed., 1994); *Women and Children in Health Care: An Unequal Majority* (New York: Oxford University Press, 1993); *Disability, Difference, Discrimination*, coauthored with Anita Silvers and David Wasserman (Lanham, Md.: Rowman & Littlefield, 1998); *Genes, Women, Equality* (New York: Oxford University Press, 2000); and *Genetics in the Clinic*, coedited with Victor McKusick, Angela Scheuerle, and Timothy Aspinwall (St. Louis, Mo.: Mosby, 2001). Currently she is Visiting Professor Emerita at Stanford University, where she is working on a book about ethical issues in women's health care across the life span.

SHELIA A. M. MCLEAN, LL.B., M.LITT., PH.D., LL.D. (EDIN), LL.D. (ABERTAY, DUNDEE), F.R.S.E., F.R.C.P.E. (EDIN), F.R.S.A., is the first holder of the International Bar Association Chair of Law and Ethics in Medicine at Glasgow University, Scotland, and is Director of the Institute of Law and Ethics in Medicine at Glasgow University. She has acted as a consultant to the World Health Organization and the Council of Europe, as well as a number of individual nation-states. She has acted as an expert reviewer for many of the major grant-awarding bodies and other similar organizations. She has published extensively in the area of medical law, is on the editorial board of a number of national and international journals, and is regularly consulted by the media on matters of medical law and ethics. Her publications include *Legal Issues in Medicine*, editor (Aldershot, UK: Ashgate Publishing Co., 1981); *Medicine, Morals and the Law*, coauthor with G. Maher (Aldershot, UK: Gower, 1983, repr. 1985); *Human Rights: From Rhetoric to Reality*, coeditor with Tom Campbell, et al. (Oxford: Basil Blackwell, 1986); *The Legal Relevance of Gender: Some Aspects of Sex-Based Discrimination*, coeditor with Noreen Burrows (London: Macmillan, 1988); *Legal Issues in Human Reproduction*, editor (Aldershot, UK: Ashgate Publishing Co., 1989); *A Patient's Right to Know: Information Disclosure, the Doctor and the Law* (Aldershot, UK: Dartmouth Publishing Co., 1990, paperback 1995); *Law Reform and Human Reproduction*, editor (Aldershot, UK: Dartmouth Publishing Co., 1992); *Compensation for Damage: An International Perspective*, editor (Aldershot, UK: Dartmouth Publishing Co., 1993); *Law Reform and Personal Injury Litigation*, editor (Aldershot, UK: Dartmouth Publishing Co., 1995); *Contemporary Issues in Law, Medicine and Ethics*, editor (Aldershot, UK: Dartmouth Publishing Co., 1996); *Ethics and the Law in Intensive Care*, coeditor with N. Pace (Oxford: Oxford University Press, 1996); *Death, Dying and the Law*, editor (Aldershot, UK: Dartmouth Publishing Co., 1996); *The Case for Physician Assisted Suicide*, coauthor with Alison Britton (London: Pandora Press, 1997); *Old Law, New Medicine* (London: Pandora/Rivers Oram Press, 1999); *Medical Law and Ethics* (Aldershot, UK: Ashgate Publishing Co., 2002); *The Genome Project and Gene Therapy*, editor (Aldershot, UK: Ashgate Publishing Co., 2003); and *Legal and Ethical Aspects of Health Care*, coauthor with John Kenyon Mason, forthcoming (London: Greenwich Medical Media Press, 2003).

RUTH B. PURTILO, PH.D., is the C.C. and Mabel L. Criss Professor at the School of Medicine and Director of the Center for Health Policy and Ethics, Creighton University Medical Center, Omaha, Nebraska. She holds a secondary appointment as Professor in the Department of Physical Therapy, School of Pharmacy and Health Professions. She graduated with a Ph.D. degree from Harvard University in 1979 and prior to that earned a Master of Theological Studies Degree from Harvard Divinity School in 1975. She is a Fellow of the Hastings

Center, New York. She has a sustained research focus on the ethics of disability, long-term care, and rehabilitation. She has published many scholarly articles and six books, including *Ethical Dimensions in the Health Professions*, 4th ed. (Philadelphia, Pa: W. B. Saunders, 2003). She was an area editor for the *Encyclopedia of Bioethics*, rev. ed. (New York: Macmillan, 1995) and is currently codirector with Henk ten Have of an International Dialogue on Ethical Foundations of Palliative Care for Alzheimer's Disease sponsored by the Greenwall Foundation, New York. Their coedited book from that dialogue, *Ethical Foundations of Palliative Care for Alzheimer's Disease* is in press (Baltimore, Md.: Johns Hopkins Press, forthcoming).

DAVID SCHLESSINGER, PH.D., is Chief of the Laboratory of Genetics at the National Institute on Aging (NIA). He received his Ph.D. with Dr. James D. Watson at Harvard University in 1960. After a postdoctoral period with Dr. Jacques Monod at the Pasteur Institute in Paris, he joined the faculty of the Washington University School of Medicine in St. Louis, where he was Professor of Molecular Microbiology, Professor of Genetics, and Professor of Medicine, and established the first Human Genome Center. Since joining the NIA in 1997, he has established groups to examine the genetic basis of aging-associated traits and diseases, as well as their basis in early human development. His research has led to over three hundred published reports in the areas of antibiotic mechanisms, recombinant NA formation and metabolism, genome structure, and inherited human diseases. He has served as President of the American Society of Microbiology and is the current President of the Human Genome Organization in America. He has also served on numerous government panels and the Council of the American Cancer Society, with formal recognitions that have included receipt of the Eli Lilly Award in Microbiology, a Josiah Macy Fellowship, and M.D. honoris causa degrees from Uppsala University and the University of Cagliari.

BRENT WATERS, D. MIN., D.PHIL., is Director of the Center for Ethics and Values and Assistant Professor of Christian Social Ethics at Garrett-Evangelical Theological Seminary. He is the author of *Reproductive Technology: Towards a Theology of Procreative Stewardship* (Cleveland, Ohio: Pilgrim Press, 2001); *Dying and*

Death: A Resource for Christian Reflection (Cleveland, Ohio: United Church Press, 1996); and *Pastoral Genetics: Theology and Care at the Beginning of Life,* coauthor with Ronald Cole-Turner (Cleveland, Ohio: Pilgrim Press, 1996) and *God and the Embryo: Religious Voices on Stem Cells and Cloning,* ed. (Washington, D.C.: Georgetown University Press, forthcoming). He served previously as the Director of the Center for Business, Religion and Public Life, Pittsburgh Theological Seminary. Waters is a graduate of the School of Theology at Claremont (D.Min.), and the University of Oxford (D.Phil.).

INDEX

Abortion, xiii, 10, 25–27, 38, 42, 102–3, 198, 205–7, 264–66
Access; equality of, 257; and ethics discourse, 10; to health care, 168; to research data, 71; restriction of, 69, 71; to technology, xii, 122. *See also* Health care
Accountability, 39–40, 68, 70–73
Adenovirus, 62, 64
Advanced Cell Technology, 265–66
Ageing; and associated disease, 145; and embryology, xii, 147–51; impact of genetics and genomics on views of, xii, 144–62; tissues and organs, 148, 150–51
Alderson, Priscilla, 210
Alexander the Great, 145
Algene Biotechnologies Corp., 58
Algorithms, 65–66
Alpert, Joel J., 220
American Medical Association, 184
American Society of Bioethics and Humanities, 194
Amgen Inc., 57, 60–61
Anderson, French, 121–23
Annas, George J., 37–38, 40, 44
Argument, 10, 120–23
Aristotle, 104, 144–45
Arney, W. R., 89
Artificial intelligence, 65–66
Asilomar, California, 5
Association of British Insurers, 46
AstraZeneca PLC, 58
Augustine, Saint, 127, 131, 134

Augustus, 128
Autonomy, 10, 37, 45, 90–91, 95–98, 111, 171–72, 192, 204. *See also* Individual; Person

Bachelard, Gaston, 98
Basic Local Assignment Search Tool (BLAST), 66
Baxter, Richard, 129, 131, 134
Bayer AG, 62
Bayley, Carol, xiii, 176
Beckett, Samuel, 145
Bell, John, 186
Belmont Report, 258
Bergen, B. J., 89
Bioengineering, 253, 260
Bioethics and Biomedical ethics. *See* Ethics
Bioinformatics, 64–66, 91, 255
Biology, ix, 43, 134
Biomedicine, 98, 131, 134
Biotechnology, 1, 5–10, 53, 57, 63, 84, 192, 258, 267
Black, J., 40
Blackmun, Justice Harry A., 264
Brahams, D., 38
Brehany, John, 1
Bridges, Calvin, 2
Bristol Myer Squibb Co., 58
British Medical Association, 41
Brodeur, Dennis, xiii, 164
Brown, Louise, 264
Burden and benefit. *See* Risk
Bushnell, Horace, 130, 134

Business, 53–73. *See also* Companies; Investment

Cancer; breast, 101, 103, 111, 113, 177–78, 181, 184, 187; breast and ovarian, 26–29, 177; classification, 57; colorectal, 177, 180; drugs, 62; and the EGF receptor, 58; and gene HER-2, 58; glioblastoma, 59; gynecological, 104; life-threatening, 168, 189; and molecular variations in tumors, 57; treatments, 256, 261; and tumor growth, 58, 64; and tumor suppressor, 62. *See also* Disease
Caplan, A. L., 96
Care; appropriate, 183; continuum of, 167; cost-effective, 68; and cure, 58, 168; decisions, 174; duty of, 44; home, 156; patient, 166; primary, 179; quality, 179; right, 166. *See also* Health care; Long term care
Caregiving, 102–3
Category mistakes, 87
Celera Genomics Group, 60–62, 65–66, 255
Cell(s); and beta-secretase, 57; biology, 2; of an individual, 255; loss of, 149; membrane, 150; mortal and immortal, xii, 148–49, 151; nucleus of, 2; pathological, 57; types, 261. *See also* Stem cell research
Cell Genesys Inc., 62, 66
Center for Health Care Ethics, ix, xiv, 1
Chadwick, Ruth, xiii, 186
Chromosome(s); abnormalities, 198, 201, 207; anomalies, 197; and chromosomal arrangements, 105; and formation of gametes, 2; and genetic mapping, 147; genes located on the, 83; and linked disease, 3; and neural tube defects, 201, 205; problems, 200; testing, 199; and the X chromosome, 150, 201
Ciba-Geigy AG, 59
Clayton, Ellen Wright, x, 23
Clinton, President Bill, 221, 262–63
Clonaid, 253
Cloning; ban of, 265; and Dolly, 11, 71; human, 1, 123, 260; human embryos, 253–54, 265; literature, 123; and patents, 265; positional, 150; reproductive, 192, 254, 265–66; techniques, 122, 253–54. *See also* Somatic cell nuclear transfer; Therapeutic cloning
Code of Federal Regulations (U.S.), 258
Collins, Francis, 4, 255
Commercialization and commodification, 7, 55, 67, 72–73, 257
Committee on Drugs of the American Academy of Pediatrics, 220–23
Committee on the Ethics of Gene Therapy (Britain), 6
Common good, xiii, 11, 68, 162, 170, 209
Common Rule, 258
Community, 68–70, 96, 139–40, 167, 171. *See also* Society
Companies; biopharmaceutical, 64; biotechnology, 59, 62, 71, 261; cloning, 253; gene therapy, xi, 62; genomics, 53–73; insurance, 97; pharmaceutical, 60–61, 71,

177, 261. *See also* Business; Investment
Competition, 59–60, 69
Conditions. *See* Diseases and conditions
Confidentiality, x, xiii, 6–7, 41–44, 73, 108, 182–83, 192, 209
Conflict of interest, 258
Congress (U.S.), 4, 7, 23, 46, 221, 254, 265–66
Consensus, 8, 9, 73
Consent, xiii, 6, 9–10, 117, 169, 171–73, 183, 191–94, 203–5, 229, 257–58, 261–62
Consumers, 111–13
Council of Bioethics (U.S.), 260
Council on Ethical and Judicial Affairs of the AMA, 190
Crick, ix, 3, 4, 7, 35

Dana-Farber Cancer Institute, 60
Daniels, Norman, 170
Danish Council of Ethics, 43, 45
Darwin, Charles, 131
De Silva, Ashanti, 216
Death(s); in an asthma study, 259; and chronic disease, 153; early, 56; of a gene therapy patient, xiv, 9, 257–59, 260; and life threatening disease, 156; and organ transplants, 261–62
Decision making/makers, xi, 28, 38–39, 92, 113, 116, 170
Decode Genetics Inc., 58
Deoxyribonucleic acid (DNA); abnormalities of, 80; and alleles, 153; and amino acid chains, 3; chips, 66; and chromosomes, 255; cloned, 147; codes, 84, 90; cultural meaning of, 91; debate, 84; double helix structure of, ix, 2, 3, 65; gene as a section of, 2, 256; human, 35; language of, 48; letters of, 255; molecule, 37, 55; and nucleotide base pairs, 3; and polymerase chain reaction (PCR), 3–4; recombinant (rDNA), ix, 3; samples, 55, 188; sequence and sequencing, 55, 132; the structure of, ix, 3, 7; tests, 153–54; variations, 66; and Watson and Crick, 3
Department of Energy (U.S.), 4
Department of Health and Human Services (U.S.), 218–19, 223, 233, 258, 261
Destiny; common lineage and, 127; and competing destinies, 133–40; constructing our, 126; and grace, 135–39
Diagnosis, 4, 56, 59, 254. *See also* Preimplantation and Prenatal diagnosis
Dignity; and chronic disease, 158; culture and, ix, xi; and genetic engineering, xi; and genetic research, xi, 116–24; of human persons, 120; as a moral constraint, 118, 123; violation of, 122–24. *See also* Autonomy; Individual; Person
Dilemma(s), ix, 10, 68–69, 129
Disability/Disabled, xiii, 10, 32, 93, 113, 153–54, 162, 193, 209–10
Discrimination, 10, 25, 30, 32, 56, 73, 162, 183, 210, 257
Disease(s); alleviation of, 96; classification of, 88; chronic, xii, 153–62, 167; concepts of, 83, 90; control of, 90; cure of, 9, 165; and debility, 162, 261; definition of, 170; diagnosis and prediction

of, 28, 171; diagnosis and treatment of, ix, 55–57; elimination of, 95; environmentally associated, 256; eradication of, 169; etiology of, 30; gene-caused, 56, 256; and gene HER-2, 58; heterogeneous, 186; iatrogenic, 188; and illness, 165; infectious, xiii, 94, 17–71, 179; inherited, 24; and intervention, 56; management of, 180; model of, 92, 170; molecular model of, xiii, 171; multifactorial, xiii, 113, 165, 172; nature of, 164; pathway, 60; prediction and prevention, 56, 165, 171, 256; predisposition to, 55; prognosis of, 80; rare, 179; role proteins play in causing, 60; susceptibility to, 66; understanding of, 186. *See also* Cancer; Genetic disease; Prenatal screening; Screening; Testing

Diseases and conditions; achondroplasia, 101; AIDS, 57, 156, 168, 230; ALS, 261; Alzheimer's, 10, 28, 57, 146, 177–78, 188, 256, 261; amyloid plaques, 57; amyotrophic lateral sclerosis, 156; Angelman syndrome, 101; arthritis, 156, 261; cardiovascular, 28, 168, 256, 261–62; cystic fibrosis, 27, 28, 55, 101–2, 106, 132, 134, 154–56, 177; dementia, 133; diabetes, 58, 153, 177, 256, 261, 262; Down syndrome, 26, 101, 197–99, 200, 205–7, 210; early onset, 147; ectodermal dysplasias, 150; familial adenomatous polyposis, 180; Fanconi anemia, 254; Gaucher's, 57; Hansen's, 156; hemochromatosis, 27, 56, 179; hemophilia, 132; hepatitis, 172; HIV positive children, 217, 225, 232–36; HIV protease enzyme, 57, 60; Huntington's, 26, 29–30, 44, 177, 261; hypertension, 28; liver, 132, 257; Lou Gehrig's, 153; migraine, 58; multiple sclerosis, 256; muscular dystrophy, 55, 80, 156; obesity, 256; osteoporosis, 146; Parkinson's, 260–61; pneumonia, 28; polio, 169; Prader-Willi syndrome, 101; premature ovarian failure, 147; psoriasis, 58; renal failure, 149; schizophrenia, 58; sever combined immunodeficiency (SCID), 236; sickle cell anemia, 24, 25, 132, 162, 177; small pox, 169; spinal cord injuries, 261; Steinert's, 80; Tay-Sachs, 24, 118–19; thalassemia, 24, 187; trisomy-13, 198; trisomy-18, 198; trisomy-21, 197; tuberculosis, 172; Turner syndrome, 102; Werner's syndrome, 146; X-linked, 101, 149, 201. *See also* Cancer; Chromosome; Prenatal screening; Screening; Testing

Disorder(s); autosomal dominant, 180; genetic, 25; late onset, 26, 183; recessive, 24; single gene, 186

Dolly, 11, 71, 261

Donaldson, Liam, 265

Drug(s); *Activase*, 57; and chemotherapeutic agents, 57; and children 216–37; development, 65, 71, 261; discovery tools, 65; distributing new, 64; dosage, 188; gene based, 57–70; gene therapy, 58; genetic targets

for, 64; *Hereceptin*, 58; ONYX-015, 62; and protease inhibitors, 57; research, 57; response to, 190; small molecule, 64; therapies, 257; toxicity, 189, 190; *Zyprexa*, 58
Duty, 120, 129

Edwards, Patrick, 264
Efficacy, xii, 122, 260
Eli Lilly & Co., 58, 62
Embryo(s); cells of the, 149; and cloning, 266; concept of the, 192; creation of, 132; destruction of, 254, 260–65; development of, 148; fertilized human, 266; human, 118; and human life, 253, 264; and in vitro fertilization, 10, 119, 132; and the NIH *Report of the Human Embryo Research Panel*, 264; and preimplantation genetic diagnosis, 201; protection of, 264; spare, 261, 265; and the status of, 135, 265–66; and the UK *Human Fertilisation and Embryology Act*, 265. *See also* Ageing and embryology; Cloning; Fetal tissue; Fetus; In vitro fertilization; Therapeutic cloning
Embryology, 147–51
Embryonic stem-cell research, x, xiv, 11, 253–54, 260, 261–67. *See also* Cloning; Fetal tissue; Fetus; Stem cell research; Therapeutic cloning
Engelhardt, H. Tristram, 126, 127, 140
Enhancement, 9, 57, 121–22
Enlightenment, 126–27
EnsEMBL, 66

Environment, 7, 53, 164, 165, 171–75
Environmental Genome Project, 256
Environmental Genome Project (EGP), 56
Equality; concepts of, 104; and egalitarian analysis, xi, 101–13; gender, xi; genetics and, x, 31, 113; and justice, 10; and policy, 32, 257; between races, 104; social, 113; as a subset of justice, 101, 104–6, 113. *See also* Gender equality; Justice
Ethics/ethical; analysis, challenges, debate, discourse, framework, issues, and literature, ix, xi, 7–10, 54, 80, 117, 121–23, 154, 162, 165–75, 188–95, 257, 260, 267; applied, 194–95; and bioethics and biomedical ethics, 84, 89, 94–98, 162–75, 188, 192; and commitment to the individual patient, 44; and commitment of medicine, 169; and codes, 40, 129; concerns, 164–65; and consequentialist argument, 121; and culture, 38, 267; and deontological moral arguments, 45, 116, 107, 120–23; and genetics, ix, 1, 7–11, 82, 83; and interpretative ethical methodology, 97; and moral imperative, 167; and pain management, 176; and patents, 63; and policy/politics, xiii, xiv, 37, 261, 265–66; and prenatal screening and diagnosis, 202–10; principles, xiii, 9, 110, 167, 171–72, 192, 205; and principle of beneficence, 108, 167; and principle of cooperation, 206–7,

263; and principle of do no harm, 108, 167, 183; and principle of use of resources, 167; and prudence, 73, 162; and responsible parenthood, 207–9; and social issues and principles, 97, 129; theories, 192, 103; western, 37. *See also* Autonomy; Confidentiality; Consent; Dignity; Duty; Freedom; Justice; Philosophy; Organizational ethics; Religion; Research; Research ethics; Respect; Responsibility; Rights; Values
Eugenics, 1, 7, 9, 10, 24–25, 47, 101
European Bioinformatics Institute, 65
European Convention on Human Rights, 39, 42
European Molecular Biology Laboratory, 65
Evolution, 10, 53, 73, 82
Experiment(s); upon human beings, 10; human gene therapy, 257, 259; and protocols, 165

Fetal tissue, 261, 264
Fetus; abnormalities of the, 197; development of the, 148–49; and disorders of the, 93; respect for the life of the, xiii, 10, 202–3; status of the, 39; therapy for the, 201. *See also* Embryo; Genetic Testing
Fletcher, J. C., 42
Fogle, T., 82
Food and Drug Administration (U.S.), 40, 62, 132, 217–18, 221, 230, 235, 258
Ford, Norman M., xiii, 197

Foucault, Michel, 87
Franklin, U., 47
Freedom, 7, 8, 10, 98, 172

Garavaglia, Pia, 210
Gearhart, John, 261
Gelsinger, Jesse, 9, 64, 217, 235, 257, 258–59
GenBank, 65
Gender; differences in genetics, xi, 101–4; equality, xi, 101, 105–11; and treatment of disease, 103
Gene(s); defective, 132–33, 256; disease-causing, 58, 70; expression and function, 3, 43, 55, 65–66, 118–19, 149; and gender, xi, 101–3; mutation, 53, 58, 80, 132, 178, 181–83; as private property, 69; sequences and sequencing, 3, 55–56, 63–64, 181; transfer protocol, xiv, 8. *See also* Genetics; Genomics; Sequencing
Gene Therapy Advisory Committee (Britain), 6
Genentech Inc., 57, 58, 61, 69
Genetic application(s), 116–23, 131
Genetic biotechnology, 9, 11, 12
Genetic cause(s), 121, 147
Genetic civilization, xi, 94–96
Genetic(s) code, 3, 54–55, 169, 170
Genetic counseling, xiii, 10, 91, 106–10, 181–83, 203–5, 257
Genetic disease, xii, 9, 55, 96, 153–62, 165, 177, 198
Genetic engineering, 5, 8, 11, 63, 72, 117, 121–23, 126, 140
Genetic essentialism, 83, 88
Genetic information, xii, 4, 7, 10, 23, 36, 41–46, 64–67, 81–82, 90–97, 164, 166, 171–75, 188, 191–93, 255–57. *See also* Information; Privacy

INDEX

Genetic intervention(s), 121, 164, 175
Genetic knowledge, 80–84, 89, 97, 131, 134, 173–74, 179
Genetic labels; and the ethical use of labels, 154–59; and long-term care policies, xii, 153–62
Genetic medicine, xiii, 7, 9, 10, 56, 68, 71, 164–75, 176, 181, 256
Genetic privacy, x, 10, 37, 41. *See also* Genetic information; Information; Privacy
Genetic research, 1, 2, 5–6, 9, 53, 55–57, 69, 71, 80–82, 115, 135, 165, 257–58
Genetic screening, xiii, 10, 91, 94–96, 185–95, 257
Genetic technology, x, xi, xii, 1, 6, 40, 65–66, 81–84, 88–90, 126–40, 168, 256
Genetic testing, 9–10, 41, 46, 56, 72, 83, 96–97, 103–7, 111–13, 173, 177–83, 190–91, 257
Genetic (gene) therapy; alternatives to, 6; capacity of, 42; categories of, 9; as a competing service, 173; effective, 256; ethics of, 9; experiment(al), 64, 257, 259; germ-line, 9, 10, 121, 164; human, 144; protocols, 8, 116–17, 257; and research with children, xiii, 216–37; somatic cell, 8, 9, 121. *See also* Drugs; Research with children; Therapy; Treatment
Genetic Therapy Inc., 59
Geneticization, xi, 80, 83–90, 96–98
Genetics; clinical, xi, 2, 4, 54, 83–86, 90–96, 179, 186, 255; ethical, legal, and social implications of, x, 23, 35; and employment, 46; and gene inheritance, 53; health outcomes, 54; human, 1, 2, 35–36, 151; impact of, 81, 84, 101, 144–62, 164–65, 194; meaning and model of, 98, 113; revolution, ix, x, 35–52, 54, 72
Genome; individual, 1–2, 94; informatics, 65; mining the human, 54; map of the human, ix, 1, 4, 56, 65, 83, 84, 147, 255, 256. *See also* Human Genome Project
Genomics; and bioinformatics, 65; business of, 53–54; human, xiv, 53, 151, 255, 267; impact of, 144–62; and preimplantation genetic diagnosis, 254–55; research and safety in , 67, 70, 73, 113, 259; technologies and therapies, 66, 72. *See also* Stem cell research
Genotype, 3, 4, 191
Genovo, 258
Genset, 58
Gentechnology, 53–98
Genzyme Corp., 57, 62, 66
Geron Corp., 261
Giardiello, F. M., 181
Gillam, Lyn, 210
Glantz, Leonard H., 37, 223
Glaxo Wellcome PLC, 58
Goodey, Chris, 205
Gostin, L. O., 48
Governor's Commission on Human Genetic Technologies (Nebraska, United States), 157
Grace, xii, 126–40
Grant, George, 139
Green, Ronald, 265
Grodin, Michael A., 220, 223
Groves, J., 156

Gustavus Adolphus College, 10
Guttmacher, Alan, 179

Haldane, J. B. S., 7
Hambrecht & Quist, 61
Hare, R. M., 118,
Harris poll, 46
Hayflick, L., 149
Health; concepts of, 9, 39, 83, 88, 90, 92, 170; and gene based outcomes, 23, 55, 59; outcomes, xi, 53, 54, 55, 67, 68, 72, 168; policy, 92, 93, 153; population, 54, 167; promotion, 54, 56, 95, 96
Health care; access to, 168, 172; delivery systems, 5, 164–75, 191, 256; and pharmacogenomics, xiii, 186–95; and screening, 186–95; transform(ing), 37, 90, 93, 133, 160, 166–67, 256. *See also* Access; Care; Managed care; Patient, Public health
Health insurance, 6, 10, 39, 46, 56, 72–73, 83, 97, 103, 162, 174
Health maintenance organizations (HMOs), 183
Hedgecoe, A., 86–87
Heller, Jan, xi, 116
Heseltine, William, 60
Heyd, David, 118, 120
House of Commons Select Committee on Science and Technology, 41
Human being(s), 35, 68, 116, 119
Human Gene Therapy Subcommittee (U.S.), 257
Human Genome Project, 4, 11, 35–48, 59, 65–66, 81–84, 144, 164–65, 173, 177, 186, 237; *Ethical, Legal, and Social Implications* of, 4, 11, 36, 165.
See also Genetics; Genome; Genomics
Human Genome Sciences Inc., 60, 61, 64
Human Genome Sequencing Consortium, 255
Huxley, Aldous, 7, 33

Imagination, 7, 11, 41, 84
In vitro fertilization, 119, 132, 255, 264, 265
Incyte Pharmaceuticals Inc., 61, 64
Individual(s), 37–38, 41, 45, 96–97, 118, 167, 171. *See also* Autonomy; Person
Infertility, 28, 261, 266
Information; access to/disclosure of, 44, 49, 64–65, 69–70, 81, 107–11, 192; accuracy of, 10, 61, 64, 70; commercial, 61, 66, 70; genetic, genomic, and genotype, 48, 162, 191; health and medical, 23, 30, 37, 46, 65, 70, 93, 182, 194; private, 31, 45, 168. *See also* Genetic information; Genetic privacy; Privacy
Inheritance and laws of heredity, 2–3, 53, 72, 190
Institute for Genomics Research, 60
Institute for Health Care Improvement, 166
Integration Panel for the U.S. Department of Defense Breast Cancer Research Program, 111
Integrity, xi, 37, 53–55, 67–68, 71–72
Investment, xi, 53–55, 59–63, 67, 73. *See also* Companies

Janofsky, Jeffrey, 219, 231
Jefferson, Thomas, 69

Johns Hopkins University, 259, 261
Johnson & Johnson, 58
Juengst, Eric, 152
Justice; gender, xi, 10, 257; and genetics, 42, 101, 218; and labels, 161–62; principle of, 42, 104, 167, 173, 193; promoting, 113, 167. *See also* Equality; Ethics; Gender Equality; Rights

Kass, Leon, 8
Kavka, Gregory, S., 118
Kennedy, I., 39
Kere, Juha, 150
Kevles, D. J., 47
Kinderlerer, J., 39, 40
Knowledge, 36, 38, 44–45, 81, 89–90, 126, 176, 180–81. *See also* Genetic knowledge; Genetic information; Information
Ko, Minoru, 148–49

Labels. *See* Genetic labels; Long term care
Law(s), ix–x, 5–7, 12, 24, 35–49. *See also* Patent(s); Regulation(s)
Lawsuit, 64, 258
Lederberg, Joshua, 8
Legal order, 45–49
Levine, Robert, 223
Life, 48, 138, 151, 165, 167, 169, 173, 175
Life sciences, 84, 192, 253, 255, 259–60, 266–67
Lindee, M. S., 83, 90
Lippman, Abby, 84, 90
Long term care; and disabilities, 155–56; and health care, 166–67; policies, 153–62. *See also* Care; Genetic labels
Longley, D., 39, 40
Lysaught, M. Therese, xiii, 216

Magill, Gerard, xiv, 1, 53, 253
Mahowald, Mary Briody, xi, 101
Managed Care, 72, 166, 167. *See also* Health care
Market, 6, 53–55, 60–61, 62, 67
Martin, George, 146
Mayo Clinic, 184
McGill University, 84
McLean, Sheila, x, 35
Meaning, 10, 97, 126, 257
Medical error, 166, 190–91. *See also* Health care
Medicalization, 86–90, 97
Medicine, 54–57, 67, 87–88, 95, 166–68, 172, 256
Men, 101–2
Mendel, Gregor, 2
Merck & Co., 58
Mergers, 62–63
Meslin, Eric M., 259
Metaphor, 8, 69, 91
Mice; Doogie, 132, 133, 134; Tabby, 150
Microsoft, 62
Millennium Pharmaceuticals Inc., 61, 62, 64
Minorities, 10, 102, 113
Molecular biology, 5, 35, 87, 90, 97
Molecular medicine, 1, 3, 7, 9, 53, 84, 171–72, 256–57
Monsanto, 62
Morality. *See* Ethics
Morgan, T. H., 2
Muller, Hermann, 3, 7

NASA, 53
Nash, Molly and Adam, 253–55, 260, 267
National Academy of Sciences (U.S.), 260

National Bioethics Advisory Commission (U.S.), 122, 234, 259, 262, 265
National Cancer Institute (U.S.), 256
National Center for Biotechnology Information (U.S.), 65
National Commission for the Protection of Human Subjects of Biomedical and Behavioral Research (U.S.), 218–19, 222
National Human Genome Research Institute (U.S.), 65–66
National Institute of Environmental Health Sciences (U.S.), 56, 256
National Institutes of Health (U.S.), 4, 5, 27, 60, 221, 225, 229, 232, 235, 258–59, 262
National Institutes of Health Gene Transfer Safety Assessment Board (U.S.), 259
National Office of Human Research Oversight (U.S.), 259
Nature, 10, 69, 72, 126, 127, 131
Nelkin, D., 83, 90
Newell, Christopher, 210
Niermeijer, M. F., 86
Nijmegen University, 80
Nirenberg, Marshall, 8
Normal/Normalcy, 10, 94, 97, 121, 133
Novartis AG, 59, 60–61

Office for Human Research Protections (U.S.), 258
Office for Protection from Research Risks (U.S.), 223–24, 234–37, 258
O'Neill, Onora, 194
Onyx Pharmaceuticals Inc., 62

Organizational ethics, xiii, 53–54, 59, 63, 67, 162, 164–75. *See also* Ethics

Pain management, 176
Palmer, Julie Gage, 9
Pangea Systems Inc., 65
Paradigm(s), 55–56, 59, 71, 171, 183, 186, 195, 264
Parfit, Derek, 118
Patent(s); applications and filing, 65–66, 70; business, xi, 53–55, 63, 66–70; and *Diamond v. Chakrabarty*, 63; and genetics, 6, 11, 69, 257; and intellectual property rights, 6, 39, 47, 63, 66–70; mechanism of, 67, 72; and prior art, 69–70; and *Regents of the University of California v. Genentech Inc.*, 69
Patent and Trademark office (U.S.), 6
Patient(s), 91–92, 97, 121, 153–62, 166–68, 176, 178, 181, 186, 190
Patient safety, 253–54, 257–60
Pauling, Linus, 25
PE Biosystems Group, 61
Pellegrino, Edmund D., 37
Perker Elmer Corp., 60
Person(s), xi, 11, 69, 116–19, 120–23, 165, 172. *See also* Autonomy; Individual
Pfizer Inc., 58, 62
Pharmaceutical(s), 4, 31, 71, 186–95
Pharmacia & Upjohn, 60
Pharmacogenomics, xiii, 1, 31, 186–95. *See also* Genetic screening; Genetic testing; Screening; Testing

Phenotype, 3, 57, 82
Philosophy, 45, 87, 97
Physician education, xiii, 176–84. *See also* Information; Knowledge
Pilia, Giuseppe, 147
Policy, x, 1, 6–11, 23–24, 67–69, 155, 253, 255, 261, 166
Preimplantation genetic screening/diagnosis, 119, 132, 134, 201, 254–57
Prenatal care, 25, 27
Prenatal screening/testing/diagnosis; and abortion, xiii, 25–26, 103; and amniocentesis, 27, 102, 199, 200–203; and autonomy, 182; and anxiety of pregnant women, 197–98; and chorionic villus sampling, 27, 102, 200; and cystic fibrosis, 106; ethical evaluation of, 197–210; and fetal congenital malformations, 197; and fetal hemolysis, 201; and maternal serum screening, 198; and morally responsible parenthood, 207–9; and nuchal translucency scan, 198–99; and preimplantation/preconception genetic diagnosis, 201; and sensitivity to disabilities, 209–10; therapeutic benefits of, 201–2; and ultrasound, 199. *See also* Disease; Genetic disease; Reproduction; Screening; Testing
President's Commission for the Study of Ethical Problems in Medicine and Biomedical and Behavioral Research (U.S.), 5, 8, 258
Prevention, 9, 56, 95, 256
Princeton University, 132

Privacy; and autonomy, 31, 41, 43–44, 48–49; genetic, 7, 47, 191; and *Griswold v. Connecticut*, 42; and the *Human Rights Act*, 37, 39; reproductive, 10, 264; right, 38, 39, 43, 44; and *Roe v. Wade*, 42
Professionals/ism; and caregiving, 39, 102, 194; cure oriented, 156; and ethics, 47, 95, 191; and the genetics professional, 107; and interdisciplinary teams, 159–61; and patient relationship, 158, 170–71
Profit, 54, 59, 63, 67, 68, 73
Progress, 45, 47, 69, 253
Protection of human subjects, 258, 259; *Ethical and Policy Issues in International Research: Clinical Trials in Developing Countries*, 259; *Ethical and Policy Issues in Research Involving Human Participants*, 259. *See also* Research ethics
Protein(s), 2, 3, 57, 60, 66, 146, 150, 256–57
Proteomics, 256–57
Prudence. *See* Ethics
Public health, xiii, 167, 171, 172. *See also* Health care
Purtilo, Ruth, xii, 153

Race, 169, 257
Ramsey, Paul, 8, 10
Recombinant DNA (rDNA), 3, 6
Recombinant DNA Advisory Committee, 5, 8
Recombinant DNA Advisory Committee (U.S.), 5, 8, 225–26, 229–36, 259

Regulation(s); *Americans with Disabilities Act*, 7, 30, 158; *Diamond v. Chakrabarty*, 63; and the executive branch, 7; *Family and Medical Leave Act*, 30; and federal agencies, 7; *Genetic Privacy Bill*, 37; *Griswold v. Connecticut*, 42; *Human Rights Act*, 37, 39; and law, 39–41; *Mount Isa Mines v. Pusey*, 45; *Patent and Trademark Act*, 7; *Roe v. Wade*, 42; *Skinner v. Oklahoma*, 42; *Tarasoff v. Regents of the University of California*, 42

Reilly, P. R., 46

Religion; and the Christian tradition, 32, 128, 130; and ethics, 10, 257, 265; and the Protestant tradition, 11, 128–29; and the Roman Catholic tradition, 11, 122, 123; and western traditions, 116

Reproduction; artificial means of, 40, 192; and choice(s), 36–39, 90; and genetic testing, 24–27; and reproductive planning, 25, 101–2, 135, 147. *See also* Cloning; Embryo; Prenatal screening

Reproductive Genetics Institute, 254

Research; consent for, 6, 257; genomics, 36, 65, 55, 58–59, 61, 68, 153, 194, 253–54; government sponsored (funded), 63, 67, 71, 72; medical, 9, 67, 69, 70, 255, 258, 264, 266; protocols and therapies, 54, 133, 257, 258; publicly funded, 31; and safety, 264, 265

Research ethics, 216, 219, 220, 221, 231, 258, 259. *See also* Protection of human subjects

Research with children, xiii, 216–37; and federal regulations and policy, 218–24; and human gene transfer (HIV protocol), 224–37; and *Policy and Guidelines on the Inclusion of Children as Participants in Research Involving Human Subjects*, 221

Resources, 68, 170

Respect, 10, 96, 108, 202–3. *See also* Autonomy; Dignity

Responsibility, x, 10, 39, 90–98. *See also* Individual; Person

Rhone-Poulenc SA, 58

Rifkin, Jeremy, 8

Rights, 36, 39, 42, 45, 48–49, 69, 120. *See also* Ethics; Justice

Risk(s); and benefit analysis, 9, 40, 42, 73, 101–2, 105, 188, 193, 227–30, 258, 261; of disease, 56, 169, 177; factors, 31, 177, 187, 190; financial, 59, 60–63; genetic, 9, 10, 25, 257; health, 102, 177, 194, 198

Robertson, John, 135

Roche, Ellen, 259

Roche Holding, 58, 62

Roche, Patricia, 37

Rome, 128

Roslin Institute, Edinburgh, 11, 261

Ross, Lainie Friedman, 108

Rothman, Barbara Katz, 84

Rothstein, M., 46

Roundworm, 255

Saint Louis University, ix, xiv, 1

Sandoz AG, 59

Sanger Center, London, 65
Schlessinger, David, xii, 144
Schwartz, Thomas, 118
Science, x, 7, 9, 11–12, 35–38, 53–54, 65, 253, 257, 260
Screening, xiii, 10, 24–28, 56–57, 62, 186–87, 197, 254, 257. *See also* Genetic screening; Prenatal screening
Searle, Judith, 197
Sen, Amartya, 104
Sequencing; and algorithms, 65; and annotation processes, 59, 65; gene sequencing machines, 60, 61; the human genome, ix, 60, 144, 179; and microarrays (chips), 149; and pattern recognition, 65
Shakespeare, William, 145
Shirkey, H. S., 220
Single nucleotide polymorphism (SNP); database, 70; technology, 55, 58; testing for, 186
Society, x–xi, xiv, 37–38, 83, 90–94, 98, 104, 123, 140, 255, 259–60, 267. *See also* Community
Somatic cell, 8, 171
Somatic cell nuclear transfer, 122, 261
Srivastava, Anand, 150
Stanford University, 184
Starfield, Barbara, 219, 231
Stem cell research, 1, 53, 151, 192, 253–55, 259–64. *See also* Cloning; Embryonic stem-cell research; Therapeutic cloning
Steptoe, Robert, 264
Stewardship, xi, 53, 54, 68, 71–73
Sturtevant, Alfred, 3
Suffering, xii, 95–96, 135–36, 156, 260, 264

Supreme Court (U.S.), 6, 264, 266
Suter, M. S., 41
Symptoms, 28, 80
Systemix Inc., 59, 61

Ten Have, Henk A. M. J., xi, 80
Tenet Healthcare Corporation, ix
Testing, xiii, 10, 24, 28, 153–54, 64, 68, 180, 182, 186, 190, 196, 261. *See also* Genetic testing
Therapeutic cloning, x, xiv, 253, 254, 260–61, 265–67. *See also* Cloning
Therapeutic interventions and procedures, 40, 165, 170, 174
Therapy (therapies), 4, 183, 201, 253, 256, 260. *See also* Genetic therapy
Thompson, James, 261
Tissue(s), 37, 148–51, 257
Traits; behavioral, 135; biological, 255; deleterious, 134; genetic, 2, 9, 53, 133, 136, 260; and intelligence, 121; multifactorial, 190; polygenic, 3
Transplant, 254, 261
Treatment, 9, 56, 158–59, 179, 190, 256
Trial(s); clinical, xiv, 132, 188, 234–37, 256; gene therapy, 253, 257, 258; phases of, 3, 62. *See also* Research
Tuskegee experiment, 5

UNESCO Declaration of the Responsibilities of the Present Generations towards Future Generations, 42
United Kingdom Parliament, 265
Universal Declaration of Human Rights, 42

University of California, 69, 184
University of Minnesota, 132
University of Pennsylvania, 217, 258
University of Pennsylvania's Institute for Human Gene Therapy, 233
University of Wisconsin-Madison, 261
U.S. Ethics Advisory Board, 264
U.S. Patent and Trademark Office, 63
U.S. Supreme Court, 63

Value(s); approaches to, 45, 47, 54, 67, 120–21, 126, 194, 267; deontological approach to, 72, 124; of human life, 253–54, 259, 260, 261; of medical knowledge, 80, 88–89; neutrality, 94, 95, 96; prescriptivity, 96–97, 110, 257; and society, 11, 70, 120, 135, 175, 253
Van Dijck, José, 84, 98
Variability and variation, 56, 186, 255

Varmus, Harold, 233, 236
Vascular Genetics Inc., 64
Vectors for gene therapy, 58, 257
Ventor, J. Craig, 60, 255
Verdi, Giuseppe, 145
Virus, 62, 257

Walters, LeRoy, 9
Warner-Lambert Company, 62
Warnock, Mary, 264
Warnock Report, 264
Waters, Brent, xi, 126
Watson, James D., ix, 3, 4, 7, 35
Wellcome Trust (Britain), 4, 60
Wertz, Dorothy C., 42, 110, 210
White House, 258
Wilkie, T., 48
Wilmut, Ian, 261
Windeyer, J., 45
Women, 101–5, 108, 197–98. *See also* Autonomy; Gender equality; Privacy
World Health Organization, 169
Wrongful birth litigation, 25